MANNED SYSTEMS DESIGN
Methods, Equipment, and Applications

NATO CONFERENCE SERIES

I Ecology
II Systems Science
III Human Factors
IV Marine Sciences
V Air-Sea Interactions
VI Materials Science

III HUMAN FACTORS

MANNED SYSTEMS DESIGN
Methods, Equipment, and Applications

Edited by

J. Moraal
Institute for Perception TNO
Soesterberg, The Netherlands

and

K.-F. Kraiss
Forschungsinstitut für Anthropotechnik
Werthhoven, West Germany

Published in cooperation with NATO Scientific Affairs Division

PLENUM PRESS · NEW YORK AND LONDON

Library of Congress Cataloging in Publication Data

Main entry under title:

Manned systems design.

(NATO conference series. III, Human factors; v. 17)
Proceedings of a conference held September 22-25, 1980, in Freiburg, West Germany, sponsored by the Special Programme Panel on Human Factors of the Scientific Affairs Division of NATO.
Bibliography: p.
Includes index.
1. Human engineering—Congresses. 2. Systems engineering—Congresses. I. Moraal, J., 1934—   II. Kraiss, K. F. III. North Atlantic Treaty Organization. Special Programme Panel on Human Factors. IV. Series.
TA166.M36   620.8'2   81-10732
ISBN-13: 978-1-4613-3308-1   e-ISBN-13: 978-1-4613-3306-7
DOI: 10.1007/978-1-4613-3306-7

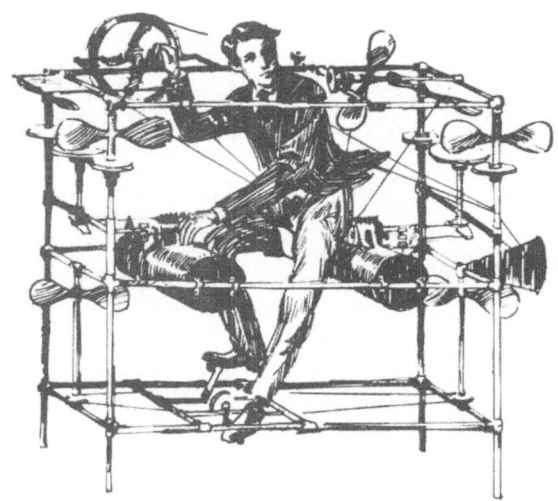

Proceedings of a conference on Manned Systems Design: New
Methods and Equipment, held September 22-25, 1980, in
Freiburg, West Germany

© 1981 Plenum Press, New York
Softcover reprint of the hardcover 1st edition 1981
A Division of Plenum Publishing Corporation
233 Spring Street, New York, N.Y. 10013

PREFACE

This volume contains the proceedings of a conference held in
Freiburg, West Germany, September 22-25, 1980, entitled "Manned
Systems Design, New Methods and Equipment".

The conference was sponsored by the Special Programme Panel
on Human Factors of the Scientific Affairs Division of NATO, and
supported by Panel VIII, AC/243, on "Human and Biomedical Sciences".
Their sponsorship and support are gratefully acknowledged.

The contributions in the book are grouped according to the
main themes of the conference with special emphasis on analytical
approaches, measurement of performance, and simulator design and
evaluation.

The design of manned systems covers many and highly diversified
areas. Therefore, a conference under the general title of "Manned
Systems Design" is rather ambitious in itself. However, scientists
and engineers engaged in the design of manned systems very often are
confronted with problems that can be solved only by having several
disciplines working together. So it was felt that knowledge about
newly developed methods and equipment, applicable in the design
process, is of common and increasing interest for all those who are
engaged in the design of manned systems, from the earliest conceptual design phases until operation under real circumstances.
This seems to be particularly true in view of restricted resources
of manpower and energy.

The organization of a conference and the publication of a book
can only be done with active co-operation of a great number of
people. It is not possible to mention all of them. However, we
like to acknowledge the invaluable help of Lt. Col. P.J. Houtzagers,
who acted as administrative director to the conference, Irene Klein
and Henny Gebbink, who both did all the work in preparation of the
conference and the manuscripts, and Konrad Steinheuer by whose
technical help the conference ran so smoothly.

We hope that the contents of the book will not only contribute to a better understanding of Manned Systems Design but particularly to a transfer of ideas and methods into practice.

J. Moraal
K.-F. Kraiss

CONTENTS

PART III: MEASUREMENT OF PERFORMANCE

PART IV: SIMULATOR DESIGN AND EVALUATION

PART I
CONCEPTS AND STRATEGIES

METHODS: PAST APPROACHES, CURRENT TRENDS AND FUTURE REQUIREMENTS

Donald A. Topmiller

INTRODUCTION AND HISTORICAL ANTECEDENTS

The field of Human Factors Engineering in the United States evolved out of demands from the increasing complexity of military weapon systems in World War II. The early workers in the field such as Paul Fitts in the Air Force, Adelbert Ford in the Navy and a little later John Weisz in the Army, were all trained as experimental psychologists, hence the methods applied during these years were transitioned almost directly out of empirical applied psychology. This was formally recognized in 1949 with one of the first published texts in the field (Chapanis et al., 1949).

During the early and mid 1950s, the application of the tools of the experimental psychologist to man-machine design problems continued to flourish. We also saw, during this period, an attempt to extract from the existing experimental literature, human performance limits, given certain man-machine interface characteristics, environmental conditions, and task environments. An excellent early effort to do this was reflected in the publishing of the "Tufts Handbook" or the Handbook of Human Engineering Data in 1951. This was the authoritive reference document for Human Engineering Data for over ten years. In the early sixties the three military services convened the Joint Army-Navy-Air Force Steering Committee to publish the Human Engineering Guide to Equipment Design. The first edition was published in 1963 and the revised second edition in 1972. The respective services also published numerous engineering handbooks of human engineering design principles and criteria.

The underlying assumption during the "Handbook Era" was that design engineers and system engineers would apply the appropriate human engineering data to manned systems design problems. Frequently, this was found not to be the case for several reasons. First, many design engineers were not aware of the existence of the handbook data, or if they were, there was very little incentive to use the data in their design trade-off analyses. (This was often the fault of the government program management since no requirement for the application of human factors was included in the contract.) Secondly, if the need was recognized, the design engineer could not apply the data since the problem or intended operational use of the system did not correspond to the conditions under which the data were experimentally derived. Finally, an integrating structure for the data base which was systems and problem specific did not exist. It was frequently found that in order to solve a specific design problem, a physical simulation, or at least a functional mock-up, of the manned system design was required. Depending on the significance of the design issue and the expense involved, the solution may or may not have been assured by physical simulators.

This situation was probably characteristic of the human factors state-of-the-art in the mid 1960s. The experimentalist, qua human engineer, found that experimental methodology used in problem specific cases was often extremely expensive and may or may not address the operational or system variables critical to the design issue. We also saw during the same period an explosive growth in digital computer technology. This technological development was exploited by the operations research community and led to the rapid computerization of large multivariate systems simulations as well as the coincident development of computer simulation languages. The first psychologist to recognize, and eventually implement, the power of the computer simulation to man-machine system design problems was Arthur Siegel from Applied Psychological Services, who along with Jay Wolf, a mathematician, under the Office of Naval Research contract developed the first 2-operator model and in 1969 published the first book in man-machine simulation models (Siegel and Wolf, 1969). Certainly this development contributed a quantum jump in human factors methodology and eventually led to several man-machine systems design and analysis techniques such as CAFES (Computer Aided Function-Allocation Evaluation System) in the Navy, with its associated submodels, and SAINT (Systems Analysis Integrated Network of Tasks) in the Air Force. Human factors engineering recognized that in order to get man-machine design principles integrated into system design and development programs, it would have to borrow heavily from the systems simulation and operations research communities.

In fact, in the mid-sixties our Laboratory sponsored a contractual survey with Air Research Incorporated to evaluate the extent to which human factors (man-machine interface) parameters had been incorporated into military Operations Research (OR) studies (Schwartz et al., 1967). A total of 250 studies were reviewed covering a period from World War II until 1965. Detailed analysis was made on 20 of

the most representative studies to evaluate the extent to which hu-
man factors variables were incorporated into the study, `and the de-
gree to which the model was sensitive to variations in the human fac-
tor parameter values. The detail review included:
- Description of the system modeled.
- Discussion of the problem treated and objectives of the analysis.
- Identification of the OR techniques used. (Hopefully a "classical"
  OR model.)
- Identification of the human factors parameters relevant to the mo-
  del of the systems.
- Discussion of the role and treatment of human factors parameters
  found in the model.
- Discussion of the sensitivity of the model to changes in values of
  the human factors parameters.

   From the analysis ensuing from the above objectives, the study
lists the following conclusions:
- There is a strong tendency for OR investigators to concentrate too
  heavily on the model rather than the natural system problems.
- The relative sensitivity of various OR techniques to variations in
  human factor variables were difficult to assess for the models sur-
  veyed because:
  a. The majority of OR analyses do not analytically treat human per-
     formance parameters (affecting system performance), since they
     seem imbedded in the terms of various OR models, but precise re-
     lations among the parameters and OR model terms are not demon-
     strated.
  b. In the studies which treat human performance variables, the pa-
     rameters considered and systems, or contexts, modeled are too
     diverse to permit direct and quantitative comparison.
  c. The lack of standard measure of human performance in systems
     serves to increase the complexity of the problem.

   In retrospect this is a lesson in irony since Siegel and Wolf
at that time had not received wide publicity in their model development efforts, but must have recognized these deficiencies which stimulated their use of monte-carle techniques for developing the original 2-man model

   At the turn of the decade into the seventies, we saw these major trends in man-machine design methods and techniques development.
First, with the stage set by the technological breakthrough in computer power many human factors groups in industry and government either developed their own computer-controlled engineering design simulation capability for performing single-operator and multi-operator
(crew) real-time mission-based simulation or they piggy-backed studies on simulation facilities developed and controlled by hardware
engineering groups. Many of these facilities were used to investigate
advanced cockpit designs, advanced control-display concepts and/or
measurement of pilot/crew workload. It was fortuitous that this high-

technology capability in man-machine design physical engineering simulation filled a unique requirement, since during the seventies, the national defense policy was one of enhancing existing weapon system capability by hardware/software modification and upgrade rather than investing in new weapons per se. This was particularly true of the decisions not to produce the B-1 and instead upgrade the avionics (electronic warfare and navigation) capability of the 20 year old B-52. Our Laboratory developed the SACDEF (Strategic Air Command Design Evaluation Facility) to evaluate the operator/crew performance aspects of the "improved" avionics systems. Our Laboratory also developed a computer-based multi-operator command/control simulation facility on which many simulation of BUIC (Back-Up Interceptor Control), AWACS (Advanced Warning And Control System) and RPV (Remotely Piloted Vehicle) surveillance and weapons direction mission were conducted. This program is summarized in a chapter by the author in Tsokos and Thrall (1979). In fact, in our BUIC simulations we attempted for the first time to integrate computer-simulation with multi-operator/multi-task real-time physical simulation. Where we had the capability to conduct the physical simulation of the operational BUIC "active tracking" tasks, we did so. Those operational tasks which could not be physically simulated, by virtue of computer and display limitations, we computer-simulated using a Siegel-Wolf model. This enables us to combine the respective powers and advantages of both simulation disciplines into one "hybrid" technique. It was indeed interesting and gratifying to find that the independent variables we were manipulating in the experimental simulation such as radar track trail length and penetrator velocity had significant effects on task performance measures such as "tracker initiation time" and were not washed-out by the potential propagation of sampling error in the monte-carlo process in the Siegel-Wolf model. Unfortunately, for a variety of reasons, we have not followed through on this "hybrid" technique to develop it to its highest potential as a manned systems simulation and design tool.

In the thirty some odd years covered in this very brief and sketchy treatment of the historical antecedents to the methods used in manned systems design, we have seen our field evolve out of the early work in applied experimental psychology through the design handbook era and becoming more interdisciplinary with strong technological influences stemming from the fields of computer and information science with overtones of operations research and systems simulation and modeling.

CURRENT TRENDS

The US Air Force Systems Command has for the past year and a half conducted a very comprehensive study of the Human Factors Engineering (HFE) field and its technology base development, application of principles, tools, methods, and design criteria throughout the research and development phases of AF Weapons Systems Acquisition. It also included extensive examination of HFE problems associated with acquiring and developing professionals in the Air Force, both civilian and military to meet the demands of advanced weapon system development, acquisition and operation.

The study was conducted primarily by a select group of senior nationally and internationally renowned human factors engineers who performed under contract to the Air Force. Certain senior military and civilian HFE worked with the contract professionals to provide needed governmental information.

Robert C. Williges and the present author worked collaboratively on one of the four task committees to perform the "Technology Assessment" (Williges and Topmiller, in press) task the results of which were relied on heavily to develop this section on Current Trends and the final section on Future Requirements.

For purposes of the "Technology Assessment" task, technology includes information, methods, and concepts/devices created for useful Human Factors Engineering (HFE) purposes. These technologies were classified into the following categories:

1.0. Reference Data Sources - These are catalogs of already collected data which maybe measures of human properties, limits, tolerances task performance capability, interface design principles, design criteria and operator-machine centered equipment, subsystem or system data.

2.0. Experimental Design Methods - are means by which data are collected to enhance the data bases, to draw inferences and to solve problems.

3.0. Human-Machine Integration Performance Metrics - are measurements used in data collection/problem solutions.

4.0. Models - are means of organizing information to represent the functioning of objects/processes being modeled by imposing formalistic rules and relationships.

5.0. Engineering Design Simulation - is concerned with the use of simulation as a design tool and is not to be confused with the production or use of training simulation.

6.0. Procedures - are rationally organized steps to aid in the production of a design (at any stage), evaluation of a design, or in extraction, extension or analysis of a design.

To idenfity specific HFE technologies falling under the six broad categories a total of one hundred and thirty-six (136) subcategories of technologies were used for collecting the data base from which the study developed its findings and conclusions. Table I lists the aggregate of technologies examined.

Table I. Listing of Current HFE Technologies

1.0. Reference Data Sources
        Scientific & Engineering Data Sources
        Professional Journal Literature
        Handbooks, Guides, Specs
           Design Handbook 1-3
           NASA Bioastronautics Data Book
           Design Handbook for Image Interpretation Equipment
           Human Engineering Guide to Equipment Design
           MIL Spec 1472
        Data Bases
           Anthropometric Source Book & Data Bank (static & dynamic)
           Old Systems Record Review
           SHERB (Sandia Human Error Rate Bank)
           HFTEMAN (Human Factors Test & Evaluation Manual)

2.0. Experimental Design Methods
     2.1. Statistical Procedures
        Univariate Methods
           Correlations
           Simple Regression
           Parametric Inferences
           Nonparametric Inferences
        Multivariate Methods
           Multiple Regression
           Polynomial Regression
           Canonical Analysis
           Principal Components
           Factor Analysis
           MANOVA (Multivariate ANalysis Of VAriance)
           Discriminate Analysis
        Pretesting Methods
        ANOVA Designs (ANalysis Of VAriance)
        DATA Reduction Designs
           Blocking Designs
           Hierarchical Designs
           Fractional-factorial Designs
           Central-composite Designs
     2.2. Tailored Methods
        Confusion Matrices
        Quasi-experimental Designs
        Response Surface Methodology
        Finite Interaction Test

3.0. Human-Machine Integration Performance Metrics
     3.1. Anthropometric Measures
           Static
           Dynamic

Table I. Listing of Current HFE Technologies (Continued)

3.2. Physiological Measures
       FFF (Flicker Fusion Frequency)
       GSR (Galvanic Skin Response)
       EKG (Electrocardiogram)
       EEG (Electroencephalogram)
       ECP (Evoked Cortical Potential)
       Eye and Eyelid Movement
       Pupillary Dilation
       Muscle Tension
       Heart Rate
       Breathing Analysis
3.3. Human-Machine-Environment Measurement Techniques
       Noise Map/Fill
       Vibration
       Impact
       Acceleration
       Noise Mapping
3.4. Subjective Opinions
       Questionnaires/Checklists/Ratings
         Open-Ended
         Multiple Choice
         Rating Scales
         Ranking Procedures
         Forced Choice
         Semantic Differential
         Critical Incidents
3.5. Automatic Recording Methods
       Event Recording
       Photography
       Audio/Video Tapes
       Motion Pictures
       OPRENDS (Operational Performance Recording and Evalua-
         tion Data System)
       Recorded Flight Data

4.0. Models
   4.1. Biomechanical Models
       Architectural Body Models
       Dynamic Dan
       Combiman
   4.2. Performance Models
       Information Theory
       Statistical Decision Theory
         TSD (Theory of Signal Detectability)
         Bayesian Decision Making
       Estimation theory

Table I. Listing of Current HFE Technologie (continued)

        Control Theory
          Quasilinear Control
          FFM (Fixed-Form Models)
          OCM (Optimal Control Models)
        Queueing Theory
   4.3. Process Models
        Short-Term Memory Models
        Visual Scanning/Detection Models
          GRC (General Research Corporation)
          MARSAM II (Multiple Airborne Reconnaissance Sensor
            Assessment Model)
          VISTRAC (VISual Target Recognition and ACquisition)
          CRESS/SCREEN (Combined Reconnaissance, Surveillance,
            SIGINET/SRI Countersurveillance Reconnaissance
            Effectiveness Evaluation)
          Autonetics Model
          Detect
          ASTCAD
          REA/BAC
        Air Traffic Control Models
        Industrial Inspection Models
        Attention/Workload Models
        Fault-Diagnosis Models
        HOS (Human Operator Simulation Model)

5.0. Engineering Design Simulation
   5.1. Human-Machine Integration Engineering Research Simulation
        General Purpose Static Aircraft Crew Station
        General Purpose Dynamic Aircraft Crew Station
        General Purpose Control Display
        General Purpose Multiperson
   5.2. Human-Machine Integration Engineering Design Simulation
        Static Mockup
        Specific Dynamic Control/Display
        Specific Crew Station
        Outside Dynamic Visual Scene
        Workplace Simulator
        Multiman Workstation
        Command and Control Simulation
        Sensor Simulation
        Computer Simulation

6.0. Procedures
   6.1. Systems Engineering Analytic & Management
      6.1.1. To Aid System Engineering Analysis
         6.1.1.1. Manual
              PERT (Program Evaluation and Review Tech-
              niques

Table I. Listing of Current HFE Technologies (Continued)

TLA-1 (Time Line Analysis-1)
CMP (Critical Path Method)
Expected Value Method
Functional Flow Diagrams
FDI (Functional Description Inventory)
Function Allocation Tradeoffs
Task Analysis
Decision Tree Analysis
Action/Information Requirements
Time Lines
Flow Process
OSD (Operational Sequence Diagrams)
Task Descriptions

6.1.1.2. Computerized
SW (Siegel-Wolf Model)
SAINT (Systems Analysis of Integrated
    Networks of Tasks)
PSM (Pilot Simulation Model)
CAPA (Computer Analysis of Personnel
    Activity)
GERT (Graphical Evaluation and Review
    Technology)
FOVEA (Field Of View Evaluation Apparatus)
WSP (Workload Simulation Program)
R&M (Reliability and Maintainability
    Model)
Sensitivity Analysis

6.1.2. To Extract From a Design
6.1.2.1. Manual
CHRT (Coordinated Human Resources Tech-
    nology)
HR/DODT (Human Resources/Design of Op-
    tion Decision Trees)
QQPRI (Qualitative and Quantitative Per-
    sonnel Requirements Information)
ISD (Instructional System Development)
CDB (Consolidated Data Base)
Comparability Analysis
Technical Order Function Evaluation
TEPPS (Technique for Establishing Per-
    sonnel Performance Standards)

6.1.2.2. Computerized
LCCIM (Life Cycle Cost Impact Model)
TRAMOD (Training Requirements Analysis
    MODel)
PAM (Personnel Availability Model)
LCOM (Logistics Composite Model)

Table I. Listing of Current HFE Technologies (continued)

                        CORELAP (Computerized RElationship LAy-
                            out Planning)
6.2. Detailed Design Procedures
    6.2.1. Manual
                Specification Compliance Summary Sheet
                Link Analysis
    6.2.2. Computerized
                HECAD (Human Engineering Computer-Aided Design)
                CATTS (Continuous Assessment of Task Time Stress)
                TBLA (Time-Based Load Analysis
                RECEP (RElative Capacity Estimating Process)
                CAFES (Computer Aided Function Allocation and
                    Evaluation System)
                DMS (Data Management System)
                FAM (Function Allocation Model)
                WAM (Workload Assessment Model)
                CAD (Computer-Aided Crew Station Design Model)
                CGS (Crew Station Geometry Evaluation Model)

The task study first conducted a workshop in January 1979 with
twenty-six key human factors engineers from government and industry
to identify technological gaps and problems. From this initial workshop, a comprehensive questionnaire was developed based on the technology categories identified in Table I, to further determine the
relative degree of use of these technologies in the Laboratory and
Systems development environments.

This questionnaire, based on the technology areas in Table I,
was used to assess the degree of utilization of the respective technologies throughout the research, development, test and evaluation
phases of the system acquisition process. The questionnaire was distributed to more than twenty (20) industrial organizations with known
human factors experience. Responses covered thirty-nine (39) specific weapon and subsystem development programs including fighter and
bomber aircraft as well as certain command and control systems covering a ten year period from the late sixties to the late seventies.
A comparable questionnaire was administered to over one hundred (100)
government human factor engineers in Air Force Laboratories or engineering development organizations.

Table II summarizes the findings of percentage use of the six
technology categories tabled across industry managers and project engineers vs. government applications and laboratory responses. It is
evident from these percentages that reference data sources receive
a high utilization rate across the board whereas models have a low
utilization rate by all respondent categories. This finding probably

Table II. Average Percent Use of General HFE Technology Categories

General HFE Technology	INDUSTRY		GOVERNMENT	
	1. Man-agers	2. Pro-jects	3. Appli-cations	4. Labor-atories
	(15)	(39)	(71)	(41)
Reference Data Source	67%	55%	46%	42%
Experimental	52%	29%	22%	51%
Human Performance Metrics	47%	30%	28%	33%
Models	13%	10%	6%	17%
Engineering Design Simulation	68%	50%	34%	39%
Procedures	26%	21%	16%	14%

reflects on the technological lag encountered in more sophisticated and quantitative mathematical modeling methods.

Table III breaks down the broad category response into selected detailed HFE technology areas by industry and government respondents. These data further illustrate that for all respondent categories that the more established technologies (Reference Data Sources, Statistical Research Methods, Anthropometric Measures and Design Simulation Methods) are the most frequently used. The newer technologies involving mathematical modeling and computerized techniques are not utilized much. This fact underscores the technological lag problem. At the bottom of the table the intercorrelations between respondent categories, demonstrate the lack of relationships between laboratory scientists' utilization of these technologies and the applications human factors engineers ($Y_{1,4}$ = .85, $r_{2,4}$ = .72, $r_{3,4}$ = .76). Whereas the correlation between technology use between managers, project personnel and government applications personnel are somewhat higher. One possible interpretation of these findings is that the technology transfer between the technology developers (laboratory scientists) and the human factors engineers (both management and project) remains a problem of some significance.

This technology transfer problem may stem from several sources, but two facets merit special concern. First, at least in the USAF Research and Development Community, there is no well defined management mechanism to insure efficient feed-forward from the laboratories to the developers nor is there a feed-back formalized procedure from the operational environment to the laboratory and development programs. Secondly, most laboratory research is conducted with basic research (6.1) or exploratory development (6.2) funding and very few programs see the "prototype" development stage (6.3). Hence, the meth-

Table III. Average Percent Use of Detailed HFE Technology Category

Detailed HFE Technology Category	INDUSTRY		GOVERNMENT	
	1. Man-agers	2. Pro-jects	3. Appli-cations	4. Labor-atories
	(15)	(39)	(71)	(41)
1.0. Reference Data Sources	67%	55%	46%	42%
2.1. Statistical Research Methods	79%	42%	34%	77%
3.1. Anthropometric Measures	18%	12%	7%	19%
3.2. Physiological Measures	80%	78%	47%	32%
3.3. Environmental Measures	44%	26%	38%	21%
3.4. Subjective Opinions	58%	45%	37%	49%
3.5. Automatic Recording Measures	71%	47%	41%	58%
4.1. Bioman Models	13%	10%	2%	4%
4.2. Performance Models	17%	13%	11%	28%
4.3. Process Models	11%	8%	4%	12%
5.1. Engineering Research Simulation	68%	43%	26%	28%
5.2. Engineering Design Simulation	68%	53%	37%	44%
6.1.1.1. Manual Systems Engineering Analysis	65%	53%	36%	37%
6.1.1.2. Computerized Systems Engineering Analysis	7%	6%	6%	6%
6.1.2.1. Manual Procedures from Design	9%	10%	15%	7%
6.1.2.2. Computerized Procedures from Design	5%	4%	6%	2%
6.2.1. Manual Detailed Design Aids	40%	33%	19%	16%
6.2.2. Computerized Detailed Design Aids	8%	4%	3%	1%

Correlations

$$r_{1,2} = .95 \qquad r_{1,4} = .85$$
$$r_{1,3} = .92 \qquad r_{2,4} = .72$$
$$r_{2,3} = .93 \qquad r_{3,4} = .76$$

odological tools involving computer-based modeling and simulation
have no formal R&D management structure to demonstrate verification,
validation and utility for developmental design application.

Based on the analysis and evaluation of the questionnaire results on current HFE technology several observations and conclusions
can be drawn. First, although the reference data sources are used in
weapon system development more than the advanced technologies of models and computerized procedures, it appears that the requirement exists for improving the package of the reference data sources in terms
of handbooks, guides and the like so they are more amenable to design use. In fact, the development of a computerized reference data
bank may be in order which would allow HFE data to be responsive to
specific design issues. A better designed reference system which
could feed the computer modeling and computer-based proceduralized
systems appears to be in order. A renewed effort should be launched
to transfer the computer-based technology into the implementing hands
of the applied human factors engineers with more enphasis on validation of these methods throughout the development and test phases.
The need exists for integration of computer-modeling and engineering
design simulation to developed hybrid techniques to exploit the respective powers of both in a cost-effective manner.

FUTURE REQUIREMENTS

To anticipate the directions of new methods and predict future
projections of needed technology, the Air Force HFE study will again
be used as a data base along with another study sponsored by the Air
Force Systems Command designed to define computer technology shortfalls to the year 2000 (COMTEC-2000) (Computer Technology Forecast
and Weapon Systems Impact Study, 1978). Only the man-machine interface technology part of this comprehensive study will be referred to
in developing future requirements.

Future Technology Assessment Study

To assess future technology projections twenty nine (29) experts from industry, government and academics were asked to write
technology projections, both near-term (3-5 years) and far term (5-
15 years) in seventeen (17) technology areas listed in Table IV.
(Some areas were covered by one or more experts.) The 17 technology
areas were collapsed into nine (9) human factors information needs
identified in Table V and analyzed across three (3) levels - human
centered, human-machine centered or human-machine-mission focused.
Most experts (54%) indicated a preponderance of future technology
needs in the area of man-machine-mission as compared to human (12%)
or human-machine (34%) applications. They also indicated information need for system operation (20%) and design principles/concepts
(29%) with design data base/handbook (15%) being the third highest

Table IV. Representative areas of future HFE technology projections

* Advanced display engineering technology
* Human factors in maintainability
* Systems research technology
* Human factors in safety engineering
* Engineering anthropometry
* Human/computer interactions
* High thermal stress
* Training analysis and simulation
* Biocybernetics
* Operator/crew workload
* Computer modeling and simulation
* Design guides and data bases
* Advanced cockpit technology
* Target detection/acquisition models
* Decision making
* Human factors in manufacturing technology
* Manpower and logistics factors in weapon
  system development

Table V. Summary of future HFE technology projections

		12%	34%	54%	
	Systems Operation	3%	6%	11%	20%
	T&E Procedures	–	1%	5%	6%
	Training Development	–	3%	4%	7%
HFE	Training Requirements	1%	4%	4%	9%
Information	Personnel Selection	–	1%	1%	2%
Needs	Personnel Requirements	1%	2%	2%	5%
	Design Principles/Concepts	4%	10%	15%	29%
	Design Database/Handbook	3%	5%	7%	15%
	MENS	–	2%	5%	7%
		H	H/M	H/M/M	
		U	A	I	
		M	C	S	
		A	H	S	
		N	I	I	
			N	O	
			E	N	

HFE LEVELS OF ANALYSIS

information need category. It is fairly obvious that most of the advanced thinkers in the human factors discipline believe that the greatest needs for future technology development are being driven by the requirement for a human-machine-mission (H-M-M) systems analytic and simulation capability. This human factors system capability is in turn being driven by high technology advanced in hardware/software computer developments creating monumental complex information processing requirements combined with mission threats becoming more challenging and evasive. H-M-M systems analysis and simulation methods must be developed to treat human, equipment and mission parameters in equivalent quantitative terms in order to isolate this respective contribution to overall systems effectiveness.

Williges and Topmiller (in press) conclude from the technology assessment that - Advances in other life sciences, engineering, and computer science along with consideration of energy supplies, future threat environments, and future weapon systems appear to provide the major impetus for future threat environments, and future weapon systems appear to provide the major impetus for future HFE technology developments. Current DOD research plans and activities reveal that these influences are driving the developments of models, computer simulation, display assessment, multi-operator systems, advanced cockpit considerations, management technology and biocybernetics. Future research is needed in maintenance design and analysis, engineering anthropology, safety system research, human/computer interfaces, and human/machine environments. Both near-term and long-term advances are required in each of these areas in order to provide the appropriate technology base for HFE.

Williges and Topmiller make the following recommendation regarding the implementation of computer-based HFE technology: Various computer-based procedures for HFE are currently available and near-term projections suggest even more improvements and developments in these procedures. But, by and large, this technology is not being implemented heavily in the design process. Several reasons, such as the unavailability of the procedures, the lack of knowledge of the procedures, and nonexistence of contract requirements to use them have been given for their sparse use. Before these various techniques can be completely evaluated, they must be integrated.

Little effort, however, has been directed toward integrating all of these approaches into a truly, computer-based HFE design methodology and providing a testbed for application and further development. Such an integrated methodology would include using computer-analytic models and engineering design simulation could be effectively used in interactive manner throughout conceptual design, fly-offs between design configurations could be accomplished by large-scale, total mission simulation, computer modeling could suggest alternative configurations for simulator evaluation, and engineering design simulation could be used as a means of specifying training requirements and follow-on trainer design.

The Human Factors Engineering Technology Assessment Study was conducted by, and participated in, by practicing HFE scientists, prac-

titioners and managers. The survey and review results certainly yield
one common theme, viz. that computer technology has driven much of
the advanced HFE technology developmental needs. It was somewhat fortuitous that about the same time the Air Force was making a self-ap-
praisal of HFE, it was also launching a major study effort to predict
future thrusts of computer technology development and the impact
these developments will have on future weapon system design concepts.
The approach used was to identify and classify the areas upon which
logical estimates of risk/benefits could be made to assess future
investment strategies and to also identify existing R&D programs which
would serve as technology drivers for the identified areas. This study
is significant since it was created and conducted under the general
direction of the computer science community and not the NFE community,
although the study directors recognized that the man-machine-interface
technology plays a significant role in making the predictions consistent with three primary objectives
1. Forcast the advancement of computer and computer-related telecom-
   munication technologies.
2. Assess the potential impact of these technology advances on the
   capabilities of existing or future weapon systems through the year
   2000.
3. Determine the policies and R&D initiative required to bring these
   technology advances to fruition and to incorporate them in future
   weapon systems capabilities.

This study was known as COMTEC-2000 for Computer Technology Forecast and Weapon System Impact Study for the next 20 years.

COMTEC-2000 MAN-MACHINE INTERFACE TECHNOLOGY SHORTFALLS

A special panel was formed as part of the overall study effort
to identify and project man-machine interface technology shortfalls.
The panel was chaired by Donald L. Monk from our Laboratory, with
members representing Flight Dynamics Laboratory, Human Resources Laboratory and the Mitre Corporation. Fig. l shows the human computer
technology shortfalls roadmap which the panel developed to define
the drivers of the technology needs which in turn determine the four
(4) man-computer interface areas of: Information Exchange, Human-Computer Symbiosis, Standards and Guidelines, and Design and Evaluation
Methodologies. These technology area mechanisms were evaluated in
terms of state-of-the-art to yield the final products for better and
more effective computer interfaces designed for man.

Fig. 2 shows the risk/payoff matrix for five (5) subareas of information exchange and Fig. 3 expands these five areas in a roadmap
which shows how current man-machine interface research in head/eye
tracking, neurophysiological/measurement (EEG), pilot workload as.
sessment, multi-function keyboard design, and voice control techniques

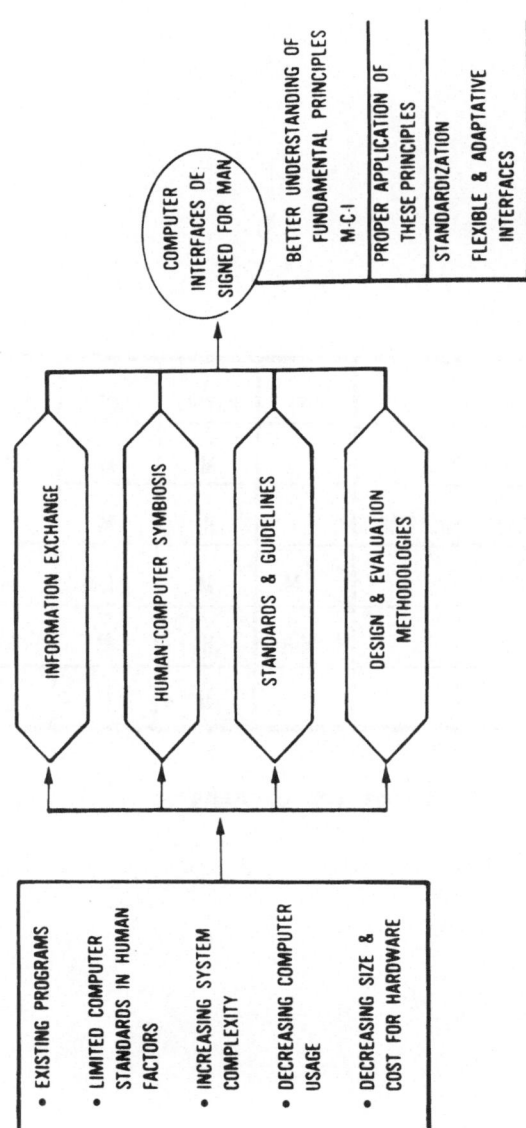

Fig. 1. Human-computer technology shortfalls roadmap

AREA	RISK	PAYOFF	COST	NEED
GRAPHIC TECHNIQUES	L	M	M	M
MULTIFUNCTIONAL INTEGRATED SYSTEMS	L	H	M	H
SELECTION/CONTROL DEVICES	L-M	M	L-M	M
NATURAL LANGUAGES	M-H	H	M	M
IDEOGRAPHS	L	M	L	L-M

Fig. 2. Information exchange

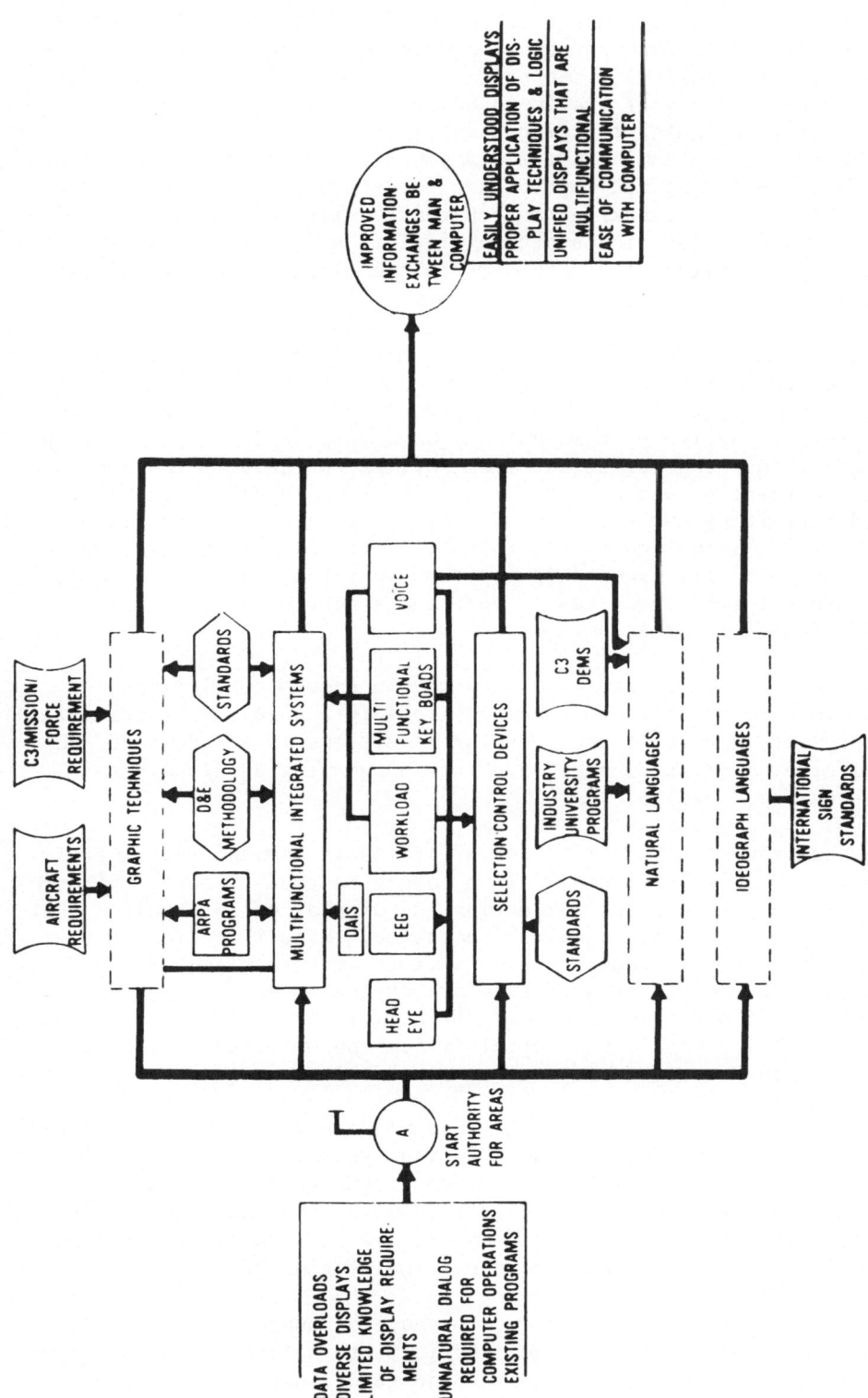

Fig. 3. Information exchange roadmap

serve as the technology base to feed the five (5) areas of improved
human-computer interface information exchange. From Fig. 2 the areas
needing higher priority on the basis of low to medium risk and cost
vs medium to high payoff are improved graphic techniques and more
operator oriented multi-functional integrated systems. Computer graphic techniques are becoming extremely powerful and present the potential for manipulating and controlling perceptual cue to enhance the
effects of size, distance and motion constancies for specific display
renditions. Multi-function systems (displays and controls) must be
designed for compatability with certain human cognitive capabilities
and limitations including short-term memory.

Figs. 4 and 5 show the risk-payoff matrix and associated roadmap respectively for the area of human-computer symbiosis. It would
appear that the potential for developing imbedding training techniques with interactive systems has not been fully exploited by the human factors and computer science communities. Principles for automatic self correcting and operator prompting techniques should be developed exploiting current knowledge of adaptive aiding techniques
and artificial intelligence algorithms. Adaptive techniques could be
used to automatically sense high operator workload periods to tempo-
rarily store information cues to pace and synchronize with operator
capacity/demand information.

One of the major thrusts in new weapon system developments is
in the command, control and communications (C^3) area. The requirement for voice control techniques for C^3 systems is an advanced HFE
technological need which emphasizes the burgeoning developments in
"human centered" design.

A requirement also exists for a "Trainable Command and Control
Information Processing System". In a modern tactical threat environment, it is not simply the collection and display of information that
must be automated, but, of equal importance, the utilization of that
data. A target nomination should be able to trigger a sequence of
semi-autonomous processes sufficient to suggest a mission profile for
the approval of a responsible duty officer. It is necessary that the
future tactical command and control systems be modified on-line in
response to unforeseeable demands of the battle situation.

Since C^3 data base systems are becoming to gargantuan, it is necessary to develop "knowledge-based fusion systems" where fusion is
not only the merging of multi-sensor data, but the interpretation of
these data achieved by their integration with other data and knowledge of a symbolic nature. The need for new ways to display and integrate these fusion systems present a challenging HFE problem.

Even given the current embryonic state of human-computer interface design it would appear urgent to initiate national and international programs to develop standards and guidelines for symbology,

AREA	RISK	PAYOFF	COST	NEED
FUNCTIONAL ALLOCATION	M	H	M-H	M
EMBEDDED TRAINING	L-M	M-H	M	M
HUMAN INTERACTION WITH AI	M-H	H	M-H	L

Fig. 4. Human-computer symbiosis.

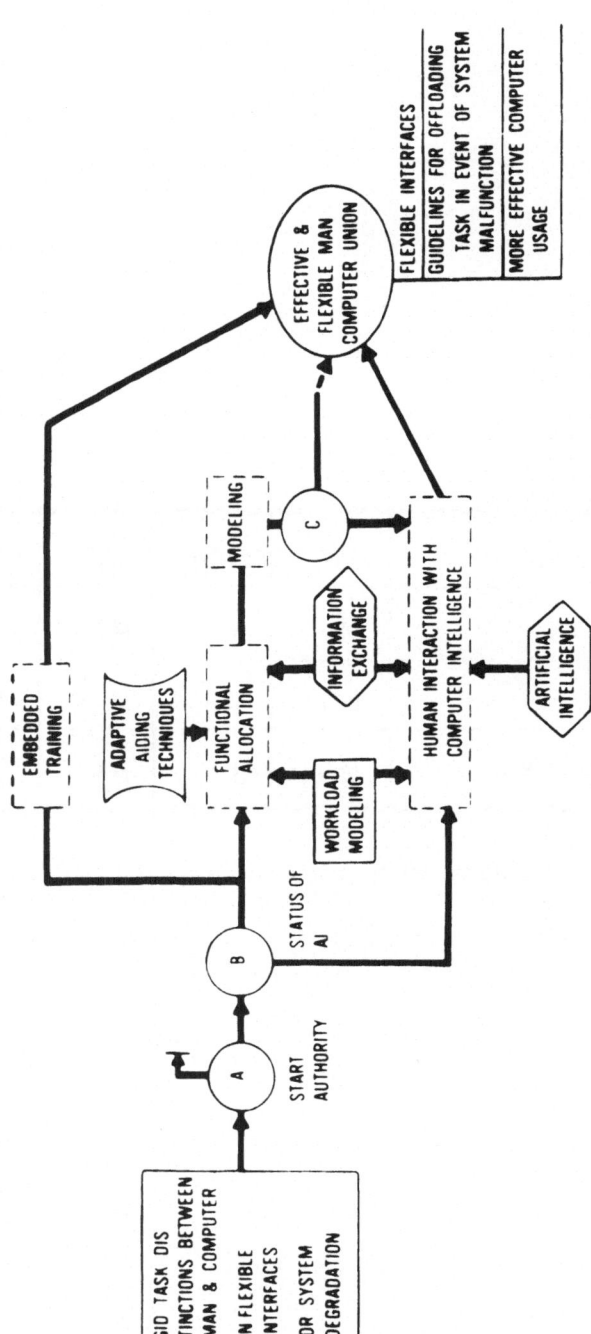

Fig. 5. Man-computer symbiosis roadmap.

display formats, command languages and equipment with Human Factors Engineers and Ergonomists playing a central role in their development. Figs. 6 and 7 again outline the risk/payoff analysis and the roadmap for developing these standards and guidelines. Certain triservice (Army, Navy and Air Force) programs including the Joint Tactical Information Display System (JTIDS), Digital Avionics Information System (DAIS) and various Command/Control programs designed for interoperability could presently use preliminary standards and guidelines in these areas. Eventually, the DOD developed standards will require international coordination and agreement for NATO joint force implementation. It is probably not too early to establish NATO committees to initiate preliminary development of international standards in the areas.

Finally, programs should be established to develop design and evaluate methodologies. A listing of eight (8) subareas to be addressed in such programs are included in Fig. 8 with the proposed roadmap for integrating these efforts in Fig. 9. Function and task taxonomies are needed along with integrated modeling approaches to quantitatively specify human-computer interface requirements to anticipate new conceptual design requirements for advanced computer-based systems with the ultimate goal of predicting overall systems performance and effectiveness well in advance of committing to a particular design configuration.

The COMTECH-2000 Summary Report concludes with the following statement: Man-machine interface will improve in response to the commercial market competition for sales to the layman of increasingly sophisticated devices. The trend will be for devices to be self-describing, assisting, the uses in their operation and maintenance. A continuing important, relatively unchanging role for men in the Air Force is forecast; and the national manpower pool from which they will be drawn will increasingly have computer experiences – it is estimated that by 1985 some 75 percent of the nations' work force will be working with computers. Man-machine dialogue will improve as better models of the data base, the uses, and his objectives are incorporated into the computer. Modalities of dialogue will range from keyboard, light-pen, and touch, through tablet and voice, to eye-movement and electroencephalogram. The latter three are of potentially great interest to the Air Force and may not fully develop from commercial R&D alone. A more intelligent, situations-dependent use will be made of display space. Similarily, programmable manipulands (e.g. soft-copy keyboards) will be developed by the commercial sector. Finally, there is a tantalizing prospect of computer mediated translations (in a limited context) from language to language and/or from verbal to pictorial representaions (ideographs).

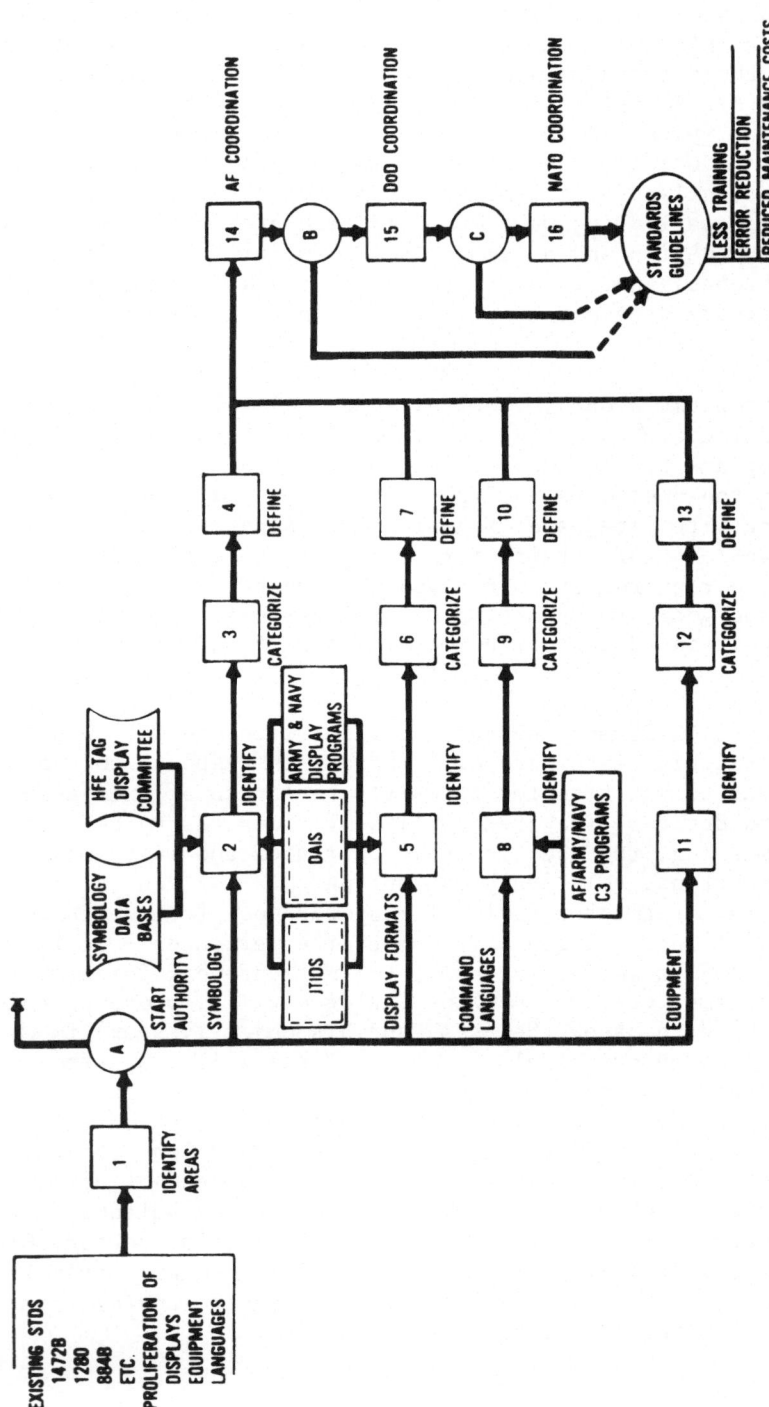

Fig. 6. Standards and guidelines roadmap.

AREA	RISK	PAYOFF	COST	NEED
SYMBOLOGY	M	H	L-M	H
DISPLAY FORMAT	M	M	L-M	M-H
EQUIPMENT	L	M	L	M
COMMAND LANGUAGES	M	H	L-M	H

Fig. 7. Standards and guidelines.

AREA	RISK	PAYOFF	COST	NEED
FUNCTION TAXONOMY	L	M-H	L	M-H
PROBLEM/APPLICATION TAXONOMY	L	M-H	M	M-H
GENERIC TASK TAXONOMY	M-H	H	M-H	H
GENERIC MODELS	M-H	H	M-H	H
GENERIC H.O. MODELING	M	M-H	M	M-H
HUMAN FACTORS METRICS	M	M-H	M-H	M-H
DESIGN GUIDE	M	H	M	H
SIMULATION/MODELING TESTBEDS	M	H	H	H

Fig. 8. Design and evaluation methodologies.

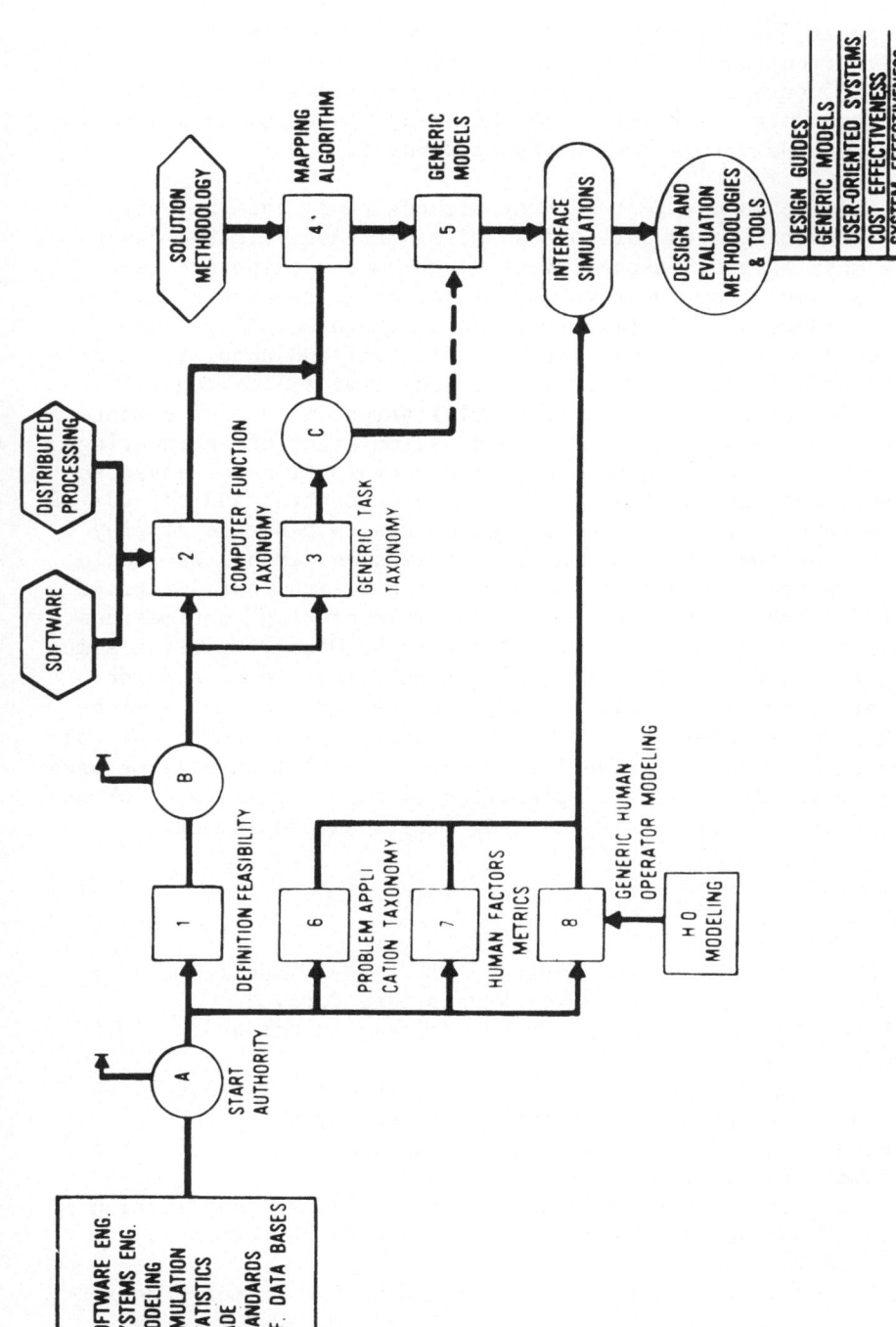

Fig. 9. Design and evaluation methodology roadmap.

SUMMARY AND CONCLUSIONS

This paper has attempted to review some of the historical development, current state-of-the technology and practices as well as estimates of future technology directions and trends in methods, techniques and data bases for man-machine interface design and their interrelationships with system design considerations.

Past, and to some extent current methods have, and are, using primarily the established data bases and manual analytic and design tools. The more advanced technologies which employ computer-based analysis and simulation await full application to the design process pending acceptance and technical upgrade of the practicing human factors engineers working in the system development and acquisition process. It is anticipated that within the next five (5) to ten (10) years we will see an upsurge in the exploitation and use of computer-based simulations and modeling, a rapid assimulation of engineering design simulation employing man-in-the-loop evaluations of advanced man-machine interface concepts including voice control and the use of adaptive neurophysiological control techniques to generate display information requirements as a function of mission demand. We should also see an increased emphasis on man-machine design considerations in overall "cost-of-ownership" and "life-cycle-costing" estimations earlier in the design and development sequence. Overall mission simulation capability will be an increasing requirement for system development programs. It is also anticipated that greater use will be made of "hybrid" simulation methods where the combined powers of computer-simulation and man-in-the-loop physical simulation will be used to anticipate the increasing complexities of operational tasks which are being driven by more sophisticated threats and equipment.

REFERENCES

Chapanis, A., Garner, W.R. and Morgan, C.T., 1949, "Applied Experimental Psychology", John Wiley & Sons, New York.

Computer Technology Forecast and Weapon Systems Impact Study, 1978, AFSC TR-78-03, Vol. I & II.

"Handbook of Human Engineering Data", 1951, 2nd ed., Tufts College Institute for Applied Experimental Psychology.

"Human Engineering Guide to Equipment Design", 1972, rev. ed., U.S. Government Printing Office.

Schwartz, M.A., Leuba, H.R. and Maltese, J., 1967, The Application of Operations Research Models to Man-Machine Performance, Unpublished Technical Report.

Siegel, A.I. and Wolf, J.J., 1969, "Man-Machine Simulation Models - Psychosocial and Performance Interaction", Wiley-Interscience, New York.

Tsokos, C.P. and Thrall, R.M., 1979, "Decision Information", Academic Press, New York.

Williges, R.C. and Topmiller, D.A., Human Factors Engineering Technology Assessment, AFSC TR, in press.

A RESEARCH PROGRAM FOR THE PROPER INCLUSION OF HUMAN RESOURCES IN

THE DESIGN PROCESS

Ben Ostrofsky

PURPOSE

The fundamental purpose of this research was to clarify the decision structure and methodology for the design of high technology, large scale systems with particular emphasis on the proper integration of human factors and their associated metrics. This effort was initiated after many years of study and development of a methodology (Ostrofsky, 1977a) that successfully integrates the necessary decisions to efficiently utilize resources in order to meet a design need. A morphology of design emerged from which economical applications can readily be achieved while properly integrating the constituent disciplines.

Of major concern, however, was the proper integration of human factors into emerging systems so that performance criteria for these systems would properly reflect the influence of these factors. Traditionally, equipment designers appear to be reluctant to accept this "soft" data as input to the design of equipment, particularly when there is difficulty in meeting the more easily identified hardware performance requirements. Hence the improper inclusion of human factors in equipment affects results, not only in the achievement of operational performance, but also in the maintenance of this performance. What is needed, then, is a methodology that allows for the explicit inclusion of these factors into the design process. Prior study (Ostrofsky, 1977a; Asimow, 1962) has resulted in such a process, and, while the U.S. Air Force has had similar procedures (AFSCM, 1966), they have experienced difficulty in properly integrating human factors into their systems.

OVERVIEW OF THE DESIGN MORPHOLOGY

The design morphology can be considered as the form or structure of the process required to establish and meet the defined needs.
The definition of such a structure can provide a means for more
effective planning, especially when the morphology is defined to a
depth which "structures" the form and content of the respective decisions at each step in the design process.

Asimow (1962) defined the life cycle of a technological system
to have the phases shown in Fig. 1.

The relationship of the design process to the production-consumption cycle satisfies the notion of relating designer planner activities to user needs. The cycle activities are of particular interest to the system designer and also appear to be one of the main
patterns in the socioecological system which encompasses the resulting design. Since the elements of the Production-Consumption Cycle
relate to user activities, it becomes necessary for the designer to
understand the nature of these activities and then relate existing
technology to them so that an effective solution to the problem e-
merges.

In attempting to plan for the most efficient use of resources
the design phases are adapted for the designer. The Primary Design
Phases of Fig. 1 must consider the Production-Consumption Cycle, and,
in fact, the success of the design can often be considered as a function of the degree to which the requirements of this cycle are included.

PRIMARY DESIGN PHASES   PRODUCTION-CONSUMPTION PHASES

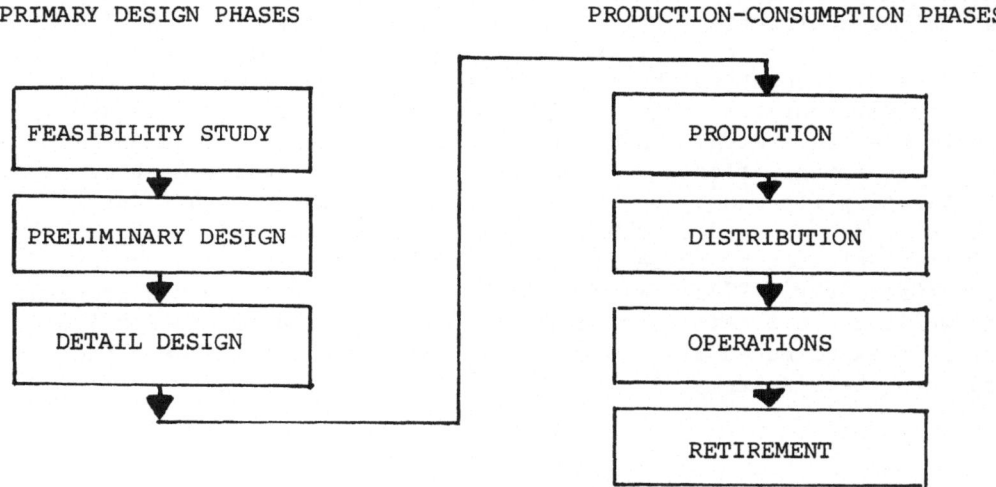

Fig. 1. Phases in the design project life.

A project begins with the Feasibility Study in order to establish a solid base from which to build ideas in depth while progress-
ing through the design process. The Feasibility Study has for its
purpose the achievement of a set of useful solutions to the design
problem. This implies, of course, that a problem has been defined,
so that a major set of activities in this phase is the establishment
and verification of need and the explicit definition of the problem
to be solved by the design. The Feasibility Study not only serves as
the keystone around which all subsequent steps are made, but it actually accomplishes several of the initial steps (see Fig. 2), including the synthesis of solutions and the screening of these solutions.
Hence the set of solutions that emerges contains the alternatives to
be compared and evaluated. If no solutions emerge from the screening
then the problem solution is "not feasible". Consequently, the completed study provides the problem needs, an identification of the design problem, and a set of useful solutions.

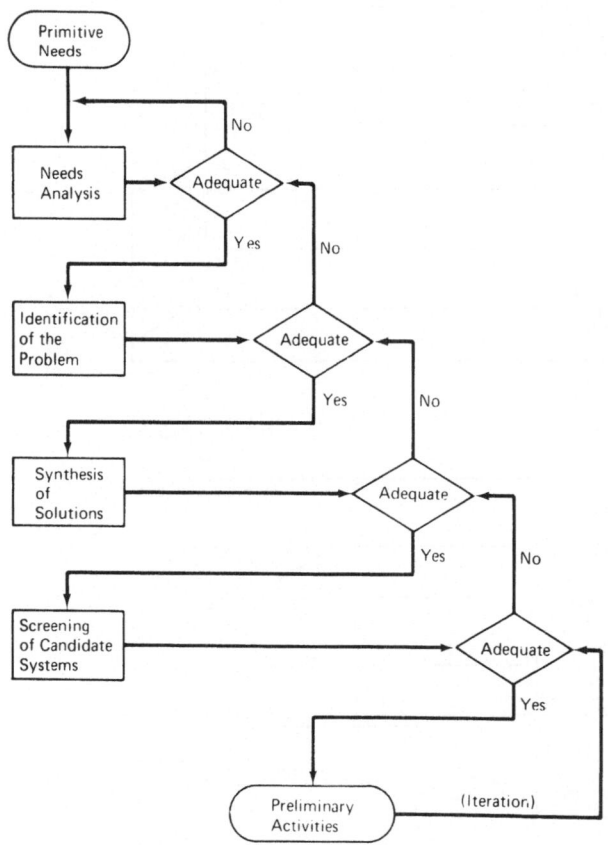

Fig. 2. The feasibility study.

Given the Feasibility Study results in the form of a set of useful solutions, the activities involved with choosing the "best" alternatives become the purpose of the preliminary activities. "Best" must
be defined in terms of criteria delineated explicitly, and to do this,
criteria must be related to parameters and attributes of the alternatives; laboratory or field testing occurs to verify relationships of
the parameters to the criteria; value must be assigned to the performance of each criterion for the respective alternative, and the "best"
alternative selected. Note that "best" implies a choice from among

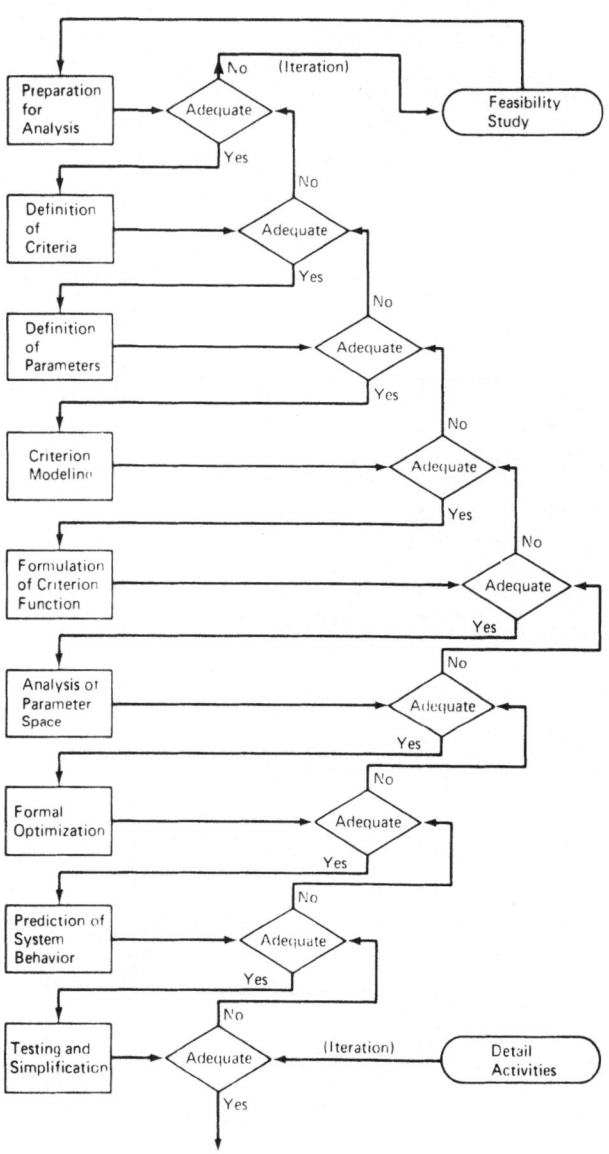

Fig. 3. The preliminary design.

those alternatives defined. Analytical aids to the identification of
a theoretic optimum alternative have been described (Ostrofsky, 1977a).
Hence strong, analytical indications are made to the designer of the
adequacy of his choice. Fig. 3 shows the preliminary design decisions.

Once the alternative has been chosen and tested, the problems of
implementation commence. For hardware systems the detail activities
must emerge with a complete description of a tested and producible design so that it can accomplish the requirements of the Production Consumption phases, in appropriate quantities, and in a planned manner.
Product-oriented systems require much detailed planning and preparation to be accomplished prior to implementation, and hence, the Detailed Phase (see Fig. 4) includes this planning.

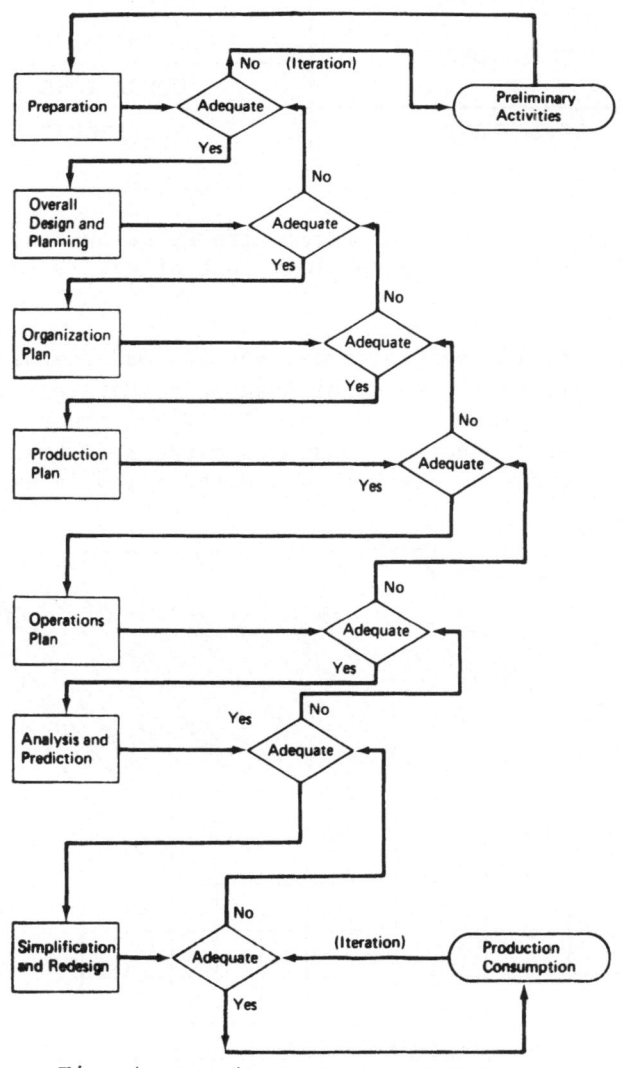

Fig. 4. Detail design activities.

SCOPE OF THE RESEARCH

The investigation started with a study to relate the design morphology to the U.S. Department of Defense Life Cycle (see Fig. 5).

	AIR FORCE LIFE CYCLE PHASES	DESIGN MORPHOLOGY PHASES
DESIGN	CONCEPTUAL	FEASIBILITY STUDY
DESIGN	VALIDATION	PRELIMINARY ACTIVITIES
DESIGN	FULL SCALE DEVELOPMENT	DETAILED ACTIVITIES
PRODUCTION CONSUMPTION	PRODUCTION	PRODUCTION
PRODUCTION CONSUMPTION	OPERATIONS	DISTRIBUTION
PRODUCTION CONSUMPTION	OPERATIONS	OPERATIONS
PRODUCTION CONSUMPTION	RETIREMENT	RETIREMENT

Fig. 5. A comparison of U.S. Air Force Life Cycle phases to the phases of the design morphology (Ostrofsky, 1977c, p. 20).

Having shown this relationship a comparison was made between Meister's human factor categories with those of Blanchard (Meister, 1971) (see Fig. 6). A listing was then prepared of all major, generic elements for each phase of the production-consumption cycle (Ostrofsky, 1977c, p. 63 a, b, c, d), and each element considered a low in an input-output matrix.

MEISTER'S CATEGORIES	HUMAN CAPABILITIES	BEHAVIORAL CONSIDERATIONS	PERSONNEL COST	TRAINING LEVEL	PERSONNEL PERFORMANCE	TEAM PERFORMANCE	MAN-MACHINE INTERFACE	PERSONNEL BACKGROUND	PERSONNEL READINESS	PERSONNEL QUALIFICATIONS	BASELINE DATA	OPERATING ENVIRONMENT	EXTERNAL ENVIRONMENT
TASK	X	X	X	X	X	X	X						
PERSONNEL		X	X	X				X	X	X			
EQUIPMENT	X		X			X							
ENVIRONMENT											X		

Fig. 6. The human factor categories versus Meister's classification (Ostrofsky, 1977c, p. 54).

While these studies were being accomplished, an annotated bibliography was prepared from 317 references (Ostrofsky, 1977b) and these abstracts were classified by Blanchard's human factor categories (Ostrofsky, 1977c).

Finally, to stay within the scope of resources available at that time, one element of each of Production, Distribution, Operations and Retirement was examined and developed in detail for the input-output matrix. Disciplines and relevant areas were defined for each sample element in each of Intended Input, Environment Input, Desired Output and Undesired Output and a comparison made of the references with each disciplines and/or element shown for each sampled element of the Production-Consumption Cycle, and a study made of those areas lacking references as they related to Blanchard (Blanchard, 1967) (see Fig. 7). Of interest is the result that, for the available titles and their annotation, only one related to team performance, two to external environment, and seven each to Personnel Cost and Personnel Background. Reasons for this are partially due to the limited time available for this study, but may indeed be due, in part to a lack of available publications in these areas.

At this point in the research, direction was offered toward the exercise of this morphology in a contractor environment, and when this was successfully accomplished on the Support Stand for the Emergency Power Unit of the F-16 aircraft, the methodology was then applied to a less, well structured system. After some search, the Maintenance Control Activities of the new Missile X were used. These studies are described below.

HUMAN FACTOR CATEGORY	No. of Ref.
1. Human capabilities	54
2. Behavioral considerations	19
3. Personnel cost	7
4. Training level	21
5. Personnel performance	53
6. Team performance	1
7. Man-machine interface	60
8. Personnel background	7
9. Personnel readiness	4
10. Personnel qualifications	9
11. Baseline date	38
12. Operating environment	42
13. External environment	2
	317

Fig. 7. Number of references annotated for each human factor area.

Results of the initial study indicated that the design morphology does, indeed, apply to the development of high technology systems
in that it provides an orderly and rational sequence of decisions requiring solution. Further, the study identified a significant communication problem between designers of equipment and human factor specialists.

In structuring the relationship among human factors, the design
decision structure of the morphology and the current literature, an
approach was developed for a complete human factors reference base
in the context of system design and development. From the initial
study, the human factors literature did not appear to cover the morphological steps in a uniform matter and apparent gaps appear to
exist in the literature relating human factors to the lifecycle process of a system. Hence the need for defining means of estimating
human factors effect on the design process was reinforced.

Approaching the input-output matrix in this elementary manner
served to alert the designer to the major considerations for his
equipment in the planned environment. The effort devoted to these
matrices was very limited since the nature of the equipment was relatively unsophisticated. However, should areas of concern arise during each study, an independent study activity could be accomplished
to clarify the problem definition. Such studies were not needed for
the EPU Service Stand.

A design concept is defined as a basic approach toward solving
the requirements problem, while a candidate system is a particular
alternative of the given concept. Fig. 8, the flow diagram representing the activities for the EPU Service Stand, has already defined
both the activities and the major decisions in the flow sequence.
The concept, then, is simply the delineation of the equipment functions at the level usable by the designer to define configurations.
Obviously, there are many concepts to solve a given equipment design
problem, the number growing exponentially with the number of equipment functions defined.

Candidate systems are developed by considering each equipment
function independently and exhaustively listing the alternatives for
them. A candidate system can then be defined by combining one alternative from each of the concept functions such that every function
defined would be accomplished if this combination of alternatives
could be realized. Fig. 9 shows the concept and the result of defining alternative methods for accomplishing each function and identified over 3.8 billion combinations, a number too large to consider
evaluating each one separately.

After some study the designers were able to intuitively eliminate more costly alternatives, leaving those that were obviously desirable as well as those that were questionable. Only the most unde-

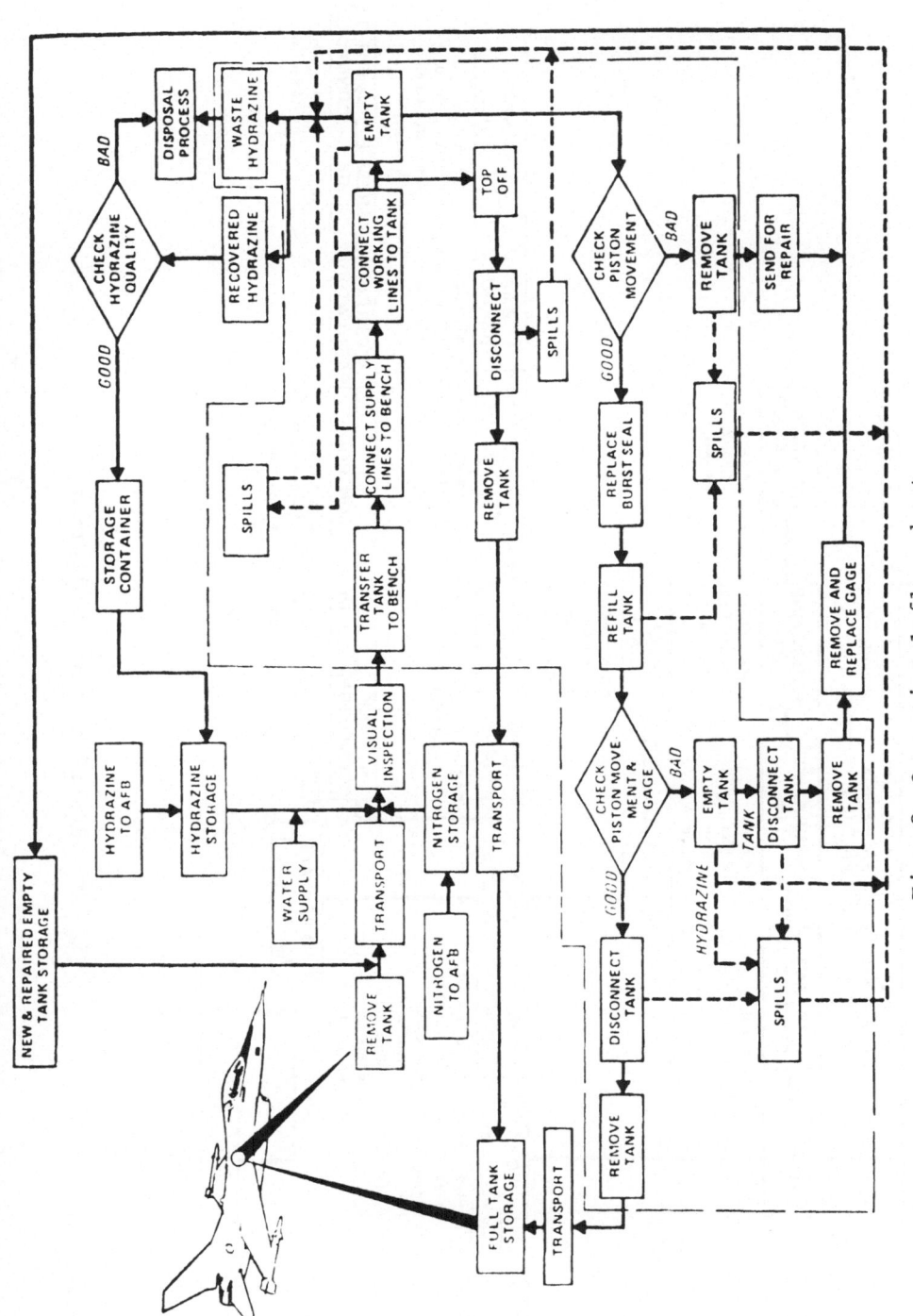

Fig. 8. Operational flow chart.

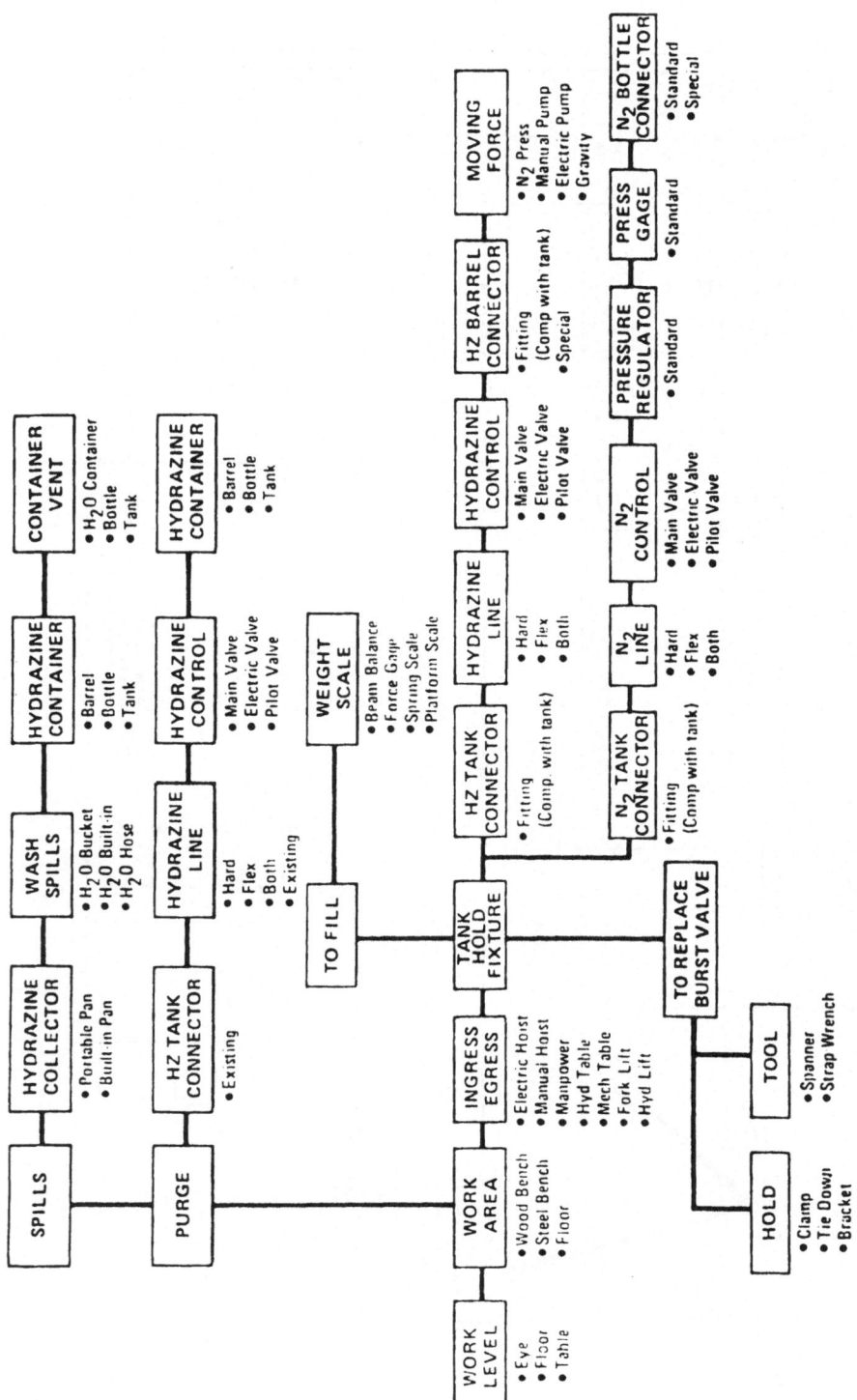

Fig. 9. Identification of alternatives.

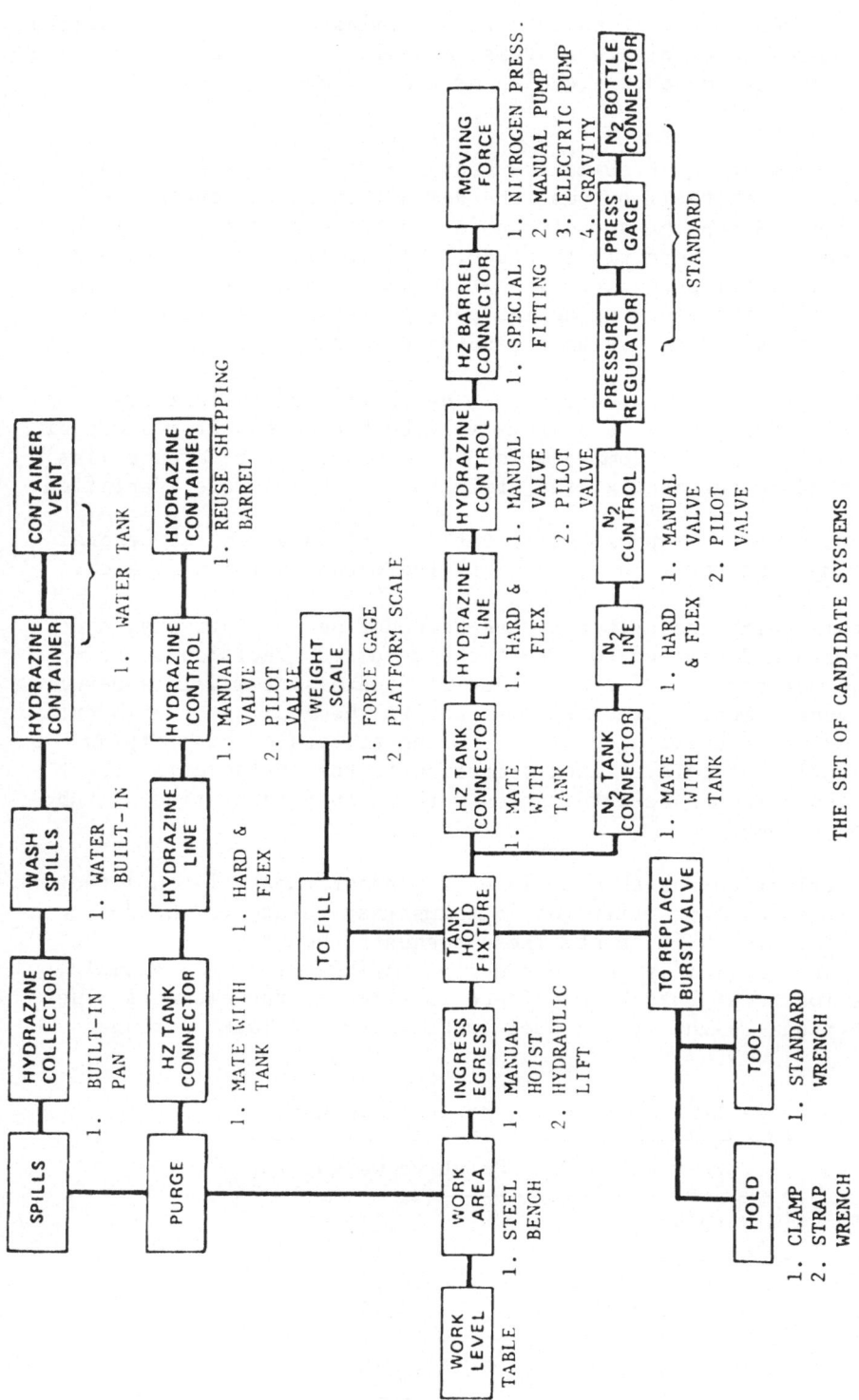

Fig. 10. The set of candidate systems.

sirable or infeasible combinations were eliminated. After this screening there were 128 candidate systems remaining, and this set provided the basis for the subsequent choice of the optimal configuration (Fig. 10).

At this point the Preliminary Design activities were initiated and these have for their purpose the definition of the optimal candidate system. Since the set of candidates to be studied has already been defined in the Feasibility Study, this Preliminary Design encompasses the activities required to define and to analyse the design space formed by the emerging design parameters and the criterion function synthesized to evaluate candidate system performance.

The initial task is the explicit definition of the criteria against which the candidate systems will be evaluated. From study of the Support Equipment Recommendation Data (SERD; see reference list) and the input-output matrix, the criteria of Table I were identified.

A survey of the support equipment design management at General Dynamics imparted the relative weights indicated in Table I below.

At this point the criteria have been defined narratively, and must be further developed to permit the required precision in comparing the performance of the various candidate systems. This development process imparts precise semantics to the meaning of each criterion, and the precision results from the process of quantifying (or modeling) that relates each criterion to the characteristics of the equipment. These characteristics were further identified as submodels and parameters.

Each criterion was then analyzed for constituent elements that would help define that criterion in terms that relate to the design equipment and its environment. These elements then serve to define the criterion explicitly for this design optimization. For example, the criterion Safety has been defined in terms of the elements Ease of Maneuvering, Weight of the Fuel Tank, Volume of Tank, Arrange-

Table I. Criteria & relative weights

Criteria $\{x_i\}$	Relative weight $\{a_i\}$
1. Safety	.318
2. Cost	.122
3. Ease of use	.174
4. Durability	.126
5. Producability	.122
6. Availability	.138
	1.000

Table II. Criteria, submodels and parameters

	1. SAFETY		2. COST	3. EASE OF USE		4. DURABILITY		5. PRO-DUCABILITY	6. AVAIL-ABILITY
	EASE OF MANEUVERING	PROBABILITY OF LEAKAGE		MANHOURS FOR SERVICING TANK	EASE OF MANEUVE MANEUVERING	NUMBER OF OPERATING CYCLES	EASE OF SHIPMENT		TIME TO MAINTAIN STAND
1. No. of connectors s	X	X							
2. Weight of tank & H-70					X				
3. Production manhours per unit			X						
4. mean cost per purchased part			X						
5. Manhours for servicing EPU tank				X					
6. Simplicity of procedures				X					
7. Readability of gauges				X					
8. Simplicity of waste disposal task			X	X					
9. Ease of shipment							X		
10. Life of F-16 program						X		X	
11. No. of purchased parts								X	
12. Time to maintain stand									X
13. Total no. of parts									
14. A/C flight hours per month					X				
15. No. of A/C per stand					X				

ment of Controls, and Probability of Leakage. An assessment was then made of the most effective manner in which to quantify each of these elements (called "parameters").

Table II shows the relationship of the parameters fo each sub-model and criterion. This arrangement provides increased visability to the accomplishment of completeness and compactness studies and is helpful toward the assurance of consistency among the parameters.

The methods described by Ostrofsky (1977a, p. 95-106) were used to model the criteria, and these models serve to attribute semantics to each criterion. Hence each is defined in terms of the relationships and assumptions delineated in the modeling exercise. Details in the development of each criterion model are presented in Ostrofsky (1978). Then, in order to prepare for the synthesis of the Criteria Function, Table III was structured. The parameters were defined from Table II and resulted from the elements used to define the criteria. Note that several of the parameters were held constant for all candidate systems (i.e.: Y2, Y7, Y10, Y14, Y15). These constants were defined in the USAF equipment performance documentation provided General Dynamics and hence no latitude was permitted at this point. However, during the subsequent analysis of the design space these constants were varied in order to observe their effects on the total criteria function.

Considerable study was accomplished by the General Dynamics engineers to arrive at meaningful values for the ranges shown in Table III. It is further observed that additional study might have resulted in additional parameters, but lack of time caused curtailment of this activity.

Table III. Range of parameters, y_k

k	y_k	minimum	maximum
1.	Number of Connectors	61	70
2.	Weight of Tank and H-70	--------110------	
3.	Production Man-hours per Unit	499 mh	935 mh
4.	Mean Cost/Purchased Part	$32.6/part	$40.2/part
5.	Man Hours for Servicing EPU Tank	2.7 mh	8 mh
6.	Simplicity of Procedures	2	10
7.	Readability of Gages	----------3------	
8.	Simplicity of Waste Disposal Task	3	7
9.	Ease of Shipment	4	15
10.	Life of F-16 program	--------300------	
11.	Number of Purchased Parts	93	103
12.	Time to maintain Stand	1.38 min/day	1.65 min/day
13.	Total Number of Parts	142	173
14.	Aircraft Flight Hours/Month	--------30------	
15.	Number of Aircraft/Stand	--------72------	

Table IV. Range of Criteria

i		x_i	$x_{i\ min}$	$x_{i\ max}$
1.	Safety		0.00039774	0.00102666
2.	Cost		$3733.00	$6449.00
3.	Ease of Use		0.002646	0.059905
4.	Durability		31.695	445.716
5.	Producability		0.537540	0.725326
6.	Availability		0.99885	0.99904

The criteria ranges must be defined to implement the particular form of the criteria function exercised in this study. In order to estimate the maximum and the minimum of each criterion, the criterion models were exercised using the appropriate value of each required parameter from Table III. Hence the submodels were used only to achieve values of the respective criterion (x_i; i = 1, ..., 6) and Table IV resulted.

In order to adequately compare the candidate systems, the relative weights, a_i, and the criteria, x_i, must be synthesized into a single function so that a figure of merit can emerge. This figure of merit, the Criteria Function represents the performance of a given candidate system when a particular alternative yields $\{y_k\}$, the set of parameters representing that particular configuration.

When the criteria x_i are examined it becomes clear that some way of handling the criterion units must be included in the function. For example, x_i, safety is measured in units of probability, volume, and weight,; x_2 is measured in inverse of dollars, x_3 is on a subjective scale, etc. Hence some method for relating the sensitivity of the unit value of x_1 with the unit value of each remaining x_i must be used. If this is accomplished improperly, the resulting combination of these criteria will not be meaningful.

A basic consideration is that the criterion function is the vehicle for comparing the values resulting from the candidate system. Hence a requirement for this function is that it should present the performance of a candidate for its parameters in units that are consistent for all criteria. Such a vehicle is obtained by identifying criterion performance as a fraction of the allowable range for that criterion:

$$\bar{x}_i = \frac{x_i - x_{i\ min}}{x_{i\ max} - x_{i\ min}} \tag{1}$$

Here x_i represents the ith criterion performance of a given candidate system, hence the numerator represents the "distance" from the minimum value of x_i that a given candidate system yields for that respective x_i, and the denominator is the range of the criterion's performance. Hence eq. (1) represents the fraction of the criterion range that a given candidate system will yield as its performance.

When this fraction is given its relative importance, a_i, the product $a_i x_i$ represents the weighted or relative value of the ith criterion. These can be added to give:

$$CF_\alpha = \sum_{i=1}^{6} a_i x_i \tag{2}$$

where CF_α = the value of the criterion function for the α candidate system.

There now exists a method for assessing the performance of the candidate system when parameter values are identified since

$$x_i = f_i \{z_i\} \tag{3}$$

and $$z_i = g_i \{y_k\} \tag{4}$$

hence $$x_i = f_i \{g_i \{y_k\}\} \tag{5}$$

From eqs. (2) and (5) (Ostrofsky, 1977a):

$$CF_\alpha = \sum_{i=1}^{6} a_i \left[\frac{f_i \{g_i \{y_k\}\} - x_{i\ min}}{x_{i\ max} - x_{i\ min}} \right] \tag{6}$$

Eq. (7) shows eq. (6) translated for this particular design problem.

Eq. (6) and its application, eq. (7) represent a simplistic approach to the structure of the criterion function. Its major limitation is the assumption of independence among the criteria (i.e. cost independent of safety, availability, etc.). One result of the development of this criterion function is the current study (Ostrofsky, 1977a, App. c) of methods for estimating the criterion interaction effects.

$$CF_\alpha = a_1 \left[\frac{(\frac{62.5}{y_2 c_1})^{\frac{1}{2}} (1 - p_1)^{y_1} - x_{1\ min}}{x_{1\ max} - x_{1\ min}} \right]$$

$$+ a_2 \left[\frac{-(y_3 c_2 + y_{11} y_4)\ (OH)\ (LC)^N - x_{2\ min}}{x_{2\ max} - x_{2\ min}} \right]$$

$$+ a_3 \left[\frac{\frac{.316\ y_6 y_7 y_8^{\frac{1}{2}}}{y_5 y_2 c_1} - x_{3\ min}}{x_{3\ max} - x_{3\ min}} \right]$$

$$+ a_4 \left[\frac{\frac{y_{10}}{y_9^2} y_{14} y_{15} (\frac{1}{k_1} + \frac{1}{k_2} + \frac{1}{k_3}) + L_1 + L_2 + L_3 - x_{4\ min}}{x_{4\ max} - x_{4\ min}} \right]$$

$$+ a_5 \left[\frac{\frac{y_{11}}{y_{13}} - x_{5\ min}}{x_{5\ max} - x_{5\ min}} \right]$$

$$+ a_6 \left[\frac{(1 - \frac{y_{12}}{T}) - x_{6\ min}}{x_{6\ max} - x_{6\ min}} \right] \tag{7}$$

The design space for this problem is a hyperspace in eleven dimensions, one for each parameter and one for the value of the criterion function. The limitations of this space are the regional constraints imposed upon the y_k, that is their respective maximum and minimum values, and the limits of CF_α, zero and one.

A sensitivity analysis was accomplished to examine the rate of change of CF_α throughout the rage of each design parameter and computer runs made to compute CF_α at various locations throughout the design space.

Table V shows where in the design space the minimum (or lowest) percent change in CF_α was located. For example, the minimum percentage change in CF_α for y, occurred when all remaining parameters were held at the values occurring at the 25% point in their respective range. The change in the criterion function (ΔCF_α) noted was -40.4%.

Table V. Minimum % change in CF_α throughout each parameter range

k	Parameter, y_k	% of Range for other Parameters	Minimum % Δ CF_α
1.	No. of Connectors	25	-40.4
3.	Production Man-Hourse per Unit	50	-15.9
4.	Mean Cost/Purchased Part	50	-1.2
5.	Man-Hours/Servicing EPU Tank	50	-4.4
6.	Simplicy of Procedures	100	-53.9
8.	Simplicity of Waste Disposal	100	-23.3
9.	Ease of Shipment	0	-16.2
11.	No. Of Purchased Parts	100	-36.5
12.	Time to Maintain Stand	75	-19.9
13.	No. of Total Parts	75	-8.6

   Table VI shows the maximum and minimum positive change in CF_α and for each parameter. This implies a potential variation of CF_α equal to their difference and this is shown in right hand column ($\Delta\%$). Examination of this column reveals that y_1, the number of connectors, can change the criterion function as much as 399.5% throughout the design space, and hence, is by far the most critical of all identified parameters to the achievement of maximum performance of the stand as identified by the criteria function, CF_α.

   Of equal interest are y_4, mean cost/purchased part, y_8, simplicity of waste disposal and y_{11}, number of purchased parts. The maximum changes in CF_α throughout their entire range in the design space are 10.9%, 25.8% and 41.9%, respectively. Hence changes in these parameters, for the respective ranges identified, have the least ef-

Table VI. Maximum variation of CF_α for each parameter

k	y_k	Maximum % Change	% Change	$\Delta\%$
1.	No. of Connectors	359.1	-40.4	399.5
3.	Production Man-Hours Unit	120.9	-15.9	136.8
4.	Mean Cost/Purchased Part	9.7	-1.2	10.9
5.	Man Hours/Servicing EPU Tank	132.3	-4.4	136.7
6.	Simplicity of Procedures	11.6	-53.9	65.5
8.	Simplicity of Waste Disposal	2.5	-23.3	25.8
9.	Ease of Shipment	145.9	-16.2	162.1
11.	No. of Purchased Parts	5.4	-36.5	41.9
12.	Time to Maintain Stand	157.1	-19.9	177.0
13.	No. of Total Parts	93.2	-8.6	101.8

fect on the criteria function. Note that y_4, the mean cost or purchased part, and y_{11}, the number of purchased parts, are the two parameters affecting change in CF the least (least sensitive parameters).

In order to proceed, however, the optimal candidate must be i-
dentified. To this end eq. (7) was programmed to compute the CF_α for
each of the 128 candidate systems (see Fig. 10) identified above.
These were ranked in descending order (see Fig. 11). Candidate number 9, identified in Fig. 12 is the configuration of the service
stand that will be developed subject to the resolution of the problems in detail design.

At this point a search was made of the design space. At the
time of this study the software was not fully developed, and hence
the achievement of the maximum CF within the design space is not
certain. However, a CF of 0.996 was achieved with this method, and
this compares with the $CF_9 = 0.859$ achieved for the best of the 128
candidates.

Table VII compares the parameters of candidate no. 9 with those
resulting from the design space search. This table can be interpreted
to represent potential growth in system performance from the configuration emerging as "best" from among the candidates considered, and
"best" for the given parametric ranges identified. The parameter values in the right hand column of Table VII may never be achieved in

RANK	CANDIDATE	CF VALUE	RANK	CANDIDATE	CF VALUE
1	9	0.859	21	33	0.603
2	1	0.829	22	20	0.589
3	25	0.779	23	49	0.581
4	17	0.747	24	18	0.570
5	10	0.670	25	19	0.562
6	11	0.667	26	81	0.559
7	12	0.663	27	89	0.554
8	13	0.636	28	21	0.538
9	2	0.636	29	44	0.506
10	4	0.634	30	60	0.501
11	65	0.628	31	14	0.496
12	3	0.628	32	52	0.494
13	28	0.625	33	15	0.493
14	41	0.624	34	77	0.489
15	27	0.617	35	66	0.484
16	26	0.616	36	16	0.479
17	5	0.613	37	64	0.477
18	29	0.613	38	67	0.477
19	57	0.611	39	69	0.476
20	73	0.611			

Fig. 11. Ranked candidate systems.

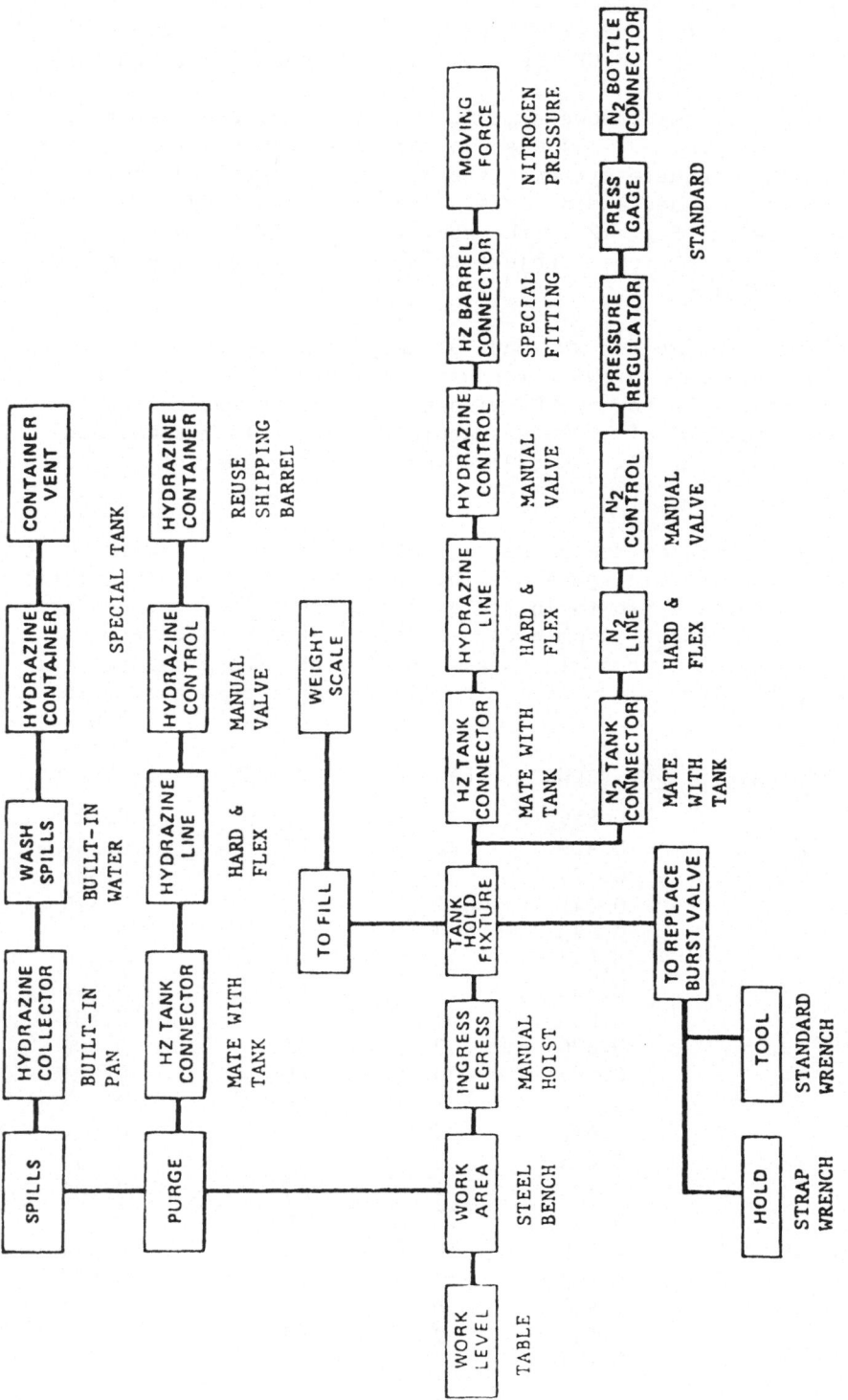

Fig. 12. Candidate system no. 9, CF_α = .859 (Rank = 1).

Table VII. Comparison of parameters from best of 128 candidates with those resulting from design space search

y_k		Candidate no. 9	Optimal From Design Space
1	Number of connectors	61	61
3	Production man-hours/unit	559	499
11	Number of purchased parts	93	103
4	Mean cost/purchased part	$ 33.70	$ 32.60
5	Man-hours for servicing	4.0	2.7
6	Simplicity of procedures	6	10
13	Total number of parts	143	142
8	Simplicity of waste disposal	7	7
9	Ease of shipment	4	4
12	Time fo maintain unit	1.38	1.42

practice, but they represent performance goals achievable from iteration in the design that change the parameters to those values shown.

To achieve the theoretic value, CF = .996:

1. production man-hours per unit must be reduced to 499 from 599.

2. the number of purchased parts can be increased from 93 to 103 while their mean unit cost must be reduced by $ 1.10 from $ 33.70 to 32.60.

3. man-hours for servicing the stand must be reduced from 4.0 m.h. to 2.7 m.h.

4. simplicity of procedures must be increased from an index of 6 to 10.

5. total number of parts should be reduced by one.

6. the time to maintain the unit might stay at its current value, but can be increased without penalty to 1.42 minutes per day.

APPLICATION OF THE DESIGN MORPHOLOGY TO AN UNSTRUCTURED PROBLEM: - MAINTENANCE CONCEPT OF MISSILE X

Having successfully demonstrated the application to a relatively well defined design problem, agreement was obtained to pursue application to an unstructured problem. In order to meet these requirements the definition of the Operational Control Center (OCC) activities for processing maintenance status change through dispatch, com-

pletion of corrective action, and post-dispatch debriefing were identified as the areas of application for the MX System. The design morphology was to be applied to the definition of optimal Fault Detection and Dispatch (FDD) activities for meeting the needs resulting
from these areas of the MX Systems. It was quickly recognized that
the magnitude of the MX System and the uncertainties assocoated with
its development necessitated extending the study through two years.
Hence final results have not yet been achieved. However, the following activities have been completed (Ostrofsky, 1979). Notice that
these are the same morphological steps accomplished on the design of
the Support Stand.

1. Basic requirements were defined; maintenance control activities
   within the OCC broadly identified.

2. A study Plan developed from the design morphology.

3. An input - output matrix for problem formulation was constructed.

4. The basic activity sequence defined.

5. Three basic scenarios developed for accomplishing maintenance:
   a. Fault Detection and Analysis in OCC
   b. Fault Detection and Analysis in the Alert Maintenance Facili-
      ty (AMF)
   c. Fault Detection and Analysis in the Strategic Missile Sup-
      port Base (SMSB)

6. Advantages and Disadvantages of each scenario were defined.

7. Alternatives to each function were identified, resulting in 60
   candidate systems for each of the three scenarios.

8. Criteria were defined by interrogating the project officers.

9. 94 Parameters, 12 submodels, and 5 criteria were defined and
   modeled.

10. Each of the 94 parameters were estimated for each of the 180
    candidate systems.

11. Criteria Function was developed and programmed.

12. Trial runs made with the parameter estimates.

HUMAN RESOURCE CONSIDERATIONS

     The "systems orientation " resulting from the morphology tends
to integrate man-machine considerations when properly approached.

Further, such an approach assures complete a priori examination of the production deployment, and operational problems.

On the F-16 project it was observed that the designer instructions (SERD) tend to identify equipment functions only, and when the morphology was used, the designer was required to consider both man and equipment functions in an integrated manner to accomplish the design objectives.

Of particular interest was the total lack of designer concern where the requirements originated. Once verified in the problem identification, however, the designers proceeded to include all requirements. Hence, human factors considerations were not discussed per se, but functional activities in maintaining the Emergency Power Unit were examined closely, and SERD requirements supplemented de facto including human factors. Hence, when designers are properly aware of operational or user problems they will include human factors effects, even to the extent of explicit quantitative modeling.

During the accomplishment of the EPU Service Stand, it was observed that the activity analysis identified and clarified requirements more quickly than normally accomplished. Further, by accomplishing the steps of the morphology, assurance was achieved of formal and complete accomplishment of each design decision, in addition to providing a formal and complete record or "audit trail" of each decision.

CONCLUSIONS

The design morphology can enhance designer performance

The morphology induces increased designer awareness of the required design tasks and requires more complete requirement definition and documentation. Since each step must be completed, a more comprehensive system study results, and better integration of "hard" and "soft" criteria is provided. Since criterion models result from both "hard" and "soft" criteria and their inputs, a more objective system optimization occurs.

Improved response to user requirements

The result of a specific consideration and its effect on resulting equipment can be readily identified. Acceptance by the user during design review was achieved more readily and definition of support needs was defined more readily.

The utility of human factors considerations in design was clearly demonstrated

This resulted from forcing a systems orientation and considering producer's and operator's environments in an integrated fashion. Further the designer's scope was broadened in meeting user needs while simultaneously increasing the technical depth to which the problem is analyzed. Also contributing to this utility is the integration of human factors criteria with other performance goals explicitly in the modeling so that emerging conclusions are the result of a totally integrated set of performance goals, both from the human factors and the technological domains.

FUTURE IMPLICATIONS OF THE DESIGN MORPHOLOGY

Rapidly increasing computer capabilities along with an increasing rate of technological innovation and application to society's needs will imply increased importance to the understanding of the generic decisions required to design a new system. While the configuration of computers will change, and the sophistication of society's use of technology will increase, the logic associated with the design and development of equipment will not change. Therefore, it will become even more important as time progresses to understand the logic of design in order to effectively take advantage of improvements in the state of the art.

REFERENCES

Air Force Systems Command Manual (AFSCM) 375-5, 1966, U.S. Government Printing Office, Washington, D.C., March.
Asimow, M., 1962, "Introduction to Design", Prentice-Hall, Inc., Englewood Cliffs, N.J.
Blanchard, R.B., 1967, Human Performance and Personnel Resource Data Store Design Guidelines, Human Factors, 6 (9), 25-34.
Meister, D., 1971, "Human Factors: Theory and Practice", John Wiley and Sons, Inc., New York, N.Y.
Ostrofsky, B., 1977a, "Design, Planning and Development Methodology", Prentice-Hall, Inc., Englewood Cliffs, N.J.
Ostrofsky, B., 1977b, An Annotated Bibliography on Morphology of Design with Inclusion of Human Factors, University of Houston, Houston, Texas, March.
Ostrofsky, B., 1977c, Morphology of Design of Aerospace Systems with Inclusion of Human Factors, University of Houston, Houston, Texas, August.
Ostrofsky, B., 1978, Application of a Structural Decision Process for Proper Inclusion of Human Resources in the Design of a Power Unit Support Stand, University of Houston, Houston, Texas, September.

Ostrofsky, B., 1979, Preliminary Activities in the Development of MX
    Maintenance Control Through the Use of a Design Morphology,
    University of Houston, Houston, Texas, September.
Support Equipment Recommendation Data (SERD), General Dynamics, Fort
    Worth Division Document No. 16PR011, Contract No. F33657-75-C-
    0310.

DEVELOPMENT AND USE OF RESEARCH METHODOLOGIES FOR COMPLEX SYSTEM/

SIMULATION EXPERIMENTATION

Robert C. Williges

INTRODUCTION

Human factors research in systems (i.e., human/machine/environ-
ment configurations) poses an important methodological challenge.
Due to the complexity of the system, the researcher must simultaneously consider a variety of interacting factors that affect several
dimensions of both individual and group performance. If these considerations are not made, the results of the research enterprise will
provide little in the way of generalizable data to aid in the design
of real systems.

Usually three general approaches are used in conducting systems
research. None of these approaches is used exclusively, and often
all three approaches are used to a certain extent in most system design activities. One approach is to make direct observations of the
operational system as a means of improving performance through modification. Often, however, it may not be possible to isolate the exact
locus of one specific system parameter through direct observations
due to lack of control over other interacting factors. Direct observations, nonetheless, still allow the researcher a means of specifying major bottlenecks in systems operations.

A second approach to systems research is to use various analytical methods of modeling and fast-time computer simulation. These
methods are particularly useful in the conceptual design stages of
new systems in which a variety of initial configurations are being
considered. Analytical methods are limited primarily by the human
performance, hardware, and environmental data bases incorporated into the modeling or computer simulation method. Obviously, an incorrect data base will lead to inaccurate solutions.

The third general approach is to use real-time (i.e., human-in-the-loop) simulation and manipulate various systems parameters according to experimental design methods. This approach provides the control that is missing in direct observations and can also be used to provide the human performance data base required by analytical methods. Traditionally, research of this type has concentrated on the performance of one operator with one display and needs to be extended to the simultaneous considerations of several factors. If approached from a generic task point-of-view, real-time simulation can also provide generalizable system design principles.

All three of these approaches to systems research incorporate similar research methodology issues. Recently, Williges (1979) attempted to summarize these issues in terms of independent and dependent variable considerations in system experimentation. Two major conclusions became quite evident in his summary. First, a variety of experimental design and analysis procedures exist, but only a limited number of them have been used in complex system research involving human performance. Second, there is only a sparse amount of methodology development research in human factors which address complex systems experimentation issues.

Given the history of various methodological shortcomings coupled with more recent advances in behavioral research methodology, it is now feasible to consider the systematic development and integration of behavioral research methodologies for complex systems/simulation experimentation. The purpose of this paper is to review several experimental design methodologies which seem most appropriate for system research. The emphasis is primarily on presenting design alternatives for efficient human performance data collection in systems/simulation experimentation. A detailed discussion of subsequent data analysis procedures is beyond the scope of this presentation.

DESIGN ALTERNATIVES FOR ECONOMICAL DATA COLLECTION

Many system research problems include additional constraints that must be considered in the experimental design. Some factors cannot be crossed in the real world and only exist as nested relationships within other factors, for example. In addition, many human factors problems require the simultaneous investigation of several factors, thereby making a completely crossed factorial design unwieldy due to the extremely large number of resulting treatment combinations, because all levels of one factor (independent variable) appear at all levels of another factor. Efficient pretest procedures are needed to reduce the large number of potentially relevant systems variables to a feasible experimental space. Often nusiance variables such as equipment variations, experimenter differences, and time limitations on an experimental session must be considered.

Several data reduction design alternatives are variable to help
solve these problems, but generally these approaches are not used extensively in human factors research. Most of these procedures are discussed to a limited extent in standard experimental design texts (e.g.
Box et al., 1978; Cochran and Cox, 1957; Hicks, 1973; Keppel, 1976;
Kirk, 1968; Myers, 1972; and Winer, 1972). Some of the most useful
procedures are defined in this section.

Single-Observation Factorials

One simple approach to reducing the number of data observations
is to eliminate replication across complex, crossed, factorial designs. The complete set of treatment combinations in the factorial
design is used, but only one observations is made in each cell of the
design. For example, a complete factorial design having six factors
each at two levels would result in 64 treatment combinations. If only one subject rather than five, for example, were observed, the data collection effort would be reduced by one-fifth. Statistical pool-
ing procedures are used in single-observation factorial designs to
obtain error terms for lower-order effects from nonsignificant higher-
order effects. Such a design is particularly useful in organizing pretesting procedures, but even then the resulting number of treatment
combinations can still become too large when many factors are included.

Hierarchical Designs

In many human factors situations the environmental constraints
do not allow various classification variables and experimenter-rela-
ted variables to be crossed with the independent variables of interest. For instance, it may not be possible to manipulate all the
treatment conditions in an experimental evaluation of various aids
provided to human inspectors on the production line in one plant due
to the set-up costs on the inspection line. Consequently, each of
two inspectors aids are set-up separately in separate plants. Nonetheless, unique characteristics of the plants which are now nested
within treatment conditions (inspection aids) may affect the research
outcomes.

Whenever factors other than subjects are nested (i.e., the levels of one factor appear only at one level of another factor) within
other factors, the design is called a hierarchical design because of
the hierarchical, or pyramidal, shape of the treatment conditions.
This category of experimental design can also greatly reduce the resulting number of treatment conditions as compared to a completely
crossed, factorial design. Hierarchical designs, however, do not allow for any evaluation of the interaction among nested variables.
So, care must be taken in defining the nesting relationships when
these designs are used purely for data reduction purposes.

Blocking Designs

Often it is not possible to obtain data from the entire factorial design all at one time, and the researcher is forced to collect
data in stages or blocks. In human factors research these blocks can
include different testing times or different experimenters collecting subsets of data. Obviously, there could be a great deal of variation among blocks, and certain aspects of the factorial design must
be confounded with these block differences. Procedures are available
whereby the experimenter can systematically confound certain components of the design with block differences while balancing the remaining aspects of the design within each block. In most cases the
investigator need only sacrifice all or part of a higher-order interaction while keeping the rest of the design balanced.

Consider an example in which the human factors specialist is
evaluating two data entry procedures (Factor A) under two levels of
pilot workload (Factor B) in two different area navigation displays
(Factor C). The tests are being conducted in an aircraft simulator,
and it is not possible to collect the data on all eight treatment
conditions resulting from the design in one flight. Only four conditions can be evaluated each flight. Fig. 1 shows how the design can
be blocked into two flights of four treatment conditions such that
the main effects and two-way interactions of the three factors will
not be confounded with possible flight differences. Only the three-
way interaction is completely confounded with the two flight effects.

At times it is desirable to make all comparisons with equal precision and not sacrifice entire effects to blocking as the three-way
interaction was in Fig. 1. Alternative procedures referred to as balanced and partially balanced incomplete block designs have been developed for this purpose when one factor consisting of several levels is being investigated. Designs of this type can include a variety of treatment levels and block sizes with the restriction that the
number of treatment levels exceeds the block size.

Fractional-Factorial Designs

When the researcher finds it impossible to collect all the data from a higher-order factorial design due to time and/or budget
constraints, some information will be lost and some effects will be
confounded. The fractional replication procedure enables the experimenter to determine systematically the nature and extent of confounding by specifying the appropriate subset of treatment combinations
to observe from the complete factorial design. By making this decision judiciously, a great deal of useful information can still be
obtained from the subset of data collected. Restrictions as to the
number of levels of each factor must be adhered to, and the designs
are limited, by and large, to factor levels including prime numbers.
Most often these designs are used for 2-, 3-, and 5-level factorial

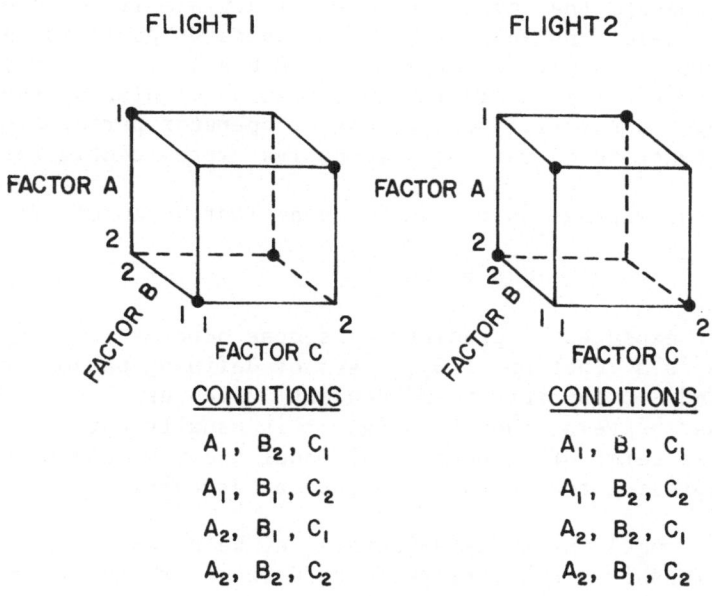

Fig. 1. Blocked 2 x 2 x 2 Factorial design.

combinations. For example, if only enough time were available on the aircraft simulator to collect data on just four of the eight treatment combinations in the previously mentioned area navigation evaluation example, the human factors specialist could choose to collect data in either flight 1 or flight 2 as shown in Fig. 1. Each flight represents a one-half replicate of the complete factorial design such that either half will allow an evaluation of the three main effects which are unconfounded from each other. Only the three-way interaction is totally lost, and the two-way interactions become aliases of the main effects in this fractional replicate.

EMPIRICAL MODEL BUILDING

For each of the design alternatives described so far it was implicitly assumed that the design would be used as a means of collecting data economically for statistical hypothesis testing. In these instances, the experimenter is concerned primarily with determining whether or not a particular factor or level of a factor had a statistically significant effect on the operator's performance in the system. Often, however, the investigator's primary aim is to determine a quantitative relationship between human operator performance and several quantitative systems parameters. Such a functional relationship is quite useful as an aid in the design of complex systems, because it allows comparative predictions of operator performance to be made as a function of various alternative system configurations.

The general expression for such a quantitative model is:

$$\eta = f(\Theta_i, X_i) \tag{1}$$

where η is the expected or predicted response outcome (dependent variable), which is a function of Θ_i, a set of defining parameters of the system, and X_i, the set of independent variables. Due to the complexity of most systems, the investigator is usually not able to define Eq. (1) in terms of a theoretical model and must use an empirical model to predict the response outcome accurately.

Empirical models have the advantages of being able to specify complex relationships in a straightforward manner which is amenable to real world applications, to require no understanding of true underlying relationships, and to use linear parameters for ease of estimation. Care, however, must be taken not to extrapolate predictions beyond the ranges of the system variables used to build the empirical model nor to use these models as a description of the underlying system process. Even though the empirical model might predict quite accurately, often it is not as parsimonious as its theoretical counterpart in terms of the number of parameters required.

One convenient form for specifying empirical models is to use polynomial expressions. Usually a second-order polynomial approximation is adequate to account for most human performance data. The general form of a complete, second-order polynomial model is:

$$Y = \beta_0 + \sum_{i=1}^{k} \beta_i X_i + \sum_{i=1}^{k} \beta_{k+i} X_i^2 + \sum_{i=1}^{k-1} \sum_{j=i+1}^{k} \beta_{2k+i} X_i X_j + \varepsilon, \tag{2}$$

where human performance, Y, is expressed in terms of an intercept value, β_0, and the weighted linear combinations of first-order terms, X_i, pure quadratic second-order terms, X_i^2, and linear interaction, second-order terms, $X_i X_j$, of the k system (independent) variables stated in terms of X's. The value ε is the estimate of error in prediction.

Sample estimates of the various β parameters specified in polynomial models such as Eq. (2) can be readily estimated in terms of standard least square regression procedures using the matrix solution:

$$b = (X'X)^{-1}X'Y, \tag{3}$$

where b is the vector of sample estimates of β, $(X'X)^{-1}$ is the inverse of the sum of squares and crossproducts matrix, and $X'Y$ is a column vector. For example, the second-order polynomial regression equation using raw score values of three system parameters (X's) would be:

$$Y' = b_0 + b_1X_1 + b_2X_2 + b_3X_3 + b_4X_1^2 + b_5X_2^2 + b_6X_3^2$$
$$+ b_7X_1X_2 + b_8X_1X_3 + b_9X_2X_3, \tag{4}$$

where Y' is the predicted value of operator performance, $b_0 - b_9$ are least square estimates of β as calculated by Eq. (3), and $X_1 - X_3$ are the three system parameters in terms of linear, quadratic, and linear interaction terms of the second-order polynomial defined by Eq. (2).

Once the experimenter has defined the appropriate empirical model, data must be collected to solve the least square estimates of the polynomial expression. At a minimum one more data point must be observed than the number of parameters estimated (b's), and the number of levels of the various system parameters (X's) must be one more than the order of the polynomial. Obviously, happenstance data or a random selection of data points could satisfy these conditions and could be used as the data matrix to solve the polynomial regression. Box et al. (1979), however, caution that happenstance data could entail inconsistencies, limited ranges of variables, semicon-founding of effects, nonsense correlations, and time dependent effects. All of these problems could be avoided through the judicious use of experimental designs.

Various designs have been developed to allow economical and efficient procedures to collect the necessary and sufficient data for generating first- and second-order polynomial regression models. These designs also allow the experimenter to take advantage of various mathematical relationships which are not possible to control in happenstance or randomly selected data. Several of these such as factorials, simplex, and central-composite designs are described by Meyers (1971). Of these design alternatives, the most flexible and most widely used is the central-composite design.

CENTRAL-COMPOSITE DESIGNS

Box and Wilson (1951) developed an experimental methodology for determining the optimal combinations of various factors present in chemical processes so as to maximize yield. They used polynomial regression models to define chemical yield relationships and developed second-order, central-composite designs as the data collection para-

digm. Subsequently, these designs have been used in a variety of chemical, engineering, agricultural, and behavioral research applications. Before reviewing some of the behavior research implications,
a brief overview of the central-composite design specification and
data analysis procedures is necessary.

Design Specification

Configuration. Consider an example in which a human factors researcher is interested in assessing the operator's driving error (Y)
in an automobile simulator as a function of frequency (X_1), velocity
(X_2), and direction (X_3) of wind gusts. A three-factor, central-composite design as shown in Fig. 2 could be used to collect the data
necessary to solve a second-order polynomial regression equation
which predicts driving error as a function of the three wind gust
conditions. The design itself is a composite of a 2^3 factorial design,
augmented by a center point and $2 \cdot 3$ additional points (star). This
composite of data points radiates from the center point, hence the
name central-composite design. In general any second-order, central-
composite design can be specified as having the following total unique data points, T,

$$T = F + 2K + 1, \tag{5}$$

where F equals the number of data points in the 2^k factorial design
and K equals the number of k factors.

The 15 specific treatment combinations for the three-factor
driving simulator example are listed below the geometric representation of the data points depicted in Fig. 2. The levels of these data
points are listed in coded form and would require linear transformations to real-world levels of the various wind gust parameters. Notice that treatment combinations 1 - 8 represent the 2^3 factorial
portion of the design; combinations 9 - 14 represent the $2 \cdot 3$ star
portion; and treatment combination 15 is the center point. It is also important to notice that each of the three wind gust factors appears at five distinct levels $(+\alpha, +1, 0, -1, -\alpha)$. When one compares
these 15 treatment combinations to the 125 treatment combinations of
the resulting 5^3 counterpart factorial design, the data collection
efficiency of central-composite designs become quickly apparent.

Obviously, the three-factor design shown in Fig. 2 can be extended to a hyperspace of any number of k factors. As the number of factors increase, a 2^{k-p} fractional design is usually substituted for
the 2^k factorial portion of the design and for the value of F in Eq.
(5). When this is done, care must be taken to choose a fractional
replicate such that all first- and second-order components are present and are not aliases of each other so that a complete second-
order polynomial regression as shown in Eq. (4) can be determined.
(See Cochran and Cox, 1957, for several examples of central-compo-

DESIGN CONFIGURATION

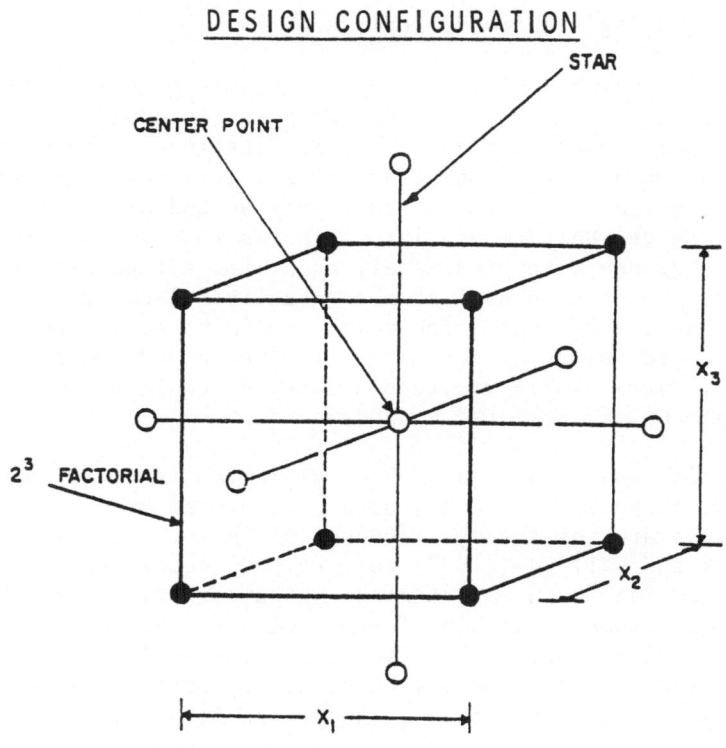

	X_1 Wind Gust Frequency	X_2 Wind Gust Velocity	X_3 Wind Gust Direction
Treatment Combination			
1	+1	+1	+1
2	+1	-1	+1
3	+1	+1	-1
4	+1	-1	-1
5	-1	+1	+1
6	-1	-1	+1
7	-1	+1	-1
8	-1	-1	-1
9	$+\alpha$	0	0
10	$-\alpha$	0	0
11	0	$+\alpha$	0
12	0	$-\alpha$	0
13	0	0	$+\alpha$
14	0	0	$-\alpha$
15	0	0	0

Fig. 2. Example of a three-factor, second-order central-composite design to evaluate automobile driving performance (Y) as a function of wind gust characteristics (X_1, X_2, X_3).

site designs using fractional replicates.)

    Replications. Usually, all replication in a central-composite
design is conducted only at the center point to enhance the data collection economy. The exact number of replications of the center point
is dependent upon certain mathematical criteria that the experimenter
wishes to consider such as uniform precision and orthogonality of regression weights. Uniform precision insures that the accuracy of the
predicted response is approximately equal for all points from the
center out to a particular distance (usually $\pm$ 1 coded units); whereas, orthogonality of regression weights eliminates any correlation
among the coded values of parameters in the second-order polynomial
regression. Myers (1971) provides a detailed explanation and mathematical treatment of these two considerations.

    Often in behavioral research, replication is conducted across
the entire design to obtain a stable estimate of error. Even though
these designs are not the most economical in terms of data collection, Clark and Williges (1973) suggest that equal replication across
the entire design is an important design alternative for behavioral
research. For example, if a different subject is observed at each
treatment condition (i.e., a between-subjects design) more degrees
of freedom for replications may be needed to obtain a stable estimate
of error. Alternatively, if each subject receives every treatment
combination (i.e., a within-subject design), then equal replications
across the design would be appropriate when collecting data on more
than one subject.

    Value of α. Fig. 2 lists the coded value of the star components
of the central-composite design as $\pm$ α. Obviously, these α's must be
translated into specific numeric values for any particular design.
The exact choice of an α value depends upon various design criteria
chosen by the experimenter. Specifically, considerations of design
rotatability, orthogonal regression weights, and blocking are usually made. (See Box and Hunter, 1957; and Myers, 1971, for a detailed
mathematical treatment of these various criteria.)

    Rotatability exists when the variance of the predicted response
is the same for all points equidistant from the center regardless of
the direction of these points. This is a convenient design quality
especially in exploratory work when the investigator is unaware of
the underlying response function relative to the orientation of the
factor axes. Box and Hunter (1957) demonstrated that for central-composite designs to be rotatable, the value of α must be:

$$\alpha = F^{\frac{1}{4}}. \tag{6}$$

    One difficulty in using second-order polynomial regression pro-
cedures for predicting human performance is that the higher-order
effects (i.e., pure quadratic effects in Eq. (2)) are intercorre-

lated unless the experimenter specifies the design appropriately. Myers (1971) demonstrated that orthogonal, second-order, central-composite designs in coded form can be described according to the following relationship,

$$\alpha = (\frac{QF}{4})^{\frac{1}{4}},$$ (7)

where

$$Q = \left[(F + 2K + C)^{\frac{1}{2}} - F^{\frac{1}{2}}\right]^2.$$ (8)

In Eq. (8), the term C represents the total number of center points. If equal replication is used, the value of C reduces to 1 and the terms in parenthesis equal the unique data points, T, in a central-composite design as specified in Eq. (5).

Under blocking conditions, subsets of the central-composite design data points are studied together to provide the investigator with additional design efficiency and flexibility. For example, blocking could be used either as a means of controlling unwanted effects (e.g., differences due to experimenters or testing days) or as an efficient way to collect data in stages. Care must be taken, however, to insure that the block mean effect is orthogonal to any first- and second-order effects due to the factors manipulated. Orthogonal blocking of central-composite designs requires the following condition be met (Box and Hunter, 1957),

$$\alpha = \left[\frac{F(2K + C_S)}{2(F + C_F)}\right]^{\frac{1}{2}},$$ (9)

where C_S equals the number of replicated center points in the star portion and C_F equals the number of center points in the factorial portion such that $C = C_S + C_F$.

In central-composite designs, the blocking is readily accomplished by separating the design into the factorial portion and the star portion. If additional blocks are required, the factorial portion can be further divided by fractional replicates as shown previously in Fig. 1. When replications occur only at the center point, they must appear in both the star and factorial portions of the design as designated in Eq. (9) by C_S and C_F, respectively. If replications are equal across the design, then the center point only appears in the block using the star portion. Consequently, C_S reduces to 1 and C_F reduces to 0 in Eq. (9).

Table I provides a summary comparison of the different values of α required for different number of factors (K) in central-composite designs both for equal replication across the design and for re-

Table I. Comparison of Total Unique Data points, α Values, and Center Point Replications for Several Rotatable, Orthogonal, and Blocked, Second-Order Central Composite Designs (after Box and Hunter, 1957; and Myers, 1971).

Experimental Factors	Total Unique Data Points	Equal Replications			Replications at Center Point (C) Only						
		Rotatable	Orthogonal	Blocked	Rotatable		Orthogonal and Rotatable		Blocked/Near Rotatable		
K	T	α	α	α	α	C	α	C	α	C(C_S + C_F)	R
2	9	1.414	1.000	1.581	1.414	5	1.414	8	1.414	6(3 + 3)	2
3	15	1.681	1.216	1.871	1.682	6	1.682	9	1.633	6(2 + 4)	3
4	25	2.000	1.414	2.121	2.000	7	2.000	12	2.000	6(2 + 4)	3
5*	27	2.000	1.546	2.345	2.000	6	2.000	10	2.000	7(1 + 6)	2
6*	45	2.378	1.724	2.550	2.378	9	2.378	15	2.366	10(2 + 8)	3
7*	79	2.828	1.885	2.739	2.828	14	2.828	22	2.828	12(4 + 8)	9

*A one-half replicate is used in the factorial portion of the central-composite design.

plication only at the center point. When replication occurs only at the center, additional considerations must be given as to the number of replications, C, to include. The values of C given in Table I for rotatable designs approximate uniform precision to a coded distance of ± 1. The values for C in orthogonal design can also be rotatable by setting α in Eq. (7) equal to the rotatable value and solving for C in Eq. (8). The values for C for the number of blocks, B, in Table I were chosen so that the value of α is nearly rotatable. Obviously, it is not always possible to have an exactly rotatable design due to the constraints of Eq. (9).

When equal replications are used in central-composite designs, the values of α differ markedly as shown in Table I. Consequently, the experimenter must often choose among rotatable, orthogonal, and blocked designs. Design constraints such as a requirement for blocking often dictate the investigator's choice; but Williges (1976) suggested using orthogonal designs whenever possible due to the ease of calculation and subsequent interpretation of the resulting polynomial regression equations.

Data Analyses. Although a detailed description of data analysis procedures is beyond the scope of this paper, a brief overview is necessary as a prerequisite to discussing some of the behavior research implications of using central-composite designs. Detailed discussions of various analysis procedures are provided by Box and Hunter (1958), Cochran and Cox (1957), Davies (1954), and Myers (1971).

Basically, two general statistical analyses are conducted on central-composite designs. First, a standard least squares, regression analysis as defined by Eq. (3) is used to determine the empirical model. Second-order central composite designs provide the necessary and sufficient data to fit complete second-order polynomial expressions as in Eq. (1) and provide enough additional data to test for higher-order effects. For example, the 15 unique data points shown in Fig. 2 along with the appropriate replications would provide the required data to fit the complete second-order polynomial stated in Eq. (4) to specify the effect of the three wind gust factors on driver performance.

The second general procedure is to conduct a subsequent analysis of variance on the polynomial regression. Suppose five different subjects drive the automobile simulator under all of the 15 different wind gust combinations shown in Fig. 2. In addition, suppose that the 15 wind gust conditions in the central-composite design were blocked into three sessions such that the first two sessions represent the two, one-half replicates of the 2^3 factorial portion of the design (i.e., conditions 1 to 8), and the third session represents the center point and 2K star components (i.e., conditions 9 to 15). A general break-down of the sources of variation and the degrees of

freedom in the analysis of variance summary table for this design
is given in Table II. Details on the analysis procedures for isolat-
ing these terms are provided by Clark and Williges (1972) and a special purpose computer program for behavior data analyses of central-
composite designs was developed by Clark et al. (1971).

Notice that the analysis of variance is broken down into two
major components dealing with variance due to regression and variance
due to residual. Regression variation is broken into single degree of
freedom components associated with each of beta weights in the second-order polynomial. Residual variation, on the other hand, can be
broken into a variety of components depending upon the central-composite design configuration. In this particular example, it can be
separated into effects due to three blocks (testing sessions), five
subjects, lack of fit, and an estimate of error. Lack of fit refers
to higher-order regression weights. If lack of fit is significant,
then some higher-order terms can account for significant portions of
variance and, in this case, a complete second-order model is not adequate to describe the functional relationship between driving performance and the three wind gust factors.

Table II. Sources and Degrees of Freedom in the Second-Order Regression Analysis of Variance Summary Table.

Source	df
Regression	(9)
X_1	1
X_2	1
X_3	1
X_1^2	1
X_2^2	1
X_3^2	1
$X_1 X_2$	1
$X_1 X_3$	1
$X_2 X_3$	1
Residual	(50)
Blocks	2
Lack of fit	3
Subjects	4
Error	41
Total	59

The ability to separate residual variation into various components so as to isolate a refined error term is an important methodological advantage of experimental design over happenstance data. In
particular, the central-composite design provides an efficient and
economical means of collecting data to satisfy a variety of design
objectives including orthogonal beta weights, blocking, and estimating lack of fit of the hypothesized empirical model. Furthermore, by
considering various behavior research constraints, the error term
can be further refined to separate between-subject effects from it.

Use of Central-Composite Designs in Behavior Research

Most human performance applications of central-composite designs
have only attempted to provide a global empirical model of operator
performance as a function of various system parameters. (See Williges and Simon, 1971; and Clark and Williges, 1973, for a general discussion of using central-composite designs in behavioral research.)
All of the applications can be divided into three classes of studies
depending upon the manner in which subjects are assigned to treatment
conditions. These three classes are within-subject, between-subject,
and mixed-factors design applications. Each is briefly reviewed as
a means of discussing some of the design modifications of basic central-composite designs used in behavior research.

Within-subject design. Often it is possible to cross subjects
with treatment conditions such that every subject receives every
treatment combination in the central-composite design shown in Fig.
2. This procedure yields a within-subject design in which the main
effect of subjects can be separated from experimental error as shown
in Table II. Within-subjects designs usually provide more sensitive
error terms as well as require fewer subjects due to repeated observations. Usually it is more convenient to replicate equally across
the central-composite design when it is a within-subject version. In
this way, the experimenter can generalize the data over more than
one subjects and avoid the situation of having repeated observations
on the same subject at the center point.

Several examples exist which used within-subjects, central-composite designs in human factors research. For instance, Williges and
North (1973) used such a design to predict the probability and latency of correct target location of video cartographic symbols as a
function of four factors including focus, density of nontarget symbols, visual angle of the observer, and TV raster lines of actual
map area. Additionally, Mills and Williges (1973) used a within-subject design in evaluating operator performance in a semiautomatic
radar surveillance system. They generated second-order empirical models of the probability of correct track initiation and track initiation time as a function of blip/scan ratio, target introduction rate,
clutter replacement probability, clutter density, and target velocity.

Recently, Williges (1980) used a within-subject, central-composite design to generate prediction equations of human operator performance in maintaining personnel records on interactive computer terminals. Specifically, he was interested in determining the effects of four parameters of the interactive computer terminals on three categories of dependent variables including the operator's overall satisfaction with the interactive system, performance in using the computer terminal, and work sampling of time allocations to various components of maintaining personnel records. The four display factors investigated in the study were system delay time (in seconds) to a computer entry, display rate of characters on the computer terminal (in characters/seconds), delay rate (in seconds) of echoing keystrokes on the display, and the buffer length of keystroke entry storage (in number of characters). Four subjects received each of the resulting 25 treatment combinations of the four-factor, central-composite design.

Table III provides a partial summary of the satisfaction ratings from the Williges (1980) study as an example of within-subject designs. The top section of Table III shows the coded values and linear transformations of the factors manipulated in the design. Notice that an α value of 1.414 was chosen to provide orthogonal, second-order polynomial regression weights as defined by Eqs. (7) and (8) and summarized in Table I.

The resulting complete second-order prediction equation for overall ratings of user satisfaction (S) and the multiple correlation (R) for the least squares regression fit are given in the center portion of Table III. Since the polynomial regression was conducted on coded values and these values are orthogonal, it is possible to make relative comparisons among the partial regression weights directly. The square of R represents the percent of variance accounted for by the polynomial regression (i.e., 54%).

A complete summary table of the analysis variance of the regression is given in the lower portion of Table III. Since the analysis was conducted on an orthogonal second-order design, the variation (SS) attributable to each partial regression weight is independent of the variation attributable to any other partial regression weights. The main effect of differences among the four subjects can be subtracted from the error terms since a within-subject design was used. Assuming no interactions of subjects with treatment conditions, the error mean square (MS) can be used to test all effects. In this particular analysis, the partial regression weights related to the linear, quadratic, and linear-by-linear components of SD and ER are all statistically significant ($p < .01$) predictors of the operator's overall rating of satisfaction with the interactive computer terminal. The test of lack of fit was not significant ($p > .25$) thereby suggesting that the second-order polynomial is an adequate empirical model for the range of variables sampled. And, finally, the test of

Table III. Partial Summary of Williges (1980) Study Using a Within-
Subject, Central-Composite Design.

Levels of Four Independent Variables				
	-1.414	-1	0	+1	+1.414
System Delay (SD)	0.10	1.55	5.05	8.55	10.00
Display Rate (DR)	240	206	125	44	10
Echo Rate (ER)	0.00	0.22	0.75	1.28	1.50
Buffer Length (BL)	1	2	4	6	7

Coded Second-Order Polynomial of
Operator's Overall Satisfaction Rating

$$S = 2.27 - 1.79SD - 0.28DR - 0.94ER + 0.19BL + 1.40SD^2$$
$$+ 0.26DR^2 + 0.85ER^2 + 0.01BL^2 + 0.30SD*DR + 0.64SD*ER$$
$$+ 0.11SD*BL + 0.05DR*ER - 0.80DR*BL - 0.27ER*BL$$
$$R = .736$$

Summary Table for Coded Regression Analysis of Variance

Source	df	SS	MS	F
Regression	(14)	(434.716)	(31.051)	9.50**
SD	1	258.047	258.047	78.99**
DR	1	6.050	6.050	1.85
ER	1	70.484	70.484	21.57**
BL	1	2.749	2.749	<1.00
SD^2	1	34.436	34.436	10.54*
DR^2	1	2.098	2.098	<1.00
ER^2	1	23.112	23.112	7.07*
BL^2	1	0.001	0.001	<1.00
SD*DR	1	5.641	5.641	1.73
SD*ER	1	26.266	26.266	8.04*
SD*BL	1	0.766	0.766	<1.00
DR*ER	1	0.141	0.141	<1.00
DR*BL	1	0.391	0.391	<1.00
ER*BL	1	4.516	4.516	1.38
Residual	(85)	(369.011)		
Lack of Fit	10	2.007	2.007	11.61*
Subjects	3	113.748	37.916	<1.00
Error	72	235.190	3.267	
	99	803.727		

*$p<.01$
**$p<.001$

the subject effect was reliable (p<.01) which underscores the advantage of removing inter-subject variability by means of the within-subject design.

To interpret the relationships shown in the second-order polynomial of Table III, a plot of the data is usually instructive. Fig. 3 depicts a perspective, transect plot of the operator's overall rating of satisfaction as a function of the two significant factors SD and ER. If additional factors need to be depicted, a series of contour and/or perspective plots would be necessary. It is also possible to superimpose contour plots of several dependent variables to determine tradeoffs among them. An alternative to graphical representation is to reduce the polynomial regression to its canonical form to facilitate interpretation of the surface described by the empirical model. Davies (1954) provides a detailed description of this procedure.

Between-subject designs. If a different subject is assigned randomly to each of the 15 different treatment combinations of the central-composite design shown in Fig. 2 for the wind gust experiment, the design would be called between-subjects. It is possible to replicate this design across either the entire design or only at the center point with the restriction that each subject is observed only once.

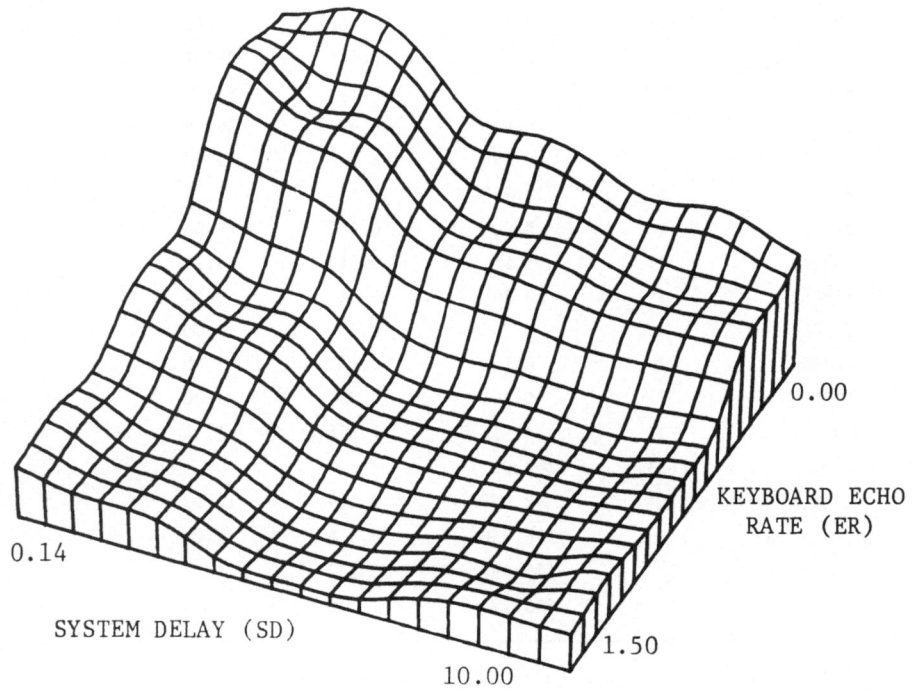

Fig. 3. Perspective plot of coded second-order polynomial of operator's overall satisfaction rating.

In certain human factors research situations the nature of the independent variables will permit only one treatment condition to be administered to each subject. For example, if the factors of the experiment represent various training conditions, then the effect of one training condition would influence subsequent performance in another training condition. Therefore, each subject should only receive a single training procedure in order to make a clearcut comparison.

Williges and Baron (1973) used such a between-subjects, central-composite design to evaluate transfer of training in a pursuit tracking task as a function of three independent variables including time between trials, number of practice trials, and difficulty of training task. They conducted the experiment both in terms of replicating only at the center point and replicating across the entire design. Although the resulting first-order polynomial prediction equation of transfer performance did not change, the statistical reliability of the partial regression weights certainly did change as a function of replication choice.

Due to individual difference, inter-subject variability often represents a large portion of the variability in behavioral research. Moreover, this variability cannot be separated from the error term of between-subject designs because subjects are nested within treatment conditions. Consequently, the F-test on the partial regression weights can be quite insensitive. As the Williges and Baron (1973) data demonstrate, the regression weights were not statistically significant with replication only at the center point due to the rather insensitive test of significance resulting from the few degrees of freedom in the error term. On the other hand, the regression weights were statistically significant when replication occurred across the entire design, because more degrees of freedom are added to the error term thereby resulting in a more sensitive test of significance. In summary, the experimenter must be cognizant of the increased variability present in between-subject designs and choose the number of required replications accordingly.

Mixed-factor design. Often it is not possible to manipulate all the factors in an experiment as either within-subject or between-subjects factors, and the resulting design is a combination of both types of factors. Such a design is called a mixed-factors design. One example of modifying a basic central-composite design to incorporate mixed-factors subject assignment was provided by Clark et al. (1973). Their approach manipulated the within-subject factors according to a central-composite design and then completely crossed the resulting treatment combinations with the between-subject factor(s). As the number of between-subject factors increases, however, this design procedure becomes quite uneconomical.

Clark (1976) proposed a more efficient mixed-factor design alternative which is constructed by initially disregarding the within-subject or between-subjects status of all factors and by determining the data points solely in accordance with central-composite design procedures. Once the unique data points are specified, subjects are assigned to· the various treatment combinations such that no subject experiences more than one level of the between-subject factor(s). An example of this version of a mixed-factor design is given in Table IV which includes three within-subject factors and one between-subjects factor.

Subsequently, Clark (1976) conducted both a theoretical and an empirical comparison of orthogonal versions of these two mixed-factor design approaches. The version shown in Table IV was comparable to the version proposed by Clark et al. (1973) in terms of design efficiency and coefficient biases as defined by Myers (1971). The empirical comparison of these two versions was conducted on a target detection study designed to investigate various parameters of visual time compression. Specifically, background noise, target configuration, target speed, and on/off ratios were manipulated. Both versions provided comparably reliable second-order prediction equations, cross validations, and mean square errors of prediction. Consequently, the more economical version shown in Table IV appears to be the more favorable alternative for constructing mixed-factors central-composite designs.

SEQUENTIAL DESIGN PROCEDURES

In most cases, system research questions cannot be answered in one single study, rather a series of studies must be conducted in an economical and strategic manner. All of the design alternatives discussed so far in this paper may appear to imply a single study. This implication is misleading, however, because all system studies require a certain amount of pretesting in order to choose the appropriate factors and to set the levels of these factors. If an investigator chooses to make a major commitment of time and effort in one large study without careful preliminary consideration and pretesting, the results could be disasterous. Box et al. (1978), for example, suggest a 25 percent general rule in which they recommend that no more than one quarter of the experimental effort should be invested in the first experimental design.

Some systematic attempts have been made to suggest overall strategies that the investigator could adopt in conducting sequential research on complex systems problems. Most of these approaches have been developed under the rubric of response surface methodology, and more recently these considerations have been extended into the behavioral research domain.

Table IV. Design Matrix for a Mixed-Factors Central-Composite Design Involving Three Within-Subject Factors (X_1, X_2, X_3) and One Between-Subjects Factor (X_4) as Proposed by Clark (1976).

Subjects	X_1	X_2	X_3	X_4
	+1	+1	+1	+1
	-1	+1	+1	+1
	+1	-1	+1	+1
S_1 and S_2	-1	-1	+1	+1
	+1	+1	-1	+1
	-1	+1	-1	+1
	+1	-1	-1	+1
	-1	-1	-1	+1
	+1	+1	+1	-1
	-1	+1	+1	-1
	+1	-1	+1	-1
S_3 and S_4	-1	-1	+1	-1
	+1	+1	-1	-1
	-1	+1	-1	-1
	+1	-1	-1	-1
	-1	-1	-1	-1
	$-\alpha$	0	0	0
	$+\alpha$	0	0	0
	0	$-\alpha$	0	0
S_5 and S_6	0	$+\alpha$	0	0
	0	0	$-\alpha$	0
	0	0	$+\alpha$	0
	0	0	0	0
S_7 and S_8	0	0	0	$-\alpha$
S_9 and S_{10}	0	0	0	$+\alpha$

Response Surface Methodology

One could view the human factors research enterprise in terms of using empirical models to generate a surface of human performance as a function of several system variables. All of these variables combined would define a response surface of human operator performance analogous to the perspective plot shown in Fig. 3. Various design strategies have been suggested as a means of exploring these response surfaces so as to reach the point of maximum performance (or minimum error) in an efficient and accurate manner. In particular, three general approaches for seeking an optimum were discussed by Cochran and Cox (1957).

One of these procedures is merely to use a random and independent selection of experimental data points (Anderson, 1953). After a specified number of trials including random treatment combinations, the particular factor combination yielding the highest overall response is regarded as the optimum. Obviously, this approach is lacking in a formal sequential design approach, but it may be the only alternative an investigator has when the systems problem involves so many factors that the use of ordered designs becomes impossible to implement. Alternatively, random selection can be supplemented either by using the set of randomly chosen treatment combinations in polynomial regression to specify an empirical model, or by using sequential design procedures to search systematically the region of optimality defined by the random selection method.

A second approach to seeking an optimum was proposed by Friedman and Savage (1947). Their approach is called the single-factor method, because it is characterized by a series of small experiments in which one factor at a time is optimized. The experimenter estimates an optimal combination of factors based on background and/or pretesting knowledge. The variable assumed to have the greatest effect on performance is optimized first in a three-level design while the other factors are held constant. Subsequently, all the other factors are optimized one at a time until an optimal configuration of all factors is achieved. Variations in this procedure can include simultaneous changes in all factors and/or the dropping of some factors in subsequent experimental rounds. The emphasis, however, is on first-order representations of each factor and little, if any, consideration is given to higher-order effects.

The third approach to sequential surface exploration, and probably the most widely used, is the method of steepest ascent proposed by Box and Wilson (1951). The three general steps followed in this approach include: (1) the use of first-order designs to determine a local slope; (2) the use of the method of steepest ascent to approach an optimum; and (3) the use of second-order designs to characterize the local region around an optimum.

A series of 2^k (or 2^{k-p}) factorial designs is used as the first-order design procedure. Following each small 2^k factorial experiment, the investigator adjusts the ranges of each factor by the method of steepest ascent which changes the levels proportional to the first-order partial regression weights. This procedure is repeated until the investigator nears an optimum point on the response surface. Second-order, central-composite designs were developed by Box and his colleagues, specifically as a means of specifying the region around an optimum. By using blocking procedures, the factorial portion of the design is used iteratively with the method of steepest ascent, and then only the star portion need be added to complete the second-order model around the optimum. In this way, response surface exploration is conducted in an extremely efficient and economical fashion.

Behavioral Research Strategies

Most of the response surface methodology procedures were designed primarily for research applications other than human performance. Indeed, most applications have been made in the chemical industry. Simon, in a series of papers (1971; 1973; 1974; 1975; 1977a; 1977b), attempted to integrate response surface methodology, along with a variety of other procedures, into a research paradigm using multifactor designs for human factors engineering experiments. He states that the basic principles of his paradigm include: avoid replicating basic experimental designs unnecessarily; avoid collecting data on higher-order effects until evidence exists as to their existance; collect and evaluate data in a sequence of progressive iterations; substitute experimenter's knowledge and analytical skills for data collection; and minimize bias effects on each individual measurement (Simon, 1973).

In the process of developing his research strategy, Simon incorporated many of the data reduction design procedures into saturated and augmented screening designs (Simon, 1973; 1977a) and methods for handling sequence effects (Simon, 1974). His strategic research paradigm is divided into five major phases which separately were intended to define the systems research problem, identify critical varables, develop a response surface, refine the equation, and verify experimental results (Simon, 1977b).

SOME UNRESOLVED ISSUES

Sequential Research Strategies

Although several approaches to sequential design procedures have been developed, no systematic effort has been made to apply these procedures in a series of systems research studies incorporating human performance. As Cochran and Cox (1958) suggest, these procedures work best when experimental errors are small and/or previous estimates of error variance are known so that individual experimenta can be kept small. This is not usually the case in behavioral research, therefore modifications of existing sequential design strategies may be necessary.

Since most complex system research problems may require the use of efficient sequential designs as opposed to a single experimental design, more emphasis needs to be given to the development of sequential design procedures for behavioral research. Several central questions need to be addressed. How can the sequential design strategy be optimized? Are there classes of systems experimentations that require different sequential design procedures? And, what are the possibilities for providing automated aiding of sequential design strategies for the experimenter?

Improved Empirical Models

Most examples of empirical models described in this paper can
be characterized as global prediction equations of operator performance across a wide range of variables as compared to empirical models used to seek an optimum and describe the region around this local optimum. When global prediction equations are made of operator
performance, the data points may be spread so widely that the resulting empirical model has marked inaccuracies in prediction. Considerations pertaining to the range of variables, type of variables, and
method of regression analysis need to be made.

Once a factor is chosen for inclusion in systems research, the
practical range must be included. Choosing this range has implications for using efficient plans of pretesting. Currently, little effort has been devoted to methods for conducting pretesting in an efficient and economical fashion. Often the system designer may only
need a functional description (i.e., mission profile) of the real
task as opposed to a detailed composite prediction equation of operator performance. What variables and range of these variables are
required to provide an accurate functional description of the mission? How must the range of variables differ when they are used for
descriptive versus predictive purposes?

Factors manipulated in the experimental design can either be
continuous, quantitative variables (e.g., time, target size, etc.)
or discrete, qualitative variables (e.g., display format, type of
image, etc.). Most complex experimental design procedures are directed primarily toward the manipulation of continuous variables,
whereas many real systems research problems involve a variety of discrete variables. Methodological extensions to incorporate continuous
as well as discrete variables in empirical models need to be considered.

The predictors in the regression equation can either be independent or correlated with each other depending on the particular design strategy used to collect the data. Namely, whenever the experimental design is not orthogonal, some of the partial regression
weights will be correlated. When the predictors are correlated, it
is possible for the resulting regression weights to be extremely un-
stable such that the weights may have the wrong sign, and/or result
in a larger mean error of point prediction. Hoerl and Kennard (1970a;
1970b) demonstrated that it is possible to correct for the biases
resulting from non-orthogonal designs through the use of ridge regression. Although the mathematics of ridge regression are well developed, the criteria used for selecting the necessary correction
factor and the exact conditions under which ridge regression is appropriate are still somewhat controversial (Rozeboom, 1979).

Time Dependent Implications

Most of the discussion in this paper has been devoted to the examination of human performance in complex systems that require event-based or steady-state behavior. In many real systems, however, the operator performance is time varying and the system itself operates on a time base. Little behavioral research has been conducted in complex systems experimentation in which time varying aspects of operator performance are considered. In fact, only a limited amount of research methodology is currently available for this purpose. Some consideration, for example, has been given to this problem in the development of quasi-experimental designs for circumstances characterized by nonrandom assignment of subjects to treatment conditions. In particular, Cook and Campbell (1976) discuss the use of interrupted time-series designs with nonequivalent control groups, nonequivalent dependent variables, and switching replications which appear to be quite useful in complex, systems research.

The general methods of time series analysis also appear to be appropriate. These methods are autocorrelations coupled with regression procedures to represent time series models of performance in both lagged and unlagged situations. (See Box and Jenkins, 1979; Montgomery and Johnson, 1976; and Ostrom, 1978, for a description of the computational details of the various time series analysis models.)

Box et al. (1978) describe the use of time series models for forecasting, feedback control, and intervention analysis. All three of these uses appear to be appropriate in system experimentation. Forecasting models can be used to predict the time-varying operator performance in the system. Feedback control applications are useful in considering the application of system research results as inputs to computer analytical modeling data bases. And, intervention analysis can be used to evaluate the effect of varying operator/team workload on system performance.

Time series models need to be evaluated as to their applicability in complex systems research directed toward time varying operator behavior. Questions dealing with the isolation of the onset and impact of multiple time series must be addressed. Describing the functional relationship of various systems parameters and an operator's time varying behavior is necessary. And, examples of the guidelines for appropriate uses of time series modeling in a systems context must be developed.

Multivariate Extensions

Analysis methods for calculating polynomial prediction equations using single (i.e., univariate) measures of operator behavior are fairly well established. Functional descriptions of multivariate ope-

rator performance are not defined even though human behavior in most
systems must be described by a variety of metrics. The degree to
which there is covariance among these dependent variables, there
will be differences between univariate and multivariate analysis procedures. Various procedures exist for clustering dependent variables.
(See Cooley and Lohnes, 1971; Harris, 1975; and Tatsuoka, 1971, for
a general description of these methods.) They rely heavily on determining a descriminant function representation of sets of variables.
The resulting dimensions can be determined either by principle components analysis which provides linear combinations of the original
variables or by various methods of factor analysis which attempt to
separate unique variance from common variance. More recently, Wooldridge (1980) recommended procedures of ridge discriminant analysis
as a means of improving this clustering procedure.

Ideally, the experimenter would like to analyze the data from
a systems experiment such that the functional relationship of various independent variables can be used to predict a multivariate combination of dependent variables. Canonical analysis is a procedure
whereby different sets of linear weights are used for the independent
and dependent variables so that the maximum multivariate correlation
(canonical correlation) is obtained (Harris, 1975; Levine, 1977; and
Tatsuoka, 1971). Even though canonical analysis specifies the maximum degree of relationship, it does not specify a functional representation of multivariate clusters and multiple independent variables.

Additional research is required to investigate the best procedure for specifying the multivariate/multifactor relationship. More
consideration needs to be given to the application of various clustering procedures as a means of combining dependent variables which can
subsequently be related to multiple independent variables through polynomial regression procedure. Once these procedures are established,
it will be possible to generate meaningful multivariate empirical
models of operator behavior in complex systems. Empirical models of
this type are of upmost importance to the system designer.

CONCLUSION

Clearly, a variety of experimental design tools are available
to the human factors investigator involved in systems/simulation experimentation. These design alternatives can be quite useful in specifying procedures for economical data collection in large scale experiments, in following sequential procedures in a series of related
research studies, and in developing empirical models of operator performance as a function of several systems variables. More applications need to be made of these approaches, and more improvements are
needed through systematic methodological research. Nonetheless, the
use of complex experimental design procedures holds promise both for

generating more generalizable system research data and for improving
human factors inputs in systems design.

REFERENCES

Anderson, R.L., 1953, Recent advances in finding best operating con-
    ditions, Journal of American Statistical Association, 48, 789-
    798.
Box, G.E.P. and Hunter, J.S., 1957, Multifactor experimental designs
    for exploring response surfaces, Annals of Mathematical Statis-
    tics, 28, 195-241.
Box, G.E.P. and Hunter, J.S., 1958, Experimental designs for the ex-
    ploration and exploitation of response surfaces, in: "Experimen-
    tal design in industry", V. Chew (ed.)., Wiley, New York, 138-
    189.
Box, G.E.P., Hunter, W.G. and Hunter, J.S., 1978, "Statistics for ex-
    perimenters: An introduction to design, data analysis, and mod-
    el building", Wiley, New York.
Box, G.E.P. and Jenkins, G.M., 1976, "Time series analysis: forecast-
    ing and control", Holden-Day, San Francisco.
Box, G.E.P. and Wilson, K.B., 1951, On the experimental attainment
    of optimum conditions, Journal of the Royal Statistical Society,
    Series B (Methodological), 13, 1-45.
Clark, C., 1976, Mixed-factors central-composite designs: A theoreti-
    cal and empirical comparison, Savoy, Illinois, University of
    Illinois, Aviation Research Laboratory, Technical Report ARL-76-
    13/AFOSR-76-6.
Clark, C., Scanlan, L.A. and Williges, R.C., 1973, Mixed-factor re-
    sponse surface methodology central-composite design considera-
    tions, in: Proceedings of the 17th annual meeting of the Human
    Factors Society, M.P. Rance, Jr., T.B. Malone (eds.), Santa Mo-
    nica, California, Human Factors Society, 281-288.
Clark, C. and Williges, R.C., 1973, Response surface methodology cen-
    tral-composite design modifications for human performance re-
    search, Human Factors, 15, 295-310.
Clark, C.E. and Williges, R.C., 1972, Central-composite response sur-
    face methodology design and analyses, Savoy, Illinois, Univer-
    sity of Illinois, Aviation Research Laboratory, Technical Re-
    port ARL-73-6/ONR-73-2/AFOSR-73-3.
Clark, C.E., Williges, R.C. and Carmer, S.G., 1971, General computer
    program for response surface methodology analyses, Savoy, Illi-
    nois, University of Illinois, Aviation Research Laboratory,
    Technical Report ARL-71-8/AFOSR-71-1.
Cochran, W.G. and Cox, G.M., 1957, "Experimental designs", Wiley,
    New York.
Cook, R.D. and Campbell, D.T., 1976, The design and conduct of quasi-
    experiments and true experiments in field settings, in: "Hand-
    book of Industrial Organizational Psychology", M. Dunnette (ed.),
    Rand-McNally, Chicago.

Cooley, W.W. and Lohnes, P.R., 1971, "Multivariate data analysis", Wiley, New York.

Davies, O.L. (ed.), 1954, "The design and analysis of industrial experiments", Chapter II: The determinants of optimum conditions, Oliver and Boyd, London, 495-578.

Friedman, M. and Savage, L.J., 1947, Planning experiments seeking maxima, in: "Techniques of statistical analysis", McGraw-Hill, New York.

Harris, R.J., 1975, "A primer of multivariate statistics", Academic Press, New York.

Hicks, C.R., 1974, "Fundamental concepts in the design of experiments", Holt, Rinehart and Winston, New York.

Hoerl, A.E. and Kennard, R.W., 1970a, Ridge regression: biased estimation for nonorthogonal problems, Technometrics, 12, 55-56.

Hoerl, A.E. and Kennard, R.W., 1970b, Ridge regression: applications to nonorthogonal problems, Technometrics, 12, 69-82.

Keppel, G., 1978, "Design and analysis: A researcher's handbook", Wiley, New York.

Kirk, R.E., 1969, "Experimental design: Procedures for the behavioral sciences", Brooks/Cole, Belmont.

Levine, M.S., 1977, "Canonical analysis and factor comparison", Sage Publications, Beverly Hills.

Mills, R.G. and Williges, R.C., 1973, Performance prediction in a single-operator surveillance system, Human Factors, 15, 337-348.

Montgomery, D.C. and Johnson, L.A., 1976, "Forecasting and time series analysis", McGraw-Hill, New York.

Myers, J.L., 1972, "Fundamentals of experimental design", Allyn and Bacon, Boston.

Myers, R.M., 1971, "Response surface methodology", Allyn and Bacon, Boston.

Ostrom, C.W., 1978, "Time series analysis: regression techniques", Sage Publications, Beverly Hills.

Rozeboom, W.W., 1979, Ridge regression: Bonanza or beguilement, Psychological Bulletin, 86, 242-249.

Simon, C.W., 1971, Considerations for the proper design and interpretation of human factors engineering experiments, Culver City, California, Hughes Aircraft Company, Technical Report No. P73-325.

Simon, C.W., 1973, Economical multifactor designs for human factors engineering experiments, Culver City, California, Hughes Aircraft Company, Technical Report No. P73-326A.

Simon, C.W., 1974, Methods for handling sequence effects in human factors engineering experiments, Culver City, California, Hughes Aircraft Company, Technical Report No. P74-415A.

Simon, C.W., 1975, Methods for improving information from "undersigned" human factors experiments, Culver City, California, Hughes Aircraft Company, Technical Report No. P75-287.

Simon, C.W., 1977a, Design, analysis, and interpretation of screening studies for human factors engineering research, Westlake Village, California, Canyon Research Group, Technical Report No. CWS-03-77.

Simon, C.W., 1977b, New Research paradigm for applied experimental
    psychology: A system approach, Westlake Village, California,
    Canyon Research Group, Technical Report No. CWS-04-77.
Tatsuoka, M.M., 1971, "Multivariate analysis: Techniques for educa-
    tional and psychological research", Wiley, New York.
Williges, R.C., 1976, Research note: modified orthogonal central-com-
    posite design, Human Factors, 18, 95-98.
Williges, R.C., 1979, Working paper on research methodologies for
    system experimentation, Blackburg, Virginia, Technical Report.
Williges, R.C., 1980, Metrics for evaluation of human-computer in-
    teraction in a personnel records task, Paper presented at IEEE
    Conference on Cybernetics and Society, Boston, Massachusetts.
Williges, R.C. and Baron, M.L., 1973, Transfer assessment using a
    between-subjects central-composite design, Human Factors, 15,
    311-319.
Williges, R.C. and North, R.A., 1973, Prediction and cross-validation
    of video cartographic symbol location performance, Human Fac-
    tors, 321-336.
Williges, R.C. and Simon, C.W., 1971, Applying response surface me-
    thodology to problems of target acquisition, Human Factors, 13,
    511-519.
Winer, B.J., 1971, "Statistical principles in experimental design",
    McGraw-Hill, New York.
Wooldridge, L., 1980, A monte carlo approach to ridge discriminant
    analysis, Unpublished Master's Thesis, University of Central
    Florida, Orlando, Florida.

PART II
ANALYTICAL APPROACHES

A REVIEW OF SIMULATION LANGUAGES AND THEIR APPLICATION TO MANNED SYSTEMS DESIGN

Bernhard Doering and Wilhelm Berheide

INTRODUCTION

Recent years have seen intensive attempts to extend the application of computer simulation to analysis, comparison, and evaluation of complex systems. The reasons for that are among other things as follows: There are now demands for effective methods for solving complex system design problems in industry, commerce, government, and society. The efficiency of the computer has increased rapidly. New sophisticated simulation languages that are more user and problem-oriented have recently been developed. Additionally, a particularly significant impetus toward the use of simulation techniques has been the trend within virtually all physical, life and social sciences to become more quantitative in their methodology.

Why is it useful to simulate systems? One of the best answers to this question comes from Teichroew et al., 1967. The use of computer simulations has become increasingly widespread in their application to the study of the behavior of systems whose states change over time. Alternatives to the use of simulation are
- mathematical analysis,
- experimentation with either the actual system or a prototype of it, or
- reliance on experience and intuition.
All approaches, including simulation, have some limitations. Mathematical analysis of complex systems is very often impossible. Experimentation with actual or pilot systems is costly and time consuming and relevant variables are not always subject to control. Intuition and experience are often the only alternatives to computer simulation available but can be very inadequate.

Simulation problems are characterized by being mathematically intractable and having resisted solution by analytical methods. The problems usually involve many variables, many parameters, functions which are not well behaved mathematically, and random variables. Thus, simulation is a technique of last resort. Yet, much effort is now devoted to "computer simulation" because it is a technique that gives answers in spite of its difficulties, costs, and time required.

THE CONCEPT OF DIGITAL COMPUTER SIMULATION

Our primary concern in this presentation is a discussion of simulation languages, i.e. languages applicable to digital computer simulation. But before we go into detail, the concepts of systems and models and the nature of simulation as it arises when digital computers are used in the study of systems should be defined and clarified. In doing so we will try to identify the constituents and relationships involved in such a study and attempt to show where and how the element of simulation enters in. We are particularly concerned with the relationship between the concepts of system, concep-

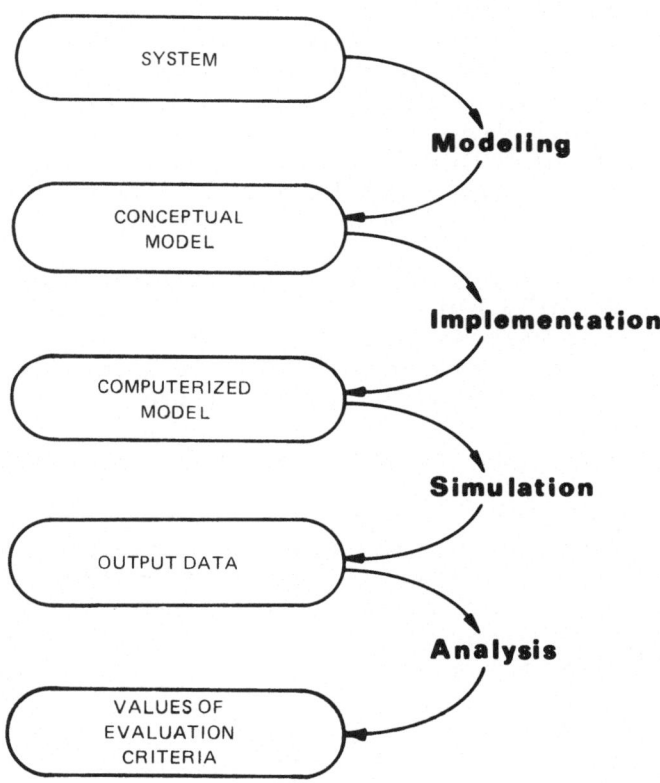

Fig. 1. The concept of digital computer simulation.

tual model, and computerized model and their relationships to the concept of digital computer simulation (Fig. 1).

The way in which simulation is treated here has its origin in the study of SYSTEMs (Fig. 1). A study usually tries to answer questions about the performance characteristics of systems. Often the need for simulation arises in the study of real systems, i.e. systems that actually exist. But frequently it arises in the study of proposed systems. Whether dealing with a real system or a proposed one, we assume that there is a definite system which is the object of interest and which is studied for one purpose or another. The system which has been selected for analysis is regarded as reality (Evans et al., 1967).

The systems that ergonomists have to deal with are man-machine systems (MMS). A MMS can be defined as a functionally organized assemblage of machines, men, and the processes by which they interact within an environment to produce some desired system outputs (Meister, 1971). A MMS is made up of machine and human elements which are regarded generally as system components (Fig. 2). There are relationships between them which can be material, energetic, or informational in nature. Relationships also exist between the system and its environment. Within a factory, for example, the environment consists of the physical room structure, loading docks, lighting, temperature, noise, etc. as well as of other MMSs which may be associated with the system considered. Usually the boundaries between a system and its environment depend on the problem to be solved and therefore can be defined only arbitrarily. Thus, from one point of view a given component may be regarded as part of the system, but from another point of view it may be regarded as part of the system environment. Another

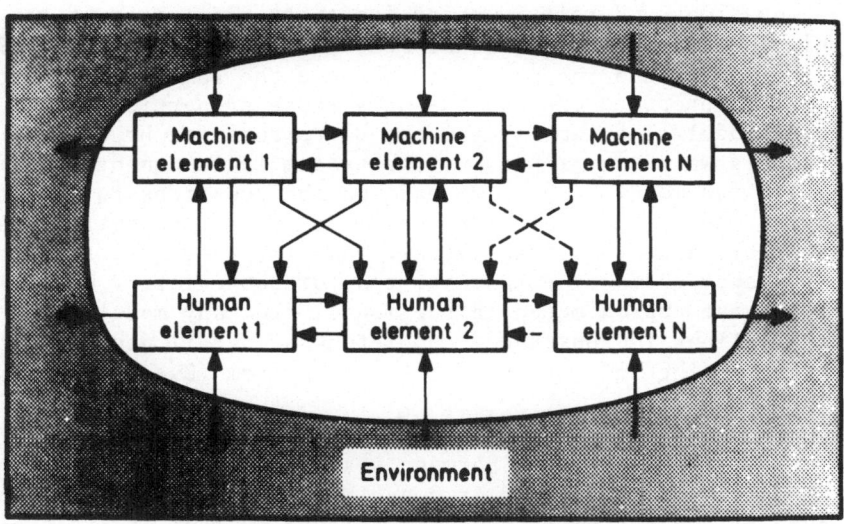

Fig. 2. Scheme of a man-machine system.

important property of system components is that they can be regarded
as systems themselves and hence as subsystems or a larger system. This
leads to a hierarchical structure of the problem area.

At any given instant in time any system component and consequently any system is in a particular condition, or state. The state can
be defined by descriptive variables of the components that serve as
tools to describe the conditions of the components at points in time
(Zeigler, 1976). The state of a system at any instant is determined
by the instantaneous state of all the components making up the system.

A description of the system state at any specific instant will
be called a state description corresponding to the state of the system at that instant (Evans et al., 1967). If a system is considered
during an interval of time it is necessary to consider many states
of the system and hence many state descriptions. Most commonly,
states of the system are considered for a chronological succession
of instants throughout the interval in question. This leads to the
construction of a state trajectory in time (Wymore, 1977), that is
a time sequence of states. A state trajectory consists of a succession of state descriptions corresponding to the states of the system
at each of a chronological succession of instants. This concept of
state trajectory will form the basis of our definition of simulation
later on.

An important concept relevant to the study of systems is the
concept of model. A model consists of a representation of the parts
of the system and of the interactions of the parts with one another
and the environment. In some cases the model may be a scaled physical object (ionic model). In other cases the model consists of mathematical equations and relationships (abstract model), or it may
be a graphical representation (visual model) (Pritsker et al., 1979).
Sometimes the system and the model are one and the same, e.g. if the
system is only a proposed one and has only those properties represented by the model. The same system can be represented by several
different models which may represent the system in fundamentally different ways. Which model is best depends on the questions to be answered with the model. With certainty any model of a real system does
not represent all aspects of the system. That is, there are features
of the real system that are either omitted, or abbreviated, or altered or otherwise approximated in the model, i.e. any model, no
matter how excellent in substance, completeness, detail etc., succeeds in representing only parts of the real system (Evans et al.,
1967).

In general, the connection between system and model is as follows: If an entity being studied is conceived of as a system, then
the structure of the system is the physical arrangement of its subsystems and components at any given time. All the changes that occur

in these subsystems, components, and the relationships among them
are called a process. Just as the structural relationships by an
iconic or visual model, the process characteristics and functional
relationships can also be revealed by using a dynamic or simulation
model. In this presentation we restrict our attention to dynamic or
simulation models that are built for use on digital computers. These
models which are called CONCEPTUAL MODELS (Schlesinger et al., 1979)
are represented by a mathematical-logical representation of a system
which can be exercised in an experimental fashion on a digital computer (Fig. 1).

Since a model is a representation of a system it is also an abstraction of a system. In MODELING a system (Fig. 1), i.e. developing that abstraction, it is necessary to decide which components of
the system have to be included in the model. To make such a decision
it is important to develop an objective for modeling that is based
on a stated problem or project goal (Pritsker et al., 1979). Reference to this objective should be made when deciding if a component
or relationship of a system is significant and, hence, should be modeled. For that, the boundaries of the system and the level of modeling detail have to be established. The success of modeling depends on how well the significant descriptive variables of the components and the relationships between them can be defined. Additionally, the desired performance measures and design alternatives to be
evaluated have to be included into the model. These can be considered
as part of the model or as input to the model. Assessments of design
alternatives in terms of the specified performance measures are considered as model outputs. Typically the assessment process requires
redefinition and redesign. Therefore, the modeling approach has to
be performed iteratively.

Once a conceptual model has been developed the next task is the
IMPLEMENTATION of the model on a digital computer (Fig. 1). With
this task the model is translated into a COMPUTERIZED MODEL (Schlesinger et al., 1979). Usually, for the implementation there are specific computers available. However, there are programs of such stature that the available resources and program objectives do justify,
and sometimes require, the acquisition of computer systems "dedicated" to that program. Given a suitable computer configuration as a
tool for simulation it is also necessary to consider carefully the
proper selection of software. Of particular importance in this selection is the language to be used in communicating with the machine.
Although a simulation model can be programmed using a general purpose
language, e.g. FORTRAN, there are distinct advantages to using a simulation language (Pritsker et al., 1979). In addition to the savings
in programming time, a simulation language also exists in model formulation by providing a set of concepts for articulating the system
description. One important part of the implementation is the verification of the computerized model. The verification task consists of
determining that the computerized model represents the conceptual

model within specified limits of accuracy, i.e. that the translated
model is executed on the computer as the modeler intended. This is
typically done by manual checking of calculations.

If the computerized model is implemented the next task is the
actual SIMULATION of the system activity or behavior with the digital computer (Fig. 1). Among modeling and simulation specialists the
word simulation means an experiment using a computerized model and
not the model itself (Gagne, 1976). Thus, we have to keep in mind
the fact, that neither the conceptual model nor the computerized
model as its substitute represents the system behavior directly. Instead, a state trajectory in time of a model provides a means of representing the activity or behavior of the system. That happens because a model is a mapping of the system, and therefore a given state
of the model represents a given state of the system. Consequently a
succession of model states represents a succession of system states.
So it is not the model itself, but rather a state trajectory of the
model that provides a description of system activity or behavior. As
mentioned before, a state trajectory in time consists of a succession of state descriptions corresponding to the states at each of a
chronological succession of instants. That means, simulation consists
of a construction in chronological order of state descriptions making
up a state trajectory. Thus, given a system and a model of that system, simulation can be defined as the use of the model to chronologi-
cally produce a state trajectory of the system model. In this application of simulation the relationship between the computerized model,
the conceptual model, and the modeled system is the following: The
computerized model causes the computer to produce output data which
is a state trajectory of the conceptual model, but which is regarded
as a state trajectory of the modeled system (Evans et al., 1967).

Thus, the model is the key product of the model building activity, as it represents the understanding of the system appropriate
to the requirements of the project. If the computerized model is assumed to be an accurate implementation of the conceptual model, then
one can discuss prediction of system behavior assumed to be based on
simulation experiments with the model itself (Gagne, 1976).

The OUTPUT DATA (Fig. 1) produced by the simulation describes
the dynamic behavior of the system considered over time. The statistical ANALYSIS of the simulation outputs is similar to the statistical analysis of the data obtained from an actual system. The main
difference is that the simulation analyst has more control over the
running of the simulation program (Pritsker et al., 1979). Thus, he
can design experiments to obtain the specific output data, i.e. VAL-
UES of EVALUATION CRITERIA, necessary to answer the pertinent questions relating to the system under study. One important task of the
analysis is the validation of the model built so far. The validation
task consists of determining that the computerized model is a sufficiently accurate representation of the actual or proposed system to

be of some use in shedding light on the specific questions which motivated the modeling project in the first place (Gagne, 1976). It is
in this validation phase that accumulated errors are discovered, and
it is here that the final acceptance of the model must be achieved.
If the computer program is found to be inadequate, it's errors are
traced back either to programming errors which were overlooked, or
to a lack of understanding of the system itself which may have resulted in an insufficient conceptual model. Validation is normally
performed in levels. Pritsker et al., 1979, recommended that a validation be performed on data inputs, model elements, subsystems, and
interface points because in simulation models, there is a correspondence between the model elements and system elements. Hence,
testing for reasonableness involves a comparison of model and system
structures and comparison of the number of times elemental decisions
or subsystem tasks are performed.

TYPES OF SIMULATION LANGUAGES

       As already mentioned, our attention will be restricted to simu-
lation on digital computers, i.e. the system of interest will be represented by a conceptual model which during experimentation is a
computer program, that represents the computerized model. Simulation
of this type is correctly called symbolic simulation using a digital
computer, or briefly, digital computer simulation. Simulation can be
conducted in real time or not. Real time computer simulation usually
is only necessary where an actual system component such as a human
functioning in real time is involved. Without real time, simulation
can be conducted in fast time, slow time, or some combination at different points of the simulation process. Concern here is only with
this kind of simulation. That means, in the simulation languages
considered here time is expanded or compressed according to the requirements of the problem. Only those digital simulation languages
which deal with pure digital simulation will be considered here, i.e.
hybrid simulation will not be dealt with.

Levels of Language Hierarchy

       The task of programming has been aided considerably by the de-
velopment of special purpose programs, assemblers, interpreters and
higher level languages which represent different levels of a language hierarchy (Fig. 3). The first stage in programming was the direct use of MACHINE LANGUAGES, i.e. 0's and 1's. However, it is time
consuming, boring and nearly impossible to program at that level.

       On the next level of the language hierarchy ASSEMBLY LANGUAGES
offer more help (Fig. 3). These languages consist of grouped alphabet
characters called mnemonics that replace the numeric instructions of
machine languages. These languages are usually machine dependent.
Assembly language programming enables us to obtain the greatest possible efficiency from a given computer, i.e. to optimize computing

speed and/or memory requirements. Sequences of assembly language
statements correspond directly to actual hardware instructions and
data-word structures in the memory of a specific machine. On the other hand, assembly languages require considerable programming effort.
If one considers the trend that computer hardware costs are decreasing while software costs are increasing, the advantage of assembly
languages have become a minor factor and therefore are hardly used
anywhere except when there are severe running time constraints (e.g.
real time programming) or if programs are to be used repeatedly many
times. Even with time constraints it might still be more effective
to use faster machines to fulfill those requirements.

HIGHER LEVEL LANGUAGEs are the next hierarchical level in computer languages (Fig. 3). They differ from lower level languages
(e.g. assembly languages) in that the instructions take the form of
fairly complex statements. They are usually machine independent,
user-oriented and for general purpose. Machine independence means
that a program written in such a language is not dependent on any
particular type of computer. Accommodation of that program to a particular type of computer is accomplished by translating the statements of the language through the compilation process into the code
of that particular machine. Higher Level Languages may be compiler
or interpreter languages. A program written in a COMPILER LANGUAGE
must be converted as a whole into machine code before execution of
the total program. The best wellknown compiler languages are ALGOL,
COBOL, FORTRAN, PL/1 and PASCAL. With an INTERPRETER LANGUAGE a program is compiled during execution converting each statement of that
particular program into machine code and executing it before going

HIERARCHY LEVEL	LANGUAGES	EXAMPLE
4	Special Purpose Languages	SIMULATION LANGUAGES
3	Higher Level Languages	FORTRAN, ALGOL. COBOL, PASCAL
2	Assembly Languages	PDP 11: MACRO 11 SIEMENS: ASSEMB
1	Machine Languages	

Fig. 3. Hierarchy of digital computer languages.

into the next statement. BASIC and APL are two examples of interpreter languages. In general, a program written in an interpreter language can also be compiled as a whole before excution. Interpreter languages are inefficient for programs which are to be executed many times, since each execution will be slowed by translation. Higher level languages are usually spoken of as user-oriented which means that the user works directly with mathematical notations or with recognizable english statements although highly stylized.

The next level of language hierarchy consists of SPECIAL PURPOSE LANGUAGEs in contrast to higher level languages which are for general purpose (Fig. 3). These languages are developed for special problem or application classes. The application class considered here is simulation and hence the special purpose languages described in the following are SIMULATION LANGUAGEs (SLs).

Relative to higher level languages SLs provide additional aids to the programmer, e.g:
- The user can express several functions in one statement.
- Debugging aids are normally available.
- SLs offer a conceptual framework for precise thinking more or less based on system theoretic concepts, i.e. the components of the system to be modeled have to be identified and described in terms of the given concept and the relations between these components have to be stated.
- SLs provide a notation for the description of dynamic models, e.g. flow diagrams, networks, or blocks.

These points may become clearer after we have explained the simulation concepts in more detail.

Discrete Simulation Languages

First, we will consider the concept of discrete event simulation. To explain this important concept in detail the standard introduction example is used: a one server queuing system. In that case a single server, e.g. a single person or device, serves some items, e.g. customers in a barber shop or cars in a wash installation. This situation is illustrated in Fig. 4. Customers arrive with unpredictable time intervals. If the server is busy when a customer arrives, the customer enters a queue. As long as there are customers waiting, the server will be in continuous operation serving the customers here

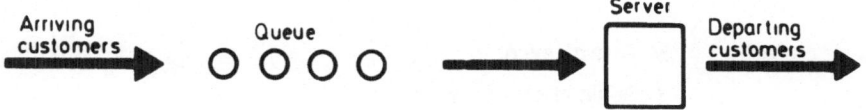

Fig. 4. Scheme of a single server queue.

on a First-In First-Out (FIFO) priority basis.

    This simple model of a queuing system is characterized by two
independent random variables:
- the time between consecutive arrivals of customers to the system
  (inter-arrival time) and
- the time required for the server to perform a service.

    The activity and event time lines of this example are shown in
Fig. 5. The upper part of the diagram shows the activities of the
queuing model on a time axis. An asterisk (*) denotes when a customer
arrives. The dashed line indicates the time period a specific customer has to wait, and the full line means serving the specific customer.
Note, that only one customer can be served at one time. The complete
line for each customer represents the total time the customer is in
the system. The middle part of the diagram illustrates the contents
of the queue over time and the lower part shows whether the state of
the server is idle or busy.

    As stated earlier, simulation is the production of a state tra-
jectory in time. In the given example there are only two variables

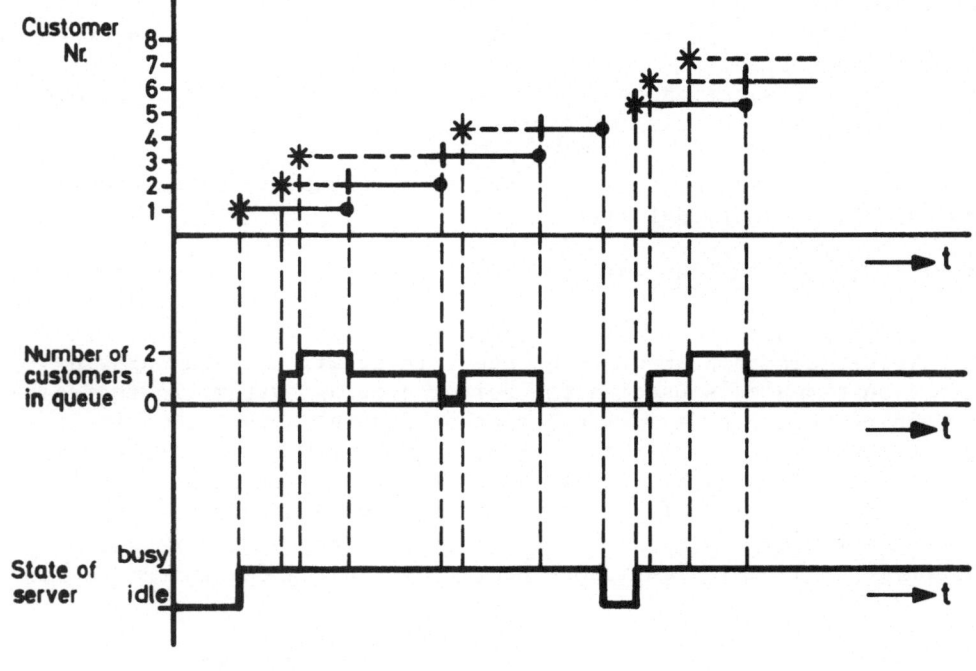

* Arrival event

\| Begin of service event

● End of service event

Fig. 5. Example of an activity and event time line.

that express the state of the model:
- the number of customers in the queue (an integer from 0 to N) and
- the status of the server (a boolean variable either idle or busy).

Note that in such a model we disregard a number of system details which are irrelevant for our purpose, such as physical movement of customers or manual operations of the server and we restrict
our attention to only these two variables. But even based upon the
outcome of this very abstract and simple model it is possible to
make important decisions if one considers how long the server is
idle, e.g. costs are incurred even when the server is doing nothing,
or how long the customers must wait, they might e.g. be dissatisfied
and go to another shop. This model, although very simple, is applicable to a variety of systems, e.g. ticket counters, bank tellers, barber shops, ships in port, etc.

The state of the described model consisting of values of the
two variables, number of customers in the queue and the status of
the server, changes at specific points in time either when the next
customer is arriving, when a specific customer is beginning to be
served, or when service for this customer has ended. These isolated
points in time at which the state changes are called EVENTs. In Fig.
5 every vertical dashed line that crosses the time line marks an e-
vent. An event is the central concept employed in so-called discrete
event simulation. We can also say that an event is an instant in
time at which an activity starts or stops.

In the described simple one-server queuing model every customer
is involved in two activities:
- the waiting activity which occurs between the arrival event and
  the begin-of-service event (represented by the dashed line in the
  upper part of Fig. 5),
- the service activity which occurs between the begin of service e-
  vent and the end-of-service event (represented by the line between
  the mark and the point in the upper part of Fig. 5).
  Note that in our example the time for the first activity is
sometimes 0, if the server is idle.

Such a time-ordered sequence of events or activities is called
a PROCESS. Even in this simple example some processes are working
in parallel. But suppose in the case of a multi-server system, there
would be a lot of activities in parallel and hence many events that
would not belong to the same process. Because on process can interact with another, e.g. upon termination of one process, another process could be stopped, there has to be a clear timing mechanism in
the computerized model that advances from one event to another on
the time line. For that timing mechanism all events have to be stored
in a schedule and the computer internally has to set the so-called
simulation clock to the time of an event, to change all states that
have to be changed at that time, and to schedule future events that
are caused by this event.

In some SLs an appropriate subroutine, that has to be provided
by the modeler, is called at these event times. The modeler describes
in this subroutine what should happen at this instant of time as described above, i.e. change the state of the system and schedule the
events that are caused by the current event. This type of SL is
classified as EVENT-ORIENTED SIMULATION LANGUAGE. These SLs provide
the user with a number of features, chief among which are the timing
routine for processing events in their proper order and some statements or subroutines to schedule or reschedule events. The most widely used event-oriented discrete SLs are GASP II (Pritsker, 1974) and
SIMSCRIPT (Kiviat et al., 1969) (Fig. 6).

Another category of discrete SLs are the PROCESS-ORIENTED SIMU-
LATION LANGUAGEs. Although most process-oriented SLs perform internally the same event algorithm as event-oriented languages, they in
addition provide the modeler with a more natural way to build a system model. This means, the modeler can think of the modeled system
more in terms of system components and their interactions. Process
simulation is based on a process network or flow chart which follows
the progress of temporary elements through the modeled system from
its arrival to its departure event. These temporary elements are
called transactions or entities.

DISCRETE SL			CONTINUOUS SL
EVENT ORIENTED	PROCESS ORIENTED	ACTIVITY ORIENTED	CSSL IV CSMP 360 ACSL MIMIC DARE DYNAMO
SIMSCRIPT GASP II	SIMULA GPSS Q-GERT	CSL	

COMBINED DISCRETE-CONTINUOUS SL	
EVENT ORIENTED	PROCESS ORIENTED
GASP IV C-SIMSCRIPT	SAINT SLAM

Fig. 6. Classification of simulation languages.

Transactions may represent any element, which is material, information or energy, expressed in quantitative units appropriate to the characteristics of interest. In the case of the server queuing model these transactions are the customers. The queue and the server as system elements represent stationary elements. The interactions between a specific transaction, system elements, and other transactions are described by a set of blocks or nodes and branches that actually form a network. At this point all that is necessary is a well defined coding process to translate these pictorial descriptions into statements that will be accepted by a process-oriented language. In this sense it is possible to speak of process-oriented SLs as being at a still higher level than event-oriented languages.

The network example of the single server queuing model in the process-oriented SL GPSS (Schriber, 1974), most widely used, would look like Fig. 7. According to the chosen example the network on the left side of the figure which represents the structure of the model is rather simple. On the right hand side the appropriate language statements can be seen that are to be entered into the computer.

As transactions, i.e. temporary elements, flow through the network, they change the state of the model, e.g. in the example the server status is changed to idle or busy and the quantity of the queue is changed. Besides the already mentioned language GPSS, the most important other SLs that use the process-oriented concept are

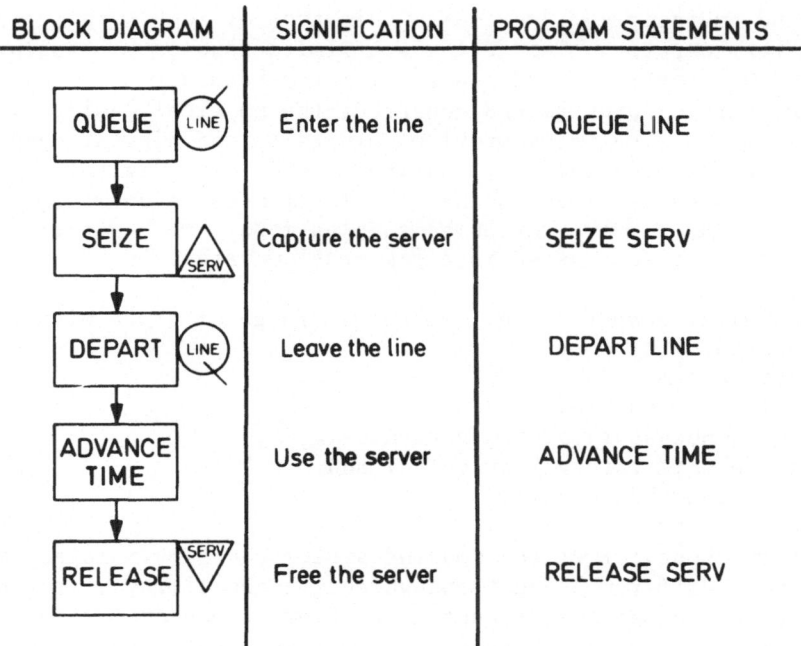

BLOCK DIAGRAM	SIGNIFICATION	PROGRAM STATEMENTS
QUEUE (LINE)	Enter the line	QUEUE LINE
SEIZE (SERV)	Capture the server	SEIZE SERV
DEPART (LINE)	Leave the line	DEPART LINE
ADVANCE TIME	Use the server	ADVANCE TIME
RELEASE (SERV)	Free the server	RELEASE SERV

Fig. 7. GPSS example of a single server queue.

SIMULA (Hills, 1973, Rohlfing, 1973) and Q-GERT (Pritsker, 1977) (Fig. 6).

For completeness, a third category of discrete event SLs, the class of ACTIVITY SCANNING-ORIENTED SIMULATION LANGUAGEs, should be mentioned briefly. With this language class the modeler describes the activities in which components of a modeled system are engaged. This simulation form concentrates on the set of conditions that determines when activities can be started or stopped. Further information about this orientation can e.g. be found in Emshoff et al., 1970. A simulation language example based on this approach is CSL developed by IBM (Fig. 6).

Continuous Simulation Languages

So far we have only been considering the concept of discrete event simulation. At this concept the state of the model changes only at discrete instants of time at which interactions between system components and transactions occur. We also made the distinction between event-orientation and process-orientation and touched the category of activity scanning orientation.

In contrast, the concept of continuous simulation uses models whose state changes continuously with time. The continuous change of state is generally represented mathematically by differential or difference equations that describe rates of change of the descriptive variables over time. Such models have long been used in the physical sciences and engineering. Previously this type of model was implemented on analog computers. The well known limitations of the analog computer, such as restricted model size, limited accuracy, and the large effort necessary to program and construct the physical analog circuitry has caused a shift by problem solvers to the digital computer and to appropriate digital SLs. Continuous models on digital computers have also been applied to the social sciences, economics, and operations research where variables such as money, men, material, information, etc. are expressed in terms of flow rates.

On a digital computer the continuous change of state with time is approximated by
- choosing some time increment delta,
- solving all the equations,
- changing the necessary status variables,
- stepping the time interval by delta, and
- solving the equations again.

Hence the behavior of the modeled system can be simulated by solving finite difference equations which, in the limit, approach the differential equations of continuous flow. It becomes clear that continuous models are also discretized in time and the values of all variables are discretized because they are changed in steps. A dis-

cretization was necessary anyway because of the limited accuracy of
a real value in a computer. The time increment is chosen according
to the resolution of the variables required by the model. There is
always a trade-off between accuracy and computer run time in selecting step size.

SLs used for implementing models with continuous state changes
are classified as CONTINUOUS SIMULATION LANGUAGEs. A continuous SL
assists the modeler by defining the format of his equation set. But
the main feature of a continuous SL is the evaluation of the equation set, to obtain specific values of the descriptive variables
in time. Sometimes, solutions of these equations can be found analytically. In many cases, however, the results can be obtained only
be integration of the derivatives. In continuous SLs, alternative
integration methods are available which can be choosen according to
the nature of the problem, accuracy requirements and computing time.

The SCi Software Commitee (SCi, 1967) stated in 1967 a standard
for continuous system simulation languages (CSSL). Along those lines
a lot of simulation Languages as CSSL I to CSSL IV and ACSL have
been produced. Other well known continuous simulation languages are
CSMP 360 and CSMP III, produced and supported by IBM, and MIMIC,
DARE and DYNAMO (Fig. 6).

To summarize, the basic major characteristics of discrete event
and continuous SLs are regarded. The crucial difference between both
types of SLs is that with the concept of discrete event simulation
the model state changes only at discrete instants in time whereas
with the concept of continuous simulation the model state changes
continuously. Another important difference is that in discrete event
simulation the major factor is the scheduling and sequencing of e-
vents, whereas in continuous simulation, differential equations are
represented. In simulation various activities proceed in parallel
with both types of concepts. In discrete event simulation only those
status variables that are affected at a particular event or by a
specific transaction that flows through the system are changed. In
continuous simulation all status variables are evaluated in parallel
at all times. The outputs of the discrete simulation are statistical
time averages and distribution histograms of descriptive variables.
For continuous simulation, graphical results in the form of continuous time trajectories are essentially obtained.

Combined Discrete and Continuous Simulation Languages

A very important concept of simulation is that of combined discrete event and continuous simulation. It can handle both characteristics concurrently. In combined discrete continuous models the state
may change discretely, continuously, or continuously with discrete
jumps superimposed. In addition there are two types of events that
can occur in combined simulation:

- time events which are scheduled to occur at specific points in time
  (already described in discrete event simulation) and
- state events which are not scheduled, but occur when the continuous part of the model reaches a particular state.

To make the simultaneous running of continuous and discrete e-
vent models meaningful, the two different model parts must be able
to interact. Combined interaction can be initiated by the two types
of events, by
- a time event which can influence its own discrete or the continuous part of the model,
- a state event which can either influence its own continuous model
  part or change parameter or values of the discrete part.

To explain these interrelations an example may be helpful. At
an unloading dock for oil tankers a discrete variable is the number
of tanker ships being unloaded. When a ship arrives for unloading
(discrete event), the number of unloaded tankers are increased. This
affects the continuous part, e.g. the amount of oil that flows into
the tank and therefore the amount of oil that is in the tank. On the
other hand, if the tank reaches a particular amount of oil (state
event), the unloading operation has to be interrupted and other e-
vents will be caused, e.g. stop the pump, switch the unloading operation to another tank, etc.

SLs used with the concept of combined discrete event and continuous simulation are classified as COMBINED DISCRETE AND CONTINUOUS
SIMULATION LANGUAGEs. SLs of this type are e.g. GASP IV, C-SIMSCRIPT,
SLAM, and SAINT (Fig. 6). It is especially this type of SLs which
is most appropriate for modeling and simulation MMSs. Therefore more
details about the combined discrete and continuous SLs will be given
in the next chapter after the criteria for selecting SLs have been
described.

Thus far, the features of the discrete event, continuous, and
combined discrete event continuous simulation concept have been considered. But what do all simulation concepts and the corresponding
SLs have in common? With every simulation concept a state trajectory
that represents the behavior of the modeled system is generated. All
SLs provide some sort of framework for building a conceptual model.
One of the main tasks of a SL is to transform a conceptual model into a computerized model, i.e. to support and to perform the implementation process. Every SL unburdens the user from programming the
time advancing mechanism and therefore from handling the concurren-
cy of activities. At least, with discrete event SLs random numbers
and variates are generated to model random processes, e.g. for systems generally composed of one or more elements that have uncertainty. Such systems have states which evolve through time in a manner
not completely predictable and are referred to as stochastic systems.
The modeling and simulation of stochastic systems requires that the

variability of elements in the system can be characterized using probability concepts. SLs also provide support for the user in evaluating and preparing results. Simulation outputs are also probabilistic and therefore statistical interpretations about them are usually required. This includes the calculations of averages, standard deviations and confidence intervals. Also graphical representation of results can be provided.

SELECTING SIMULATION LANGUAGES APPLICABLE TO MAN-MACHINE SYSTEM SIMULATION

In determining criteria for selecting simulation languages it is useful to remember the concept of digital computer simulation where four steps are outlined (Fig. 1):

- MODELING the system to get a conceptual model,
- IMPLEMENTATION of the conceptual model by transforming it into a computerized model.
- SIMULATION with this computerized model to obtain the behavior of the system in the form of output data, and
- ANALYSIS of this output data resulting in evaluation criteria.

The user of SL would prefer to concentrate his efforts only on the first and last step, i.e. to form a conceptual model and then analyze the model output data. The second step, implementation of a conceptual model, could be accomplished by a programmer and the simulation itself is performed automatically by the computer. However, there are reasons why a SL should support the user in all four steps. The criteria for selection of a language can be classified according to these steps.

Most important for the MODELING process is whether the language fits the problem, that means e.g. for combined discrete and continuous processes you can only use the appropriate combined SL. In selecting a language it is necessary to look not only at the particular problem at hand but also at what other problems might occur that need simulation. In other words the SL should be adaptable to problems that will likely arise in the future in that particular field. In most cases, a trade-off will have to be made between the ease of formulation of the problem and the problems flexibility of that simulation language. Generally, the more problem specific a simulation language is, the less flexible it is for other problems.

Also which framework a SL offers for the conceptual model is important to the modeling process. If this framework is based on systems theory, it can match those systems which were decomposed according to this theory relatively well. Such systems approaches are becoming more and more common. The elements of the system may then correspond directly to structural elements that the language offers.

There is a tendency towards that goal but such languages are not a-
vailable yet. (The reader desiring more information may consult
Oeren et al., 1979). However, some concepts available in SL are quite
useful for constructing conceptual models. One of the main reasons
why e.g. GPSS is so successful, is because its modeling concept uses
block diagrams which engineers are usually familiar with. These block
diagrams represent the structure of the system and transactions that
flow through this block diagram are equivalent to the temporary elements, i.e. matter, energy and/or information, that flow through the
system. A lot of systems can be structured with such a conceptual
framework, e.g. job shops, transportation systems, inventory systems,
and information systems. The modeler is given a tool that leads him
to think of his system in terms of the SL. This may help him to structure and classify the elements of his system and their relationships.
Other SLs, e.g. SIMULA, offer a similar concept. It is important that
the graphical construction of a conceptual model by means of block
diagrams or networks facilitates documentation of the model and provides a very good basis for discussion of the problem in inter-dis-
ciplinary teams.

Another criteria for the modeling process is the ease by which
the process can be learned in that particular language. Not only is
the described modeling concept important, but also how well the language is documented and if there are good manuals and courses available. The more that the modeling process is separated from the following implementation process the more one can confine his thinking
to the conceptual model and problems ergonomists have to deal with.

The crucial point for IMPLEMENTATION of a conceptual model on
a computer is that the selected language matches the capabilities of
the computer. If the language needs a specific compiler that is not
available for that particular machine, it can naturally be excluded
from further discussion. But even if the compiler is available one
should consider whether your present computer will have to be replaced in the future or whether the developed model may have to run
on another machine. A SL consisting of FORTRAN routines can e.g. be
transferred to almost any computer that can cope with the size of
the language.

After the conceptual model is stated, the user wants little to
do with the implementation process. If a clear "interface" between
modeling and implementation exists, it can be done by another person, a programmer or computer system specialist. The closer that
code elements that have to be fed into the computer are to a the representation in the conceptual model, the easier, quicker, more reliable and more automated this task can be done. However, this is
only possible if there are restricted elements for model building in
a particular language. On the other hand this restriction can affect the flexibility.

Again, if good documentations, manuals, and courses are available, the implementation step can be more effectively learned and
accomplished. Very important also is to what degree the language
provides error diagnostics, e.g. tracing and debugging aids. When errors occur, they should occur only on the input level of the language.
This requires a reliable translator or compiler.

How the SIMULATION itself is carried out, is not very important
from the users point of view. Decreasing computer hardware costs
makes running time and core requirements less important. In addition
to that it is hard to evaluate languages on this criteria, because
while running time can be very lengthy with one particular problem
in a particular language it might be vice versa with another problem.
In what particular orientation a language proceeds or whether the
language is a compiler or interpreter language is only important to
the extent that the modeling concept and the implementation step are
affected. Good error diagnostics have to be provided for the simulation step also. An interesting aspect for the future might be interactive simulation, where some decisions can be made by men during
the simulation process dependent on the state of the modeled system.

The last step in the digital simulation process is ANALYSIS of
data produced by simulation. For that analysis the data have to be
handled like data derived from a real experiment or test. One can
not expect a SL to also provide for the total statistical analysis.
For such purposes statistical packages such as SPSS (Nie et al.,
1975) or BIOMED (Dixon, 1968) are available. However, because in
most cases it is neither useful nor possible to store all state
changes that occur during simulation, we need some sort of data reduction, aggregation, interpretation and documentation. Therefore
all simulation languages offer certain basic statistical functions
such as calculation of means and standard deviations, and certain
formats for presenting results such as histograms and some sort of
plots. Some of these possibilities are performed automatically whereas others have to be called up in a subprogram or can be initiated
by using some symbols or codes during the implementation process.

In selecting SLs used in designing manned systems it is expe-
dient to look at those processes by which the components of an MMS
produce desired system outputs. Two kinds of processes can be distinguished: The machine processes of the machine components and the
task performance processes of the human components, i.e. of the operators and maintainers.

A machine process consists of activities in a logical or chronological order that are accomplished by the machine component of a
MMS to produce desired outputs. If the system is, for instance, an
aircraft or another type of vehicle, system states can be regarded
as outputs that in the case of the aircraft determine the flight
trajectory. The human task performance process is composed of those

tasks which an operator has to execute to affect the machine process and in this way the system outputs. There are two different categories of the task performance process operators have to perform
for affecting the machine processes.

To explain this statement in more detail let us take an aircraft system as an example. In describing the machine process of an
aircraft system we have to consider the differential equations of
flight dynamics. These are continuous in nature. Therefore the machine process is continuous in nature also. The human task performance process in an aircraft system depends on the mode of aircraft
operation. In the case of the autopilot mode the machine process is
highly automated. The pilot essentially has to monitor the machine
process and decide the correctness of the flight's progress. If necessary he modifies the flight path, for instance, by giving new desired values of the speed, the altitude, or the heading. All these
activities of the pilot during the autopilot mode have discrete outputs. Thus, the task performance process of the pilot is discrete in
nature consisting of discrete monitor and decision outputs. If the
aircraft is in the manual mode the pilot has to control the attitude
of the aircraft continuously. In this case the human task performance
process is a continuous one. That means, in the aircraft system regarded both discrete and continuous task performance processes occur. In general, one can say that in MMSs discrete and continuous
processes may appear. For problems arising with manned systems design
it is therefore useful to select a combined simulation language.

Therefore we restrict our attention now to combined discrete
and continuous SLs to the exclusion of the others. It is also not advisable to consider some languages that have been used only in academic institutions, that are not well documented and which exist only
in preliminary versions, because there is no support if problems
arise. In 1977, 18 languages for simulation of combined discrete and
continuous systems were reviewed by Oeren (1977). Some of these deal
with hybrid computer simulation which is not of interest here. Out
of the others only one has been widely used, namely GASP IV (Pritsker,
1974).

GASP IV has a lot of applications, is supported by a company,
is written in FORTRAN and is therefore portable and well-documented.
Flexibility arises from its event-orientation and that it is actually a FORTRAN Program package that allows one to insert his own specific FORTRAN routines. On the other hand no graphical approach is
available for the modeling concept.

A very popular discrete event SL SIMSCRIPT II.5 has recently
been augmented with continuous features and the new language is referred to as C-SIMSCRIPT (Delfosse, 1976). It might be comparable
to GASP IV because it has nearly the same characteristics. An advantage is the free form English-like syntax which tends to be self

documenting. On the other hand this language needs a specific compiler that is not distributed as widely as FORTRAN compilers.

A very new SL for combined discrete and continuous simulation
came out in 1979 called SLAM (Simulation Language for Alternative Modeling) (Pritsker and Pegden, 1979), that might replace GASP IV and
C-SIMSCRIPT in the future, because of its unique features. SLAM is
written in FORTRAN. It is based on GASP IV and can be used similarly, i.e. event-oriented. Additionally SLAM can be used process-oriented, i.e. it employs a network structure comprised of specialized
symbols in a manner similar to GPSS. It is possible to use both features either exclusively or concurrently. In other words to the por-
tability and flexibility of GASP IV is added a higher level process-
oriented concept. The only disadvantage that can be seen so far is
higher core space and time requirements.

A fourth SL that should be considered for manned systems design
is SAINT (Systems Analysis of Integrated Networks of Tasks)(Duket et
al, 1978; Wortman et al., 1978). This language is also based on GASP
IV and is especially designed for simulation of man machine systems.
It uses a graphical network modeling approach, whose fundamental elements are tasks, resources (equipment and/or personnel) required to
perform the tasks, and relationships among the tasks. A task can be
thought of as an aggregation of several blocks or nodes in the sense
of GPSS or SLAM. Because SAINT is a specialized language, it cannot
be as flexible for a variety of problems. Although it is based on
GASP IV, it is not possible to use all its features as can be done
in SLAM. More about this language is given in the Chubb article entitled SAINT, A Digital Simulation Language for the Study of Manned
Systems, in this publication.

USE OF COMPUTER SIMULATION IN MANNED SYSTEM DESIGN

During design of manned systems ergonomists have to deal with
various problems at different levels of detail. An extensive description of the activities ergonomists have to perform during the design
process is given e.g. by De Greene (1970), Doering (1976), Meister
(1971), Shackel (1974), Singleton (1974) and Singleton et al. (1967).
In this presentation only a short overview of the main ergonomic design steps performed during the development phases of manned systems
will be given to show where computer simulation may be applied ad-
vantageously. Later some application examples will be described.

Given that system performance requirements and system functions
have already been determined, e.g. by systems analysts, the first
problem that ergonomists are confronted with is the allocation of
system functions to personnel and machine elements of the system
(Fig. 8). The FUNCTION ALLOCATION process occurs with increasing level of detail at all stages of the system development process, i.e.

basic design steps

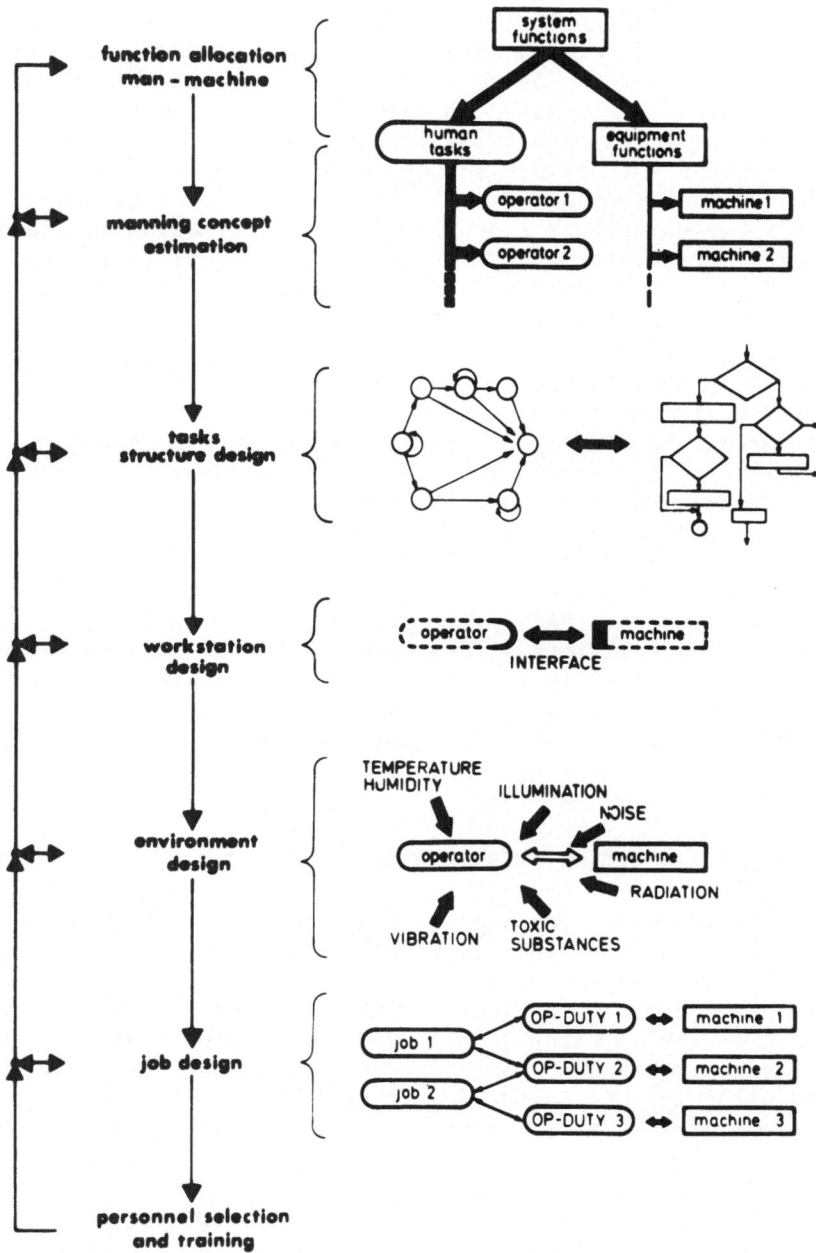

Fig. 8. Ergonomic design steps during manned system design.

from preliminary planning in the conceptual phase to final design at
the end of the design phase of a system. The further analysis of
those functions which have been assigned to machines and the design
of machine components (with the exception of man-machine interfaces)
are accomplished primarily by the engineering staff. The ergonomists
are mainly responsible for the analysis and synthesis of those tasks
that are assigned to personnel. The analysis results are the basis
of further design tasks ergonomists have to perform during manned
system design.

One of the first ergonomic design tasks is to make a preliminary ESTIMATE OF MANNING REQUIREMENTS. By this estimation the number
and types of personnel required for system operation and maintenance
are roughly determined. Starting with preliminary manning requirements a preliminary tasks structure and workstation concept will be
generated which may be changed if detailed results of the following
ergonomic design steps become available.

In DESIGNING THE TASKS STRUCTURE in detail, the logical and
functional sequence of tasks and procedures of personnel, i.e. task
performance processes, are arranged. When system personnel perform
these processes system outputs are affected. The tasks structure has
to be designed to accommodate variations in the course of events and
activities that go on in the machines and in the system environment.

DESIGN OF WORKSTATIONs here means design of man-machine interfaces, workspaces, or consoles, and the arrangement of equipment and
personnel in work areas. These designs are determined essentially by
the characteristics of those tasks that are the constituents of the
task performance processes.

The task performance processes of personnel are effected by
ENVIRONMENTAL FACTORS such as lighting, noise, temperature, humidity,
ventilation, acceleration, vibration, weightlessness, radioactivity,
pollutants, and contaminants, etc. These factors tend to reduce personnel efficiency or worse, lead to health problems. Therefore, ergonomists have to consider and control these environmental influences
in their design.

Knowledge of tasks, workstations, and the environment are necessary to determine in detail the number and types of personnel required for operation of system workstations and areas and maintenance
of system equipment which in turn is a prerequisite to JOB DESIGN.
Job design consists of grouping tasks, functions, duties and responsibilities into jobs or positions to be performed by different individuals.

For completeness it should be mentioned further that, after the
system is designed, ergonomists are involved in SELECTION AND TRAIN-
ING OF PERSONNEL and the design of job aids. Although the major a-

mount of matching between man and machine is accomplished by design of machine elements, final matching and adjustments of individual differences in personnel are effected by selection and training.

As system development proceeds from the conceptual phase and through definition and design phase, stated ergonomic design steps have to be treated in increasing detail. Additional details about system requirements and constraints also become known in this process and these in turn affect the design process. Because of the nature of this design process a general systems engineering approach for problem solving can be applied to the ergonomic design steps during system development. This approach (Fig. 9) consists of the following three major phases (Haberfellner, 1978/79):
- goal elaboration,
- solution generation, and
- solution selection.

During the first step of the GOAL ELABORATION phase the problem area is structured, i.e. the boundaries between system and environment are established. This is done by analyzing the problem situation, e.g. the missions that the system will have to fulfill. By this analysis requirements and constraints relative to missions, system performance, costs, personnel, time aspects, applicable technologies, system environment, etc. are determined. This data constitutes the informational frame for decomposing design goals down to the level of functional requirements which are the basis for the next phase.

The next phase is SOLUTION GENERATION. It consists of solution synthesis and analysis. During synthesis a sufficiently broad spectrum of alternative design concepts are generated out of which the optimal problem solution can later be selected. Analysis is hard to separate from synthesis because normally each synthesis activity is accompanied intuitively, and perhaps unconsciously, by some analysis activity. But here we are talking instead about a formal and conscious analysis performed after major and important design results have been synthesized. During this analysis each of the concepts generated is examined with regard to design goals in order to eliminate inadequate solutions and identify unsatisfactory ones which may later be improved. It is especially during analysis where computer simulation techniques can be applied.

The last phase in the problem solving approach is SOLUTION SELECTION. In this phase previously generated design concepts are evaluated with regard to their ability to fulfill design goals stipulated in the first phase. In this connection detailed design goals are regarded as evaluation criteria and assigned weighting factors that establish their relative importance. To have an objective means of evaluating functions it is desirable to specify as many criteria as possible in quantitative terms. Assuming that all quantitative data have been gathered for each criterion, it is possible to com-

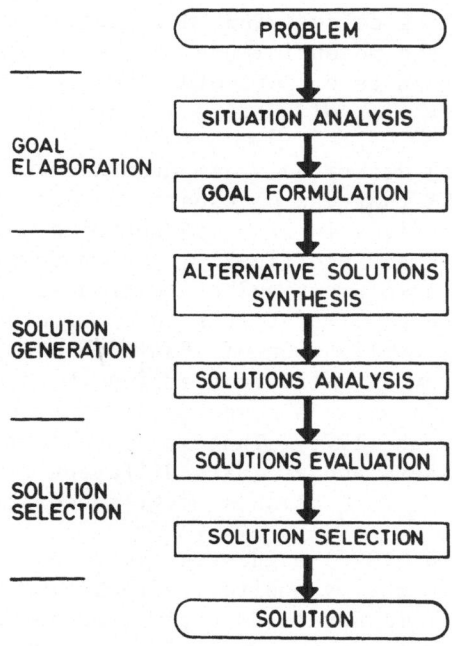

GOAL
ELABORATION

SOLUTION
GENERATION

SOLUTION
SELECTION

Fig. 9. Problem solving approach in systems engineering.

pare alternative concepts with every other on the basis of each criterion. Finally, the concept with the highest total criteria value
is selected. Detailed information on how to determine criterion values and handle weighting factors are available for instance from
Dathe (1971), McCrimmon (1968), and Zangemeister (1971).

The problem solving approach should be applied by ergonomists
to each of their design steps. Consequently, design goals have to be
considered during each design step. These design goals are the basis
for generating alternative design concepts and evaluating solution
alternatives. In this approach computer simulation is mainly applied
by ergonomists to analysis of existing or proposed tasks structure
and workstation design concepts. Applications are less often found
for the analysis of function allocation concepts, manning concepts,
work environments, and job designs. Examples of some applications
are as follows.

Adams and Reddy (1978) investigated the operational and economic effect of factors affecting the performance of industrial in-

spectors both in detecting defects and in identifying acceptable products. In the paper the potential use of the simulation language SAINT in relating the results of perceptual and behavioral research to the design of quality control systems are also discussed.

Ayoub et al. (1976) described a computerized approach with the simulation language GERT that can be used to select alternative job performance aid (JPA) designs in order to optimize human performance on given maintenance tasks. Specifically, the computer approach proposed would perform an assessment and evaluation of the effectiveness of specific JPA design alternatives in a variety of maintenance systems and the optimum application of JPAs within a given maintenance system. Another investigation concerning the assessment and evaluation of JPA formats using a computer simulation approach is given by Smillie and Ayoub (1976). That study demonstrates that SAINT can be used to assess and evaluate different combinations of JPA formats under various task demands, e.g. speed stress.

Chubb and Berisford (1977) suggested the development of models of manned system performance by monitoring real-time mission scenario simulations in order to developed techniques for applying human factors to military operations research. They employed SAINT in analyzing the navigation tasks of a strategic military aircraft system and established possible goals for such modeling exercises such as alternate design selection, force structure planning, performance deficiency identification, design evaluation, tactics assessment, training effectiveness demonstration, etc.

Doering (1973) described a computerized approach for simulating the task performance process of an operator in a man-machine system to get data about operator training requirements and to evaluate alternative workstation layouts. For describing operator tasks an extended form of a decision table was used that in addition to normal decision table data contained task description information, such as task times and required displays and controls. The implementation language was FORTRAN.

A computerized model for measuring the interaction between a primary task and some secondary loading task is used by Hann and Kuperman (1975). They intended to demonstrate how SAINT can be used to model psychological theory. Once a basic model was constructed, candidate processes could be added or selected and parameters might be varied by simply adding or altering input data. In this way test and development of theory can proceed quickly with minimum costs.

Kuperman and Seifert (1975) and Kuperman et al. (1977) developed a computer simulation model for evaluating display concepts of a digital avionics information system (DAIS). Networks developed for the model represented both multi-function switching and multi-purpose display concepts of DAIS and dedicated display and control concepts of conventional aircraft systems. The exercise of the computer model provided estimates of the nature of primary and secondary task interaction, performance prediction, and validation of modeling concepts and accuracy within the limits of available empirical data.

A computer simulation approach for analyzing occupational accidents was given by Smillie and Ayoub (1976). Used in the study was the simulation language Q-GERT for modeling the accident process in order to discover potential hazards in the occupational system. It was shown that the proposed modeling approach coupled with a computer algorithm is useful in describing and evaluating certain accident phenomena.

Wortman et al. (1975) reported about the simulation of a remotely piloted vehicle/drone control facility using SAINT. Here, the SAINT model was used in conjunction with a real-time multi-operator simulation of that facility to study the effect of system modification on system performance. In comparison to real-time man-in-the-loop simulation, SAINT has the advantage that alternative system configurations and operator procedural changes could be evaluated with little expense and effort. In another investigation carried out by Wortman et al. (1978) a computerized SAINT model was used for the assessment of human performance elements of a guided missile air defense system. The paper describes the SAINT network simulation model of the system in which the operator monitors targets and controls the assignment of fire units through the use of a visual display of flight path and parameters.

An example for analyzing function allocation and job design concepts by means of a computerized approach was separately described by Buck and Maltas (1979) and Maltas and Buck (1975). They discussed the use of computer simulation of man-machine systems in the design of large industrial processes. Simulating the processes of a hot strip mill by means of a SAINT model they investigated the effect of higher degrees of automation, different capacities of process limiting operations, and task allocation, on operator idle time. They recommend such simulation studies especially when on-site studies are too risky or costly and when laboratory studies alone would have questionable generalizability.

Further application examples of computer simulation to manned system design can be found in other articles of this publication. Many other examples for applying computer simulation are given in the books of Pritsker, 1974, 1977, and Pritsker et al., 1979, in which the simulation languages GASP IV, Q-GERT, and SLAM are described in detail.

REFERENCES

Adams, S.K. and Reddy, M.R., 1978, Potential Uses and Applications
    of SAINT to the Analysis of Quality Control Systems, in: Pro-
    ceedings of the Human Factors Society, 22nd Annual Meeting,
    184-190.
Ayoub, M.A.,Smillie, R.J. and Edsall, J.C., 1976, A Computerized Ap-
    proach for the Assessment and Evaluation of Job Performance Aids,
    in: Proceedings of the 6th Congress of the International Ergono-
    mics Association, Univ. of Maryland, The Human Factors Society,
    Santa Monica, CA, 184-190.
Buck, J.B. and Maltas, K.L., 1979, Simulation of Industrial Man-Ma-
    chine Systems, Ergonomics, 22 (7), 785-797.
Chubb, G.P. and Berisford, K.L., 1977, Manned System Modeling: SAINT
    Applied to Strategic Navigation, in: Proceedings 10th Annual Si-
    mulation Symposium, W.G. Key, S. Kumar, L. May and J.L. Morrow
    (eds.), Tampa, Florida.
Dathe, M., 1971, "Moderne Projektplannung in Technik und Wissenschaft",
    Carl Hanser Verlag, München.
De Greene, K.B. (ed.), 1970, "Systems Psychology", McGraw-Hill Book
    Company, New York, Düsseldorf.
Delfosse, C.M., 1976, Continuous Simulation and Combined Simulation
    in SIMSCRIPT II.5, CACI, Inc., Arlington, VA.
Dixon, W.J. (ed.), 1968, "BMD Biomedical Computer Programs", Univer-
    sity of California Press, Berkeley.
Doering, B., 1973, A Method to Improve the Human Operator-Machine In-
    terface Through Digital Computer Modelling of the Man-Machine
    System, in: Proceedings of the Defence Research Group Seminar
    on the Optimum Balance Between Man and Machine in Man-Machine
    Systems, Utrecht, The Netherlands.
Doering, B., 1976, Analytical Methods in Man-Machine System Develop-
    ment, in: "Introduction to Human Engineering", K.-F. Kraiss and
    J. Moraal (eds.), Verlag TÜV Rheinland GmbH, Köln.
Duket, S.D., Wortman, D.B., Seifert, D.J., Hann, R.L. and Chubb, G.P.,
    1978, Documentation for the SAINT Simulation Program, Aerospace
    Medical Research Laboratory, Aerospace Medical Division, AFSC,
    Wright-Patterson Air Force Base, OH.
Emshoff, J.R. and Sisson, R.L., 1970, "Design and Use of Computer Si-
    mulation Models", The Macmillan Company, New York.
Evans, G.W., Wallace, G.F. and Sutherland, G.L., 1967, "Simulation
    Using Digital Computers", Prentice-Hall, Inc., Englewood Cliffs,
    N.J.
Gagne, R.E., 1976, Computer Modeling and Simulation Handbook MK III,
    Simulation, may, 147-154.
Haberfellner, R., 1978/79, Problemlösungszyklus, in: "Systems Engi-
    neering", 2nd ed., W.F. Daenzer (ed.), Peter Hanstein Verlag,
    Köln, Verlag Industrielle Organisation, Zürich.
Hann, R.L. and Kuperman, G.G., 1975, SAINT·Model of a Choice Reaction
    Time Paradigm, in: Proceedings Human Factors Society, R.N. Hale,
    R.A. McKnight and J.R. Moss (eds.), 19th Annual Meeting, Dallas,

Tex., 336-341.

Hills, P.R., 1973, An Introduction Using SIMULA, Publication No. S55, Norwegian Computer Center, Oslo.

Kiviat, P.J., Villanueva, R. and Markowitz, H., 1969, "The SIMSCRIPT II Programming Language", Prentice-Hall, Inc., Englewood Cliffs.

Kuperman, G.G. and Seifert, D.J., 1975, Development of a Computer Simulation Model for Evaluating DAIS Display Concepts, in: Proceedings Human Factors Society, R.N. Hale, R.A. McKnight and J.R. Moss (eds.), 19th Annual Meeting, Dallas, Tex., 347-353.

Kuperman, G.G., Hann, R.L. and Berisford, K.M., 1977, Refinement of a Computer Simulation Model for Evaluating DAIS Display Concepts, in: Proceedings of the Human Factors Society, 21th Annual Meeting, San Francisco, Cal., Human Factors Society, Inc., Santa Monica, Ca., 305-310.

Maltas, K.L. and Buck, J.R., 1975, Simulation of a Large Man/Machine Process Control System in the Steel Industry, in: Proceedings Human Factors Society, R.N. Hale, R.A. McKnight and J.R. Moss (eds.), 19th Annual Meeting, Dallas, Tex., 193-205.

Mc Crimmon, K.R., 1968, Decision Making Among Multiple-Attribute Alternatives: A Survey and Consolidates Approach, Memorandum, RM-4823,APRA, RAND-Corporation.

Meister, D., 1971, "Human Factors: Theory and Practice", Wiley Intersciences, New York.

Nie, N.H., Hull, C.H., Jenkins, J.G. and Bent, D.H., 1975, "Statistical Package for Social Sciences (SPSS)", McGraw-Hill, New York.

Oeren, T.I., 1977, Software for Simulation of Combined Continuous and Discrete Systems: A State-of-the-Art Review, Simulation, Feb., 33-45.

Oeren, T.I. and Zeigler, B.P., 1979, Concepts for Advanced Simulation Methodologies, Simulation, March, 69-82.

Pritsker, A.A.B., 1974, "The GASP IV Simulation Language", John Wiley & Sons, New York and London.

Pritsker, A.A.B., 1977, "Modeling and Analysis Using Q-GERT Networks", John Wiley & Sons, New York and London.

Pritsker, A.A.B. and Pegden, C.D., 1979, "Introduction to Simulation and SLAM", John Wiley & Sons, New York, System Publishing Corp., West Lafayette, Ind.

Rohlfing, H., 1973, "SIMULA, Eine Einführung", Bibliografisches Institut, Mannheim.

Schlesinger, S., Crosbie, R.E., Gagne, R.E., Innis, G.S., Lahvani, C.S., Loch, J., Sylvester, R.J., Wright, R.D., Kheir, N. and Bartos, D., 1979, Terminology for Model Credibility, Simulation, March, 103-104.

Schriber, T.J., 1974, "Simulation Using GPSS", John Wiley & Sons, New York.

SCi Software Committee, 1967, The SCi Continuous-System Simulation Language, Simulation, 9, 281-303.

Shackel, B. (ed.), 1974, "Applied Ergonomics Handbook", IPC Science and Technology Press Ltd., Guilford, Surrey.

Singleton, W.T., 1974, "Man-Machine Systems", Penguin Books Ltd.,

Harmondsworth, Middlesex, England.

Singleton, W.T., Easterby, R.S. and Whitfield, D.S. (eds.), 1967, "The Human Operator in Complex Systems", Taylor & Francis Ltd., London.

Smillie, R.J. and Ayoub, M.A., 1976, A Computer Simulation Approach for Analyzing Occupational Accidents, in: Proceedings 6th Congress of the International Ergonomics Association, Maryland, USA, 226-235.

Smillie, R.J. and Ayoub, M.A., 1977, The Assessment and Evaluation of Job Performance Aid Formats Using SAINT, in: Proceedings of the Human Factors Society, A.S. Neal and R.F. Palasek (eds.), 21th Annual Meeting, San Francisco, Ca., 311-315.

Teichroew, D., Lubin, J.F. and Truitt, T.D., 1967, Discussion of Computer Simulation Techniques and Comparison of Languages, Simulation, Oct., 181-190.

Wortman, D.B., Duket, S.D., Seifert, D.J., Hann, R.L. and Chubb, G.P., 1978, Simulation Using SAINT: A User-Oriented Instruction Manual, Aerospace Medical Research Laboratory, Aerospace Medical Division, AFSC, Wright-Patterson Air Force Base, Ohio.

Wortman, D.B., Duket, S.D., Seifert, D.J., Hann, R.L. and Chubb, G.P., 1978, The SAINT User's Manual, Aerospace Medical Research Laboratory, Aerospace Medical Division, AFSC, Wright-Patterson Air Force Base, Ohio.

Wortman, D.B., Duket, S.D. and Seifert, D.J., 1975, SAINT Simulation of a Remotely Piloted Vehicle/Drone Control Facility, in: Proceedings Human Factors Society, R.N. Hale, R.A. McKnight and J.R. Moss (eds.), 19th Annual Meeting, Dallas, Tex., 342-346.

Wortman, D.B., Hixson, A.F. and Jorgensen, C.C., 1978, A SAINT Model of the AN/TSQ-73 Guided Missile Air Defence System, Proceedings of the 1978 Winter Simulation Conference, Miami Beach, Florida, 879-888.

Wymore, A.W., 1977, "A Mathematical Theory of Systems Engineering - The Elements", Robert E. Krieger Publishing Company, Huntington, New York.

Zangemeister, C., 1971, "Nutzwertanalyse in der Systemtechnik", 2nd ed., Wittmannsche Buchhandlung, München.

Zeigler, B.P., 1976, "Theory of Modelling and Simulation", John Wiley & Sons, New York.

THE HUMAN OPERATOR SIMULATOR: AN OVERVIEW

Norman E. Lane, Melvin I. Strieb, F.A. Glenn and

R.J. Wherry

INTRODUCTION

The information explosion brought about by electronics technology has rapidly and irrevocably changed the character of manned
weapon systems. Developments in microprocessors, displays and sensors have provided a capability to generate, synthesize, process and
present extraordinary quantities of data. This new capability has
not always been wisely employed in system design. The rapidity of
technology development has frequently outpaced intelligent planning
for its application to systems. The result has all too often been a
lack of attention to proper integration of new equipment and a failure to achieve an appropriate balancing of functions among the hardware, software and operator components.

As the information available to the operator increases precipi-
tously, the key question for human factors engineering is no longer
the adequacy of each individual control or display, but the success
of the design in making balanced use of the capabilities of all system components, including the operator, in managing that information.
It is in this context that new developments in computer models for
human factors must be considered.

Models which address operator performance in a total system are
still relatively rare. Pew et al. (1977), Strieb et al. (1977) and
Greening (1978) define the roles and uses of numerous operator models and systematically compare current efforts in the modeling area.
The models reviewed vary widely in assumptions, intended use, generality and a host of other factors. Most model or simulate only a

121

specific part of the operator's job, such as manual control, tracking or display reading, with different and unrelated models for each
portion of operator behavior. Recent variations on optimal control
models have given the operator some limited cognitive or decision
making capability, but are still primarily oriented toward system
controller functions (see George, 1979, for example). The full spectrum of roles which the operator must perform in modern systems, including his functions as an information processor and decision maker,
has not generally been treated in a system context.

In contrast to these highly detailed part-task operator models,
a number of techniques aimed at estimating the mission performance
of a complete system have become available in the last decade. Most
are conceptually and structurally derived from the initial work of
Siegel and Wolf (1961) and have been variously categorized by the
reviewers cited above as task-analytic, analytic prediction or time-
motion models. These models include SAINT (Wortman et al., 1978),
the Workload Assessment Model (WAM) (Edwards et al., 1977), the Timeline Analysis Program (TLA) (Anderson and Miller, 1976), and a number of proprietary models developed by equipment manufacturers for
in-house use (Asiala, 1975; Klein and Cassidy, 1972; cited by Greening, 1978). Assumptions underlying these approaches are described in
Jahns (1973) and model details are compared by Greening (1978).

Analytic-prediction models have accumulated a considerable applications history and have been of distinct value as objective techniques for preliminary screening of task allocation during early design. Linton et al. (1977), for example, use WAM to estimate operator workload in a proposed VSTOL conceptual design. All these approaches, however, depend on the analyst to estimate performance
times for each operator task; none as yet contains a detailed model
of operator behavior, although Greening's (1978) detailed analysis
indicates the beginnings of such models in several of the techniques
reviewed.

The tendency for manual and optimal control models to address
increasingly larger portions of the operator job and the progressive
inclusion of explicit operator model components in analytic prediction techniques suggest an eventual and highly desirable convergence
of the two approaches. The need to combine evaluations of overall
system performance with detailed simulations of the operator's functioning within the system has become a critical challenge to human
factors technology. The Human Operator Simulator (HOS) was designed
to provide both of these capabilities within a single modeling framework. The purpose of this paper is to introduce the method and to
describe some of its general characteristics. The material presented
here provides only an overview of HOS and its basic structure. A
more complete discussion and detailed examples are contained in
Strieb et al. (1978) and in Strieb and Wherry (1979).

HOS IN CONTEXT

What is HOS?

Simply stated, HOS is a collection of digital computer programs which allows for the simulation of a total system performing a complex mission. The name, Human Operator Simulator, is somewhat misleading. HOS simulates not only the behavior of the operator, but the operating characteristics of the system hardware and software together with any sensors, targets, emitters or other external data sources. HOS is a generalized model. It becomes a simulation of a specific operator/system performing a specific mission when the analyst provides certain kinds of information. Definitions of equipment characteristics and procedures to be followed by the operator are communicated to HOS using the Human Operator Procedures (HOPROC) Language, a simplified English/FORTRAN computer language, specifically keyed toward convenient descriptions of tasks and procedural steps.

HOS was developed intermittently over a period of more than 10 years in response to an inability to define objectively the specific impact of poor human engineering on overall mission performance. The basic difficulty was initially described by Wherry (1969) and summarized in retrospect in 1978 (Strieb et al., 1978, Appendix A):

> "I found that it was virtually impossible to get the Navy interested in correcting any single deficiency, because no single deficiency was ever so bad that it alone made the aircraft unsafe or caused unsuccessful ... mission performance. It was obvious ... that the cumulative effect of a series of minor deficiencies ... would have a major impact ... on mission success. To be able to convince others, however, would require a detailed model of the impact of various (design) features on human information absorption, processing and transmission in a task-sequencing framework to illustrate such cumulative effects."

The impetus for HOS development was thus the same as that underlying the large-scale analytic prediction models discussed above. However, the approach used in HOS differs from that of analytic prediction models in a number of important ways. Most significant is the manner in which task times are determined. In analytic prediction models, the time required for each task is provided by an analyst or sampled from time distributions. In HOS, the time to perform tasks is built up from the times required to execute detailed human performance micromodels. Although these times can be changed by the user if he wishes to modify the "standard" operator, they are internal to the model and relate to task performance in a consistent rule-based manner.

The structure of HOS provides a number of sophisticated features

that make it particularly useful for simulating complex missions.
The HOS operator is goal oriented; he will automatically eliminate
or dynamically resequence actions that have become unnecessary or
that change in priority due to events during the simulation. In addition to these internal rules for resequencing, the analyst may
also supply external formal strategies or decision rules which are
dependent upon the status of system variables at the time the decision must be made. Task-sequencing algorithms resident in HOS are
based both on initial priorities and on changes arising within the
mission; these in turn may be further modified by another feature-
-the ability to control the degree of precision (internal limits)
which the HOS operator will use to track, control or compute. Characteristics of the HOS operator can be varied to simulate performance
by operators with less training or below-average abilities. Each of
these features is described in more detail in later sections.

Assumptions Underlying the Model

There are four assumptions regarding human performance that are
basic to the HOS approach.

Assumption 1. Human behavior is predictable and goal oriented,
particularly so for a trained operator within a defined system.
Thus, a significant part of the variance of mission-relevant operator performance can be explained by appropriate predictive equations.

Assumption 2. Human behavior can be described as a series of
discrete micro-events, which can be aggregated to explain task performance. The occurrence of micro-events is controlled within HOS by
a set of micromodels, corresponding to different basic human functions. The HOS operator is capable of performing seven of these "primitive" functions:

1. Information Absorption,
2. Recall of Information,
3. Mental Calculation,
4. Decision Making,
5. Anatomy Movement,
6. Control Manipulation, and
7. Relaxation.

Each action that the operator performs is a combination of one
or more of these micro-events. Internal rules within HOS automatically determine the function combinations that make up a task, decide
on the sequence of their occurrence and compute the time required to
perform the task under the conditions in effect each time that task
is executed. Thus, any task requiring eye or hand movement, for example, must access the Anatomy Movement micromodel to find out if and
when that micro-event can occur and how long it will take to complete.
A series of such accesses comprise a complete task such as perception

of a displayed number or adjustment of a control to a specified setting. The idea that human behavior can be effectively described as
a sequence of discrete events has been popular throughout the history of behavioral research. Donders (1869) conducted a program of experiments to determine the time requirements for basic mental processes to predict performance on more complex processes. Hick (1948)
has argued that manual tracking performance can be described as a
series of rapid discrete adjustments, a condition which he attributes
to limitations imposed by the neural refractory period. Stroud (1955)
and Kristofferson (1967) advocate a "psychological moment hypothesis"
according to which human perceptual processing occurs as quantum
events paced by a rapid biological cycle. Fitts and Posner (1967) indicate that the number of repetitive stimuli that can be perceptual-
ly distinguished and the frequency at which repetitive monitoring processes can be executed are both limited to about 10 events per second.
Even complex continuous actions that last for many seconds can generally be described as a series of repetitions of a limited number of
types of discrete processes.

Assumption 3. Humans have a single channel for input and output
and a single "information processor". More particularly, although an
operator may be switching back and forth among many concurrent tasks,
only one input/output can be initiated and only one task which requires the central processor (perception, memory or calculation) can
be performed at a given moment in time. Motor actions may occur simultaneously with a central processing event, but only if they are
initiated prior to the event which requires the processor. The assumption that humans are single channel processors is, in practice,
not a constraining one. Although the basic structure of man's information processing abilities is far from resolved, operation of the
HOS model is not particularly affected by changing theories; by adjusting the time required to switch attention between events or by
modifying time to initiate an input/output, HOS can be made indistinguishable from a multi-channel or multi-processor model.

Assumption 4. Trained operators rarely make errors or forget
their procedures. HOS assumes that the operator always knows device
locations and remembers what he is to do so long as he is presented
with sufficient and accurate information. Although real operators
may on occasion make "stupid" mistakes such as skipping a key procedural step or actuating the wrong control, the philosophy of HOS
is that:

1. Such errors attributable solely to the operator (without influence of design, training or fatigue) are extremely rare;
2. the frequency distribution and points of occurrence of such errors
   are poorly understood, particularly for new systems, and modeling
   them at random can introduce serious aberrations into simulation
   outcomes;
3. basic evaluations of system effectiveness (the purpose of HOS)

should be conducted under standard error-free conditions <u>before</u>
random errors are introduced into performance.

HOS is primarily oriented toward estimating key performance
times and identifying problems in procedural organization or crewstation layout. It is, however, quite easy to cause the HOS operator
to make mistakes or fail to complete a mission. Errors in procedures
and mental calculations can be introduced, or information accuracy
degraded, leading to suboptimal outcomes. Some of the most useful applications of HOS to date have involved deliberate and systematic manipulation of designs, procedures, and parameters to determine when
performance degradation occurs with respect to time or mission outcomes (Lane et al., 1977; Lane et al., 1979). The HOS philosophy requires only that these manipulations be systematic and deliberate,
not random, and the HOS architecture and specifications for the "standard" operator are designed to encourage such forethought.

It is important to recognize that HOS is not a theory, but a
pragmatic design tool. Although the four assumptions above provide
some theoretical foundations for HOS, it is not necessary for them
to be absolutely true about human behavior (as, of course, they are
not) for HOS to be of useful validity. The important issue for HOS
is whether it provides approximations of man/machine performance
adequate for evaluation of systems and diagnosis of system problems.
As a later section will document, there is compelling evidence that
HOS does forecast human task performance in systems with surprising
fidelity, suggesting that the component models in totality are of
reasonable accuracy for prediction even if they are not direct ana-
logues of how tasks are actually accomplished. The architecture of
HOS was designed to allow for easy replacement and modification of
component models. A user dissatisfied with a particular micromodel
can, in most cases, change or replace the micromodel equations without significant changes to the operations of other micromodels.

When to Use HOS and Why

For a specific design or evaluation problem, selection of techniques should be made on the basis of need for precision vs. time-
liness of answers and on the information gained about system effectiveness for a given investment of time and money. These criteria
will frequently dictate a phased transition from approximate models
to progressively more refined ones prior to the use of high-fidelity
hardware simulators or prototypes. The HOS system is applicable over
a considerable portion of system development. By varying the level
of procedural detail, it could be used throughout development by
being configured to emulate the best characteristics of other techniques. HOS was, however, designed for certain specific types of applications and is most effective for these purposes. This section
outlines the uses for which HOS is recommended and the rationale
supporting these recommendations.

HOS is intended to fill the gap between analytic prediction models and full-scale hardware simulator. The analytic prediction techniques discussed previously are useful for preliminary function allocation, and their greatest value is as sophisticated calculators and
summarizers, combining best estimates of individual task performance
into an overall forecast of mission effectiveness at a point in the
design process at which changes to the system are still feasible.
Although HOS can be used at these very early stages, its most sensi-
ble application is as a planned supplement to analytic prediction in
the early stages and as a more detailed follow-on in later stages.
In the Navy's human factors design model sequence, use of HOS follows the use of the WAM technique, building on the preliminary task
analytic data required to exercise WAM, and HOS is also used concurrently with WAM for generating performance times on novel or complex
tasks for which satisfactory time estimates are not available. Thus,
WAM is used as a gross screening technique, focusing attention on potential problem areas which can be addressed more precisely by HOS.
HOS is then used to continue screening of problem areas, narrowing
down critical tasks and time frames to be emphasized during evaluation on manned hardware simulators.

HOS has major advantages over hardware-based simulators for detecting operability problems during the mid-range of system design
and for relatively precise initial evaluations of the success of
system component integration (Lane et al., 1977). Since hardware is
required, the practice of leaving complex design questions for resolution on dynamic hardware simulators guarantees both a major time
lag and an inability to correct identified problems without major delays and cost increases. The use of real operators on real hardware,
though necessary for ultimate validation of design integration, is
likely to leave undetected some key system deficiencies. The same
characteristics of flexibility and intuitive problem solving that
make man a superior system component also cause him to be a less-
than-ideal subject for evaluation of system procedures. Operators
selected for simulator evaluations are rarely representative in ability and as experienced as those who will eventually operate the
system. In addition, training for use of the new system may not be
constant across operators, operator motivation is uncontrolled, and
each operator will tend to use his own internal definition of satisfactory performance. Natural adaptability and desire to succeed will
cause an operator, particularly an experienced one, to perform tasks
however he can, taking shortcuts and changing procedures at will to
try to cope with task demands. A common outcome is the decision that
systems and procedures are adequate when in reality a skilled motivated operator must work around them to get the job done. The "standard" operator used by HOS is an attempt to detect potential problems in operability more directly and earlier in time than is possible with hardware simulators.

Like any model, HOS takes time and planning to use. The initial

planning for human factors evaluations must include a general idea
of how HOS will be used and should provide realistic timelines for
building the simulation months before output is required. Although
the development of the initial HOS simulation for a major system can
be time consuming, the architecture of HOS has been designed to sim-
plify this process as experience with the model is acquired. An operator or hardware procedure coded for one system can often be used
without major modification on another system. For example, a general
procedure to monitor a radar system, once developed for one simulation, is usable in every subsequent simulation that has a radar system. An acoustic signal processor simulation developed for one antisubmarine warfare (ASW) aircraft can be used without modification
for nearly all ASW systems because of equipment similarities across
platforms.

There are several system classes and conditions for which HOS
would not necessarily be the method of choice. The HOS operator is
stationary; he cannot move about without the provision of special
movement equations to the simulation. HOS has no explicit model of
communication; a system in which performance might vary significantly as a function of differences in communication times would currently be difficult to simulate with HOS. Systems which are primarily
composed of control tasks, while they can be adequately modeled with
HOS, are most efficiently approached by conventional control models
or by provision of control equations to HOS in special procedures.
HOS currently allows for only one operator to be fully modeled. Other operators in the system can be simulated with as high a degree of
detail as desired, but they are indistinguishable from hardware procedures in function, and each action and decision rule for additional operators must be provided as HOPROC code by the analyst. Although there are no prohibitive architectural problems in supporting
full HOS modeling of multiple operators, core storage restrictions
currently make this impractical.

In brief, although HOS may be used from the very beginning of
a system through the complete cycle, it is of greatest power when
alternative designs and tentative operator and equipment functions
have been identified. HOS is compatible with, and should logically
follow, large-scale analytic prediction models used as a preliminary
screening technique. HOS is a logical precursor to hardware simulator and prototype validation studies and can be used to narrow the
focus of hardware simulation toward the most important problem areas.

THE STRUCTURE OF HOS

The complete HOS system consists of three components, related
as shown in Fig. 1. The HOPROC Assembler/Loader (HAL) translates
HOPROC code into a set of general instructions to the HOS model
which describe the equipment, define the operator procedures and

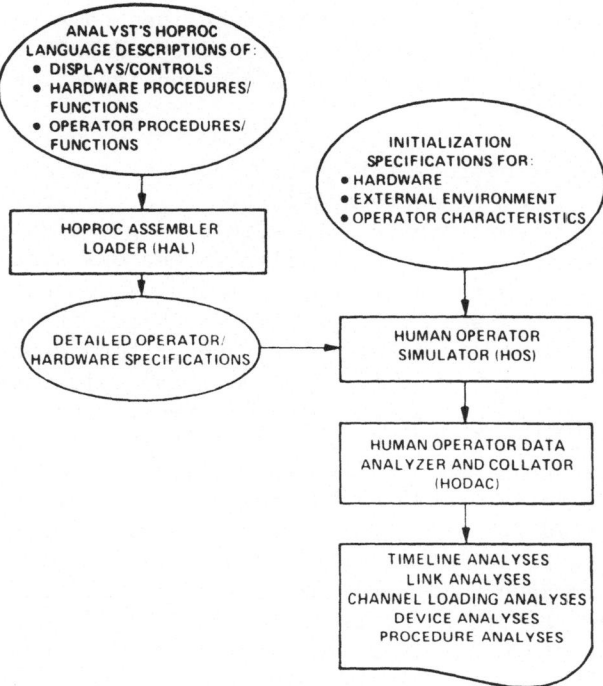

Fig. 1. HOS Program Flow.

identify the data sources available to the operator and the hardware. The Hos model itself contains general rules that govern operator behavior. When provided with the crewstation layout, the initial values of data variables and parameters describing the operator's abilities, the HOS model carries out the defined mission and produces a file containing detailed mission data. The Human Operator Data Analyzer/Collator (HODAC), a flexible report generator and statistical analysis package, analyzes this data to produce a variety of reports describing the simulation outcomes. Each of these components is described below.

The Human Operator Procedures (HOPROC) Language

The HOPROC language is used to describe the equipment that interfaces the operator to the system hardware/software, the way in which the hardware/software configuration functions and what the operator must do in order to properly use the equipment. The locations and characteristics of the displays and controls, the parameters describing the operator's capabilities and the details of the specific scenarios to be exercised are not described in HOPROC. Rather, they are entered directly into the HOS model at run time to

allow for easy modification. Thus HOPROC describes only the relatively fixed aspects of a system and its mission; the HOPROC code for a
specific configuration changes only if a new configuration alternative is to be evaluated.

HOPROC is a powerful tool for describing operator and system
performance for a number of reasons. First, it is flexible - capable
of being used for systems ranging from simple single display, single
control situations to highly complex military weapons systems. Second, it is adaptable - able to be readily modified to accommodate
new situations and new system configurations. Third, it is easy to
understand. Its combination of English for describing procedures and
FORTRAN for describing the requisite mathematical calculations make
it possible for someone unfamiliar with the specific syntactical and
grammatical constraints of HOPROC to understand the procedures that
someone else has written. Fourth, it is modular. Procedures developed
for one simulation can be readily adapted for use in other simulations or supplemented for expansion of the initial simulation. Fifth,
it is able to support simulation of the many relevant factors impact-
ing operator performance. In fact, HOPROC can allow introduction into
a simulation of any factor which is sufficiently well understood for
its influence on performance to be estimated. Most factors with reasonably clear influence have been incorporated into micromodels. Others, such as fatigue or motivation, can be simulated in HOPROC procedures if the analyst is willing to posit what the effects should be.

There are three major sections to the HOPROC inputs for a HOS
simulation. In the title declarations, the analyst identifies by name
the equipment in the operator's crewstation and describes certain of
its basic characteristics - specifically the settings associated with
discrete devices and the scale factors associated with any continuous devices. In the functions definitions, the analyst defines any
mathematical calculations needed in the simulation. These include the
mental calculations that the operator may have to perform as well as
the equations that describe equipment functions - for example, the
dynamics of the aircraft and emitter or target characteristics. In
the procedure definitions, the analyst describes the tactics that the
operator is to use, the actions that must be performed to accomplish
specific goals, and the equipment characteristics that govern response of the system to the operator's activities.

Since the functions definitions are mathematical equations, they
are written in a pseudo-FORTRAN that permits easy access to all the
key simulation variables. Since the procedures are the tasks that
either the operator or the system must perform and the circumstances
under which those activities occur, they are most readily represented
in a descriptive "natural" language. HOPROC procedures definitions
use an English-like grammatical and syntactical structure.

HOPROC uses only seven major classes of statement to describe

the tasks that the operator and system can perform. However, within
each class of statement, it is possible to vary statement structure
to allow for several hundred different operations. These are related,
as in any programming language, by straightforward rules for how
statements should be organized and punctuated. Procedural statements
can be simple actions (e.g. DEPRESS PUSHBUTTON) or complex conditio-
nals (e.g. IF THE DISTANCE-TO-TARGET IS LESS THAN THE WEAPONS-RANGE
AND THE RATE OF THE TARGET IS LESS THAN THE WEAPONS-RATE THEN LAUNCH-
THE-WEAPON).

Structure of the HOS Model

When supplied with inputs describing the characteristics of the
operator, the physical layout of the crewstation, specific characteristics of the displays and controls, and details on the mission scenario, the HOS model converts the generalized system defined by the
HOPROC procedures into a simulation of a specific operator in a specific crewstation executing a specific mission in a specific situation. The components of the HOS model that execute and manage the
simulation, and their interrelationships, are shown in Fig. 2.

Each directive given to the HOS operator is interpreted by the
Instruction Handler module. The Instruction Handler determines the
sequence of operator activities required to satisfy the directive.
In order to predict the performance times associated with each ope-

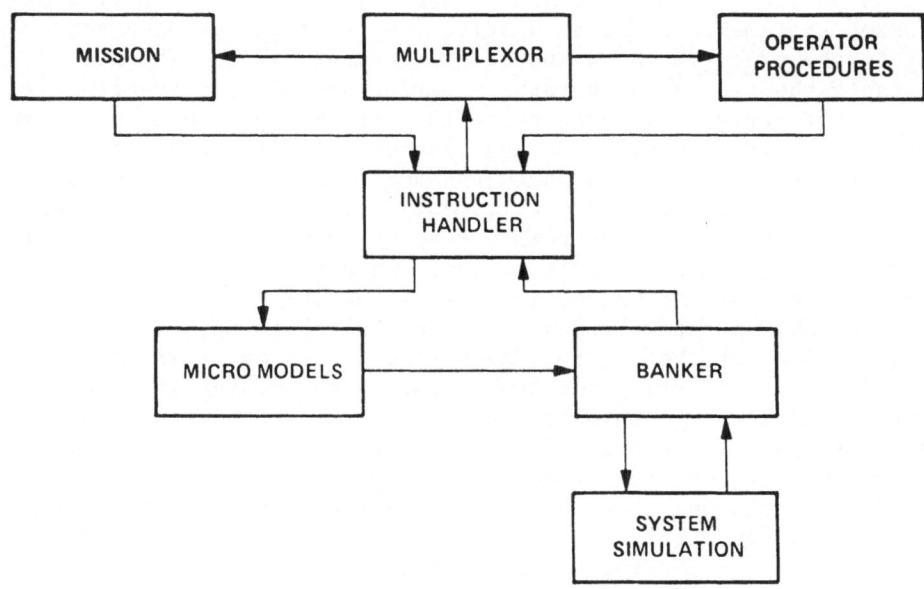

Fig. 2. HOS Simulation Structure.

rator activity, the Instruction Handler accesses one or more of the
micromodels. As each action is performed, the time charge associated
with the action is accumulated by the Banker module. The Banker periodically calls a System Simulator module to update the states of all
system variables.

HOS processes each directive in sequence beginning with the first
directive in a MISSION procedure. As statements are processed, it may
happen that the operator either cannot process a particular directive
(it may, for example, call for some resource that is unavailable) or
the directive may call for the execution of another series of directives grouped in another operator procedure. When this happens, the
Multiplexor module is called. Algorithms in the Multiplexor select
the next procedure to work on based on the relative criticalities of
all currently active procedures, how long it has been since each procedure was last worked on, etc.

The HOPROC directives that describe the operator (and equipment)
procedures are decomposed by internal rules into a series of discrete
operator actions. Thus, for example, the operator procedure statement
"TURN SWITCH ON" is decomposed into a series of actions in which the
operator turns his eyes to look at the switch, selects a hand to use
to perform the manipulation, moves the hand to the switch and turns
it on. More complex statements may require the operator to read information from a display and perform a comparison with some desired or
mentally computed value before performing an action or a series of
actions.

Although the general sequence of operator activities implied by
each HOPROC statement is deterministic, the specific sequence pertaining to the execution of a specific statement is not. The HOS
model adapts the simulated operator's performance to the specific
situation that pertains at each point in time in the simulation, just
as an actual operator would. The adaptiveness of HOS represents one
of the most important ways in which the HOS model differs from other
models. For example, the HOPROC statement "TURN SWITCH ON" describes
a specific action that the operator is to perform. Rather than simply
assigning a time charge for this action, the HOS operator goes
through the same sort of processing that an actual operator would.
Thus, if the HOS operator "knows" that the switch is on, or, after
"turning" to look at the switch, "sees" that it is on, then the suc-
ceeding actions - selecting the appropriate hand, moving the hand to
the switch, and actually turning the switch on - are not necessary.
HOS only assesses time charges for those actions that are actually
performed. In addition, the time charges that are assessed are determined by the conditions that pertain at the time the actions are required. For example, the time for moving the operator's hand and/or
eyes to a display or control are dependent on where the operator's
hands or eyes were at the time the action was required and whether
they were otherwise occupied; the time to manipulate a control is

dependent on what position the control was in initially and what type
of control it is. Thus the cumulative sequence of actions, and the
times and reasons for actions accurately represent those times that
would be obtained from an actual operator. This differs significantly
from the approach taken in some models where the probability of performing an action and the times required to carry out the actions
are drawn from distributions whose parameters are supplied by the
analyst.

Analysis of HOS Data

The basic output from HOS is a time history of simulation events.
These data can be analyzed in a variety of different ways, depending
on the purpose of the evaluation. The initial analysis usually consists of examining the sequences of operator activities to verify the
decision logic used by the simulated operator. This analysis can suggest new ways of presenting information or new types of information
that should be displayed. More importantly, codifying the operator's
procedures and decision rules in HOPROC can serve as the basis for
developing training programs that can teach operators consistent and
disciplined strategies. In addition, the Human Operator Data Analy-
zer/Collator (HODAC) program provides a series of nine standard analyses for differing aspects of system evaluation. Among the most important of these are:

• A Channel Loading Report that computes the percentage of time
  within selected intervals that each body part is occupied.

• A Link Analysis that provides data on the frequencies of usage of
  groups of displays and controls and transitions from one group to
  another.

• A Procedural Analysis that reports how often each procedure was
  executed, the time spent in each and the number of times each was
  delayed due to unavailability of system resources.

These reports enable detailed diagnosis of system problems, making possible statements such as "The operator spent a total of 4.5
seconds manipulating a particular control for an average manipulation
time of 0.45 seconds with a standard deviation of 0.15 seconds." This
diagnostic capability has been used to detect, for example, a joystick control with a slew rate inappropriate to the slew speed required by aircraft movement (Lane et al., 1979). In addition, if an
analyst has specific calculations to be performed on the data generated by a specific simulation, the mission data file can be accessed
directly using the analyst's own programs.

The HOS Micromodels

The HOS Instruction Handler has access to seven micromodels of human performance. The models and experiments on which the micro-models were based are those operator performance models that were considered to be the most appropriate for the types of problems for which HOS was originally developed. There were many situations in which data or accepted models for particular processes were unavailable. Rather than waiting until data had been collected or until generally accepted models had been formulated, provisional micromodels were adopted for these processes. HOS has been designed so as to readily accomodate other micromodels, alternative model formulations or new experimental data; thus if the need for changes to these provisional micromodels becomes necessary, new models or data can be easily incorporated.

Information Absorption. The HOS model of perception is tailored specifically to the information processing requirements of a complex man-machine interface. Whereas most contemporary theories and models for perception attempt to describe how sensory signals are processed into higher order codes, the HOS absorption model is concerned only with the changes through time in the operator's knowledge of the state of a display or control. It is therefore best to refer to it as an absorption model rather than a perception model. The same model is used to describe the absorption of information both by vision and by touch. Absorption occurs in quantum steps called micro-absorptions that are repeated until a termination criterion is satisfied. Each micro-absorption consists of an updating of the operator's estimate of the value of the device and his confidence in his knowledge of that value. Each micro-absorption results in the assessment of a time charge for the micro-absorption based on a variable parameter. Termination criteria cause the absorption process to stop when the operator's confidence is sufficiently high or when the operator cannot afford to spend any more time on the absorption process. The value that the operator believes a device to have is determined from the actual value of the device by adding an error term that is normally distributed about the actual value and whose magnitude is dependent upon an accuracy for the device as supplied by the analyst.

Information Recall. The HOS information recall model consists of two submodels - a model for short-term memory and one for long-term memory. The long-term memory model is currently limited to the recall of certain types of predetermined information. Specifically, the HOS operator is assumed to have a completely accurate and instantaneous recall of the locations of all displays and controls in his crewstation, most of their characteristics, the procedures that must be followed in carrying out a job, and the calculation processes for any computations that must be performed. These assumptions are consonant with the basic assumption of the HOS model - namely, that the operator being simulated is a trained operator who performs routine operations automatically.

The short-term memory model is more elaborate. The probability

of recalling an item of information from short-term memory, P, is
dependent on the confidence level attained during the last absorption
of the information, C, and the time that has elapsed since the last
absorption, t. The formula used to predict the probability of recall
is:

$$P = C^{\sqrt{t}}.$$

Whenever recall is attempted, a number is drawn from a uniform distribution and compared with P. If the random number is less than P, recall is considered to be successful. If the random number is greater
than P but still "close", another random number is chosen until either recall succeeds or until a number sufficiently larger than P is
selected. If the operator cannot recall the desired information, the
information absorption or mental computation micromodels are accessed.

   Mental Computation. The basic difference between the mental computation model and the information absorption model is that, in the
latter, information is absorbed from a display or control while in
the former, displayed information is used to calculate a value that
is not displayed anywhere in the crewstation. For example, a typical
mental computation associated with driving an automobile is determining how much farther one can go on a tank of gas. The computation requires the absorption of an item of information (the amount of fuel
remaining) coupled with some prior knowledge (the distance one can go
with that amount of fuel).

   When a mental calculation is required, HOS determines what information is needed in order to perform the calculation. If the HOS
operator can remember the information, the calculation is performed
at once. If he cannot remember the information, an appropriate sequence of actions is initiated to enable the operator to obtain the
data. In the above example, the displayed information required is the
amount of fuel remaining. If the operator cannot remember this, HOS
would cause him to look at the fuel gauge and read its value.

   Each mental calculation can require as many as ten different
data items. These may be the values of displays, settings of controls or the results of other mental calculations. The amount of
time required for a mental calculation is considered to be the a-
mount of time required to gather all the items of information needed
for the calculation plus some additional time to "put it all together." Because of the high potential variability in the complexity of
a function calculation, the analyst is required to supply a function
computation time for each function - HOS itself will supply the
times required to gather all the items of information needed for the
calculation.

   Errors in mental computation are assumed to be the result of
errors associated with the data used in the calculation. The calcula-

tion process itself is considered to be error-free. Thus, if the operator makes an error or obtains an inaccurate data value when either recalling the data or reading the data needed for a calculation, then the result of the calculation will be incorrect or inaccurate. It should be noted, however, that, as a result of the way in which mental calculations are described to HOS, the analyst has the ability to inject errors into the function calculation if he so chooses.

Decision-Making. A procedure is an operator task consisting of any number of steps, any step of which can invoke the execution of another procedure or any other operator action. For example, the mission in a simulation modeling a pilot's activities is a procedure that invokes a procedure for takeoff, another for cruise, another for landing, etc. Within each procedure are steps that describe the operator actions - reading displays, adjusting controls, etc. Decision-making by the operator can operate at two different levels - between procedures (inter-procedural) and within a procedure (intra-procedural).

Intra-procedural decision are of two types - decisions about what to do next (how to accomplish a procedural step) and decisions based on comparisons between system parameters (tactical decisions based on the analyst's model of operator performance). Tactical decisions are specified by the analyst in HOPROC as "IF...THEN" statements that invoke the information absorption, recall and mental computation micro-models. Decisions about how to accomplish a procedural step are generally resolved by the HOS model itself. HOS will attempt to execute each step in a procedure in sequence until it can go no further, for whatever reason. If it finds itself blocked, it will attempt to "unblock" itself. If it can, it will continue with the current procedure; if it can't, it will look for some other procedure to work on, at which point the inter-procedural queueing logic is invoked.

There are basically two types of events that will block the operator:

1. An action is required that the operator cannot perform because the action requires body resources that are busy.

2. Information or a control action is required that is not possible because a device is currently inactive (not enabled).

The latter situation is more common. When it occurs, HOS will automatically invoke a special type of procedure - an enable procedure - whose function is to activate the device that is inactive. When the first situation occurs, HOS will go off and work on another procedure until the required body part is free, by invoking the inter-procedural decision-making (procedure selection) logic.

Procedures can be invoked in any of three different ways. The analyst can specify either that:

1. The procedure is to be executed immediately and no more steps in the current procedure are to be executed until the invoked procedure has been completed,

2. The invoked procedure is to be placed on an <u>active procedure list</u> and is to be executed whenever the HOS procedure queueing logic deems it appropriate,

3. The invoked procedure is to be placed on the active procedure list and executed <u>periodically</u> until removed from the active procedure list, whenever the HOS procedure queueing logic deems it appropriate.

In situation 1, control transfers immediately to the invoked procedure and no more steps in the invoking procedure are executed until the invoked procedure is completed. Procedures placed on the active procedure list by method 3 are called <u>monitor</u> procedures in that they are usually used to cause the operator to periodically monitor a particular display or control.

The selection of the next procedure to work on (inter-procedural queueing logic) is controlled by the Multiplexor module. The Multiplexor algorithms consider both the initial criticality (priority) of the procedure, and how long it has been since the procedure was last executed. As the length of time since the procedure was last executed increases, the <u>effective criticality</u> of the procedure increases (over an initial criticality that can be set by the analyst). In addition, for monitor procedures, the effective criticality is further modified by a factor that is dependent on how close the device being monitored is to a defined set of <u>limits</u>. As the estimated value of the device approaches its limits, the effective criticality of the device increases. When it exceeds the defined limits, the effective criticality increases still more rapidly. The computed effective criticalities for each procedure on the active procedure list are compared and the procedure with the highest effective criticality is chosen as the next procedure to work on.

<u>Anatomy Movement</u>. The anatomy movement model is accessed whenever HOS determines that a body movement is required in order to accomplish the objective of an instruction. HOS automatically selects the appropriate body part, moves it to the required location, and adds to the simulation time a time that corresponds to the time the action would have taken a real operator.

The anatomy movement model consists of two micromodels — one to determine which body part to use for a particular action, the other to assign a time charge for the movement. The body part selection

micromodel is based on several common-sense principles. The first is
that the body part to be used is determined by the function to be
performed and the device being referenced. Thus, if the operator is
going to be reading data from a device, the eyes are usually the appropriate body part to use. However, there may be some devices whose
value cannot be determined visually - touching them with a hand or
foot may be more appropriate. Some devices may use two modalities -
the eyes are used to absorb information while the hands are used when
the value of the device is to be changed. HOS permits the analyst to
specify for each device the most appropriate modality for each function (reading and/or altering).

The time charge assigned for an eye movement is computed from
equations that were developed by combining data from an experiment
that involved lateral eye movements (Dodge and Cline, 1901) with data from an unpublished experiment by Wherry and Bittner that involved
both lateral and convergence movements.

The HOS model for hand movement times is based on data from a
number of sources, principally Fitts (1954), Fitts and Peterson
(1964), and Topmiller and Sharp (1965). In the Topmiller and Sharp
study, the subject moved his hand from an armrest in order to operate
controls at various locations in space. Since these data were most
relevant to potential HOS applications, they were used to describe
the ballistic travel time of hand movements. Since Fitts' movement
time model (Fitts' Law) considered movement precision, Fitts' model
is used to determine the time required for terminal positioning as
a function of device size.

The HOS model for foot movement time between control pedals describes the relationship between movement time, amplitude of movement, size of target, and required change in leg extension. The model, which is adapted from Fitts' Law, is based on data published by
Drury (1975) and by Davies and Watts (1969).

Some key characteristics of the anatomy movement micromodel
that should be noted here are:

1. Movements, like the instructions that initiate them, are executed
   serially for each body part.

2. Movements are ballistic - once initiated they cannot be stopped
   nor can another action be initiated while the movement is taking
   place.

3. Movement times are fully deterministic, based on where a body
   part is and where it is being moved to - there is no variability
   over a given distance (though variability could easily be introduced).

4. If a movement cannot be performed, an interrupt will be generated
   enabling the operator to select another procedure from the active
   procedure list for execution.

   Control Manipulation. Times associated with control manipulations
are highly variable because of the diverse types of controls used in
different operator situations. Consequently, HOS allows the user to
supply the parameters of the equations that determine the time associated with a control manipulation. In addition, there are a set of
"packaged" calculations that compute control manipulation times for
two basic control types – discrete controls and continuous rotary controls. For discrete controls, the analyst is required to supply a time
that represents the time required to move the control through a single
setting. If a control manipulation results in a movement through several settings, the time assigned will be the time required for a single setting multiplied by the number of settings. The time charge for
continuous rotary controls is based on data presented by Karger and
Bayha (1966).

   Unlike the actions discussed previously, control manipulations,
once initiated, can proceed in parallel with other actions. Thus, the
operator can be performing several control manipulations concurrently,
as well as processing information, recalling information, moving other parts of the anatomy, etc.

Operator Variability

   As described above, most of the equations that govern the operator micromodels in HOS are fully deterministic. This is consistent
with two of the premises in the HOS model – that the operator is a
trained operator and that performance variations observed in experiments on individual operators are largely the result of situational
differences, as opposed to differences in basic performance parameters. However, there are clearly differences in operator performance
– both between operators and for the same operator under different
operational conditions. By changing some of the HOS operator parameters the analyst can examine the effects that these have on total
system performance. In addition, by modifying the equations described
above, one can readily describe an operator with a different performance profile.

EVIDENCE FOR THE MODEL

General Issues of Model Validity

   In discussing the issue of validation, it is useful to distinguish between simulation languages and simulation models. Most general purpose simulation systems (e.g. GPSS, SIMSCRIPT, GASP and DYNA-
MO) are simulation languages. That is, they provide a method by which

an analyst can describe a system to a simulation program. The program itself typically makes no assumptions about the system being modeled. It simply exercises the model described to it in the simulation language according to the rules provided by the analyst. The validity of such simulation systems is dependent on whether the simulation language enables the modeler to faithfully represent the characteristics of the system that he wishes to model and on the fidelity with which the modeler represents the system in the first place. Thus, with simulation languages there are two issues that impact upon validity - the ability of the analyst and the ability of the simulation language to capture the analyst's model.

Simulation models, on the other hand, are simulation systems that have an explicit model of the system concerned. For example, some military engagement models contain specific assumptions about rules of engagement, dispositions of forces, and engagement statistics. The user supplies data on probable force levels from which the model predicts engagement outcomes. Dynamic simulators, e.g. cockpit simulators, are also examples of a simulation model. They are a "representation of reality", reflecting the dynamics of an actual system to the extent possible within the constraints of the simulator and requiring, as user inputs, only a set of initialization (and control) parameters to perform their function.

Simulation models may require only simple inputs, such as the values of certain model parameters, or very complex inputs - instead of just force levels, a tactical engagement model could require rules of engagement as well. In the case of complex inputs, these data may be expressed in the form of simulation language. Thus, in addition to the issue of the validity of the model's internal representation of the system, there can be the additional two validity issues associated with simulation languages. Each of these filters on reality could result in loss of simulation fidelity through the process outlines in Fig. 3.

Since HOS is a simulation system that contains both a model (the HOS model of the operator) and a language (HOPROC) any discussion of the validity of HOS must consider three things:

1. The validity of the HOS model of the operator.

2. The ability of the HOPROC language to describe the relevant processes associated with simulating a given system, and

3. The extend to which a specific analyst's model, described in the HOPROC language, accurately represents a given system.

These are three clearly separate issues and each must be "valid" for the results of the model to be "valid". That is, the HOS operator model must be a sufficiently accurate model of how human beings

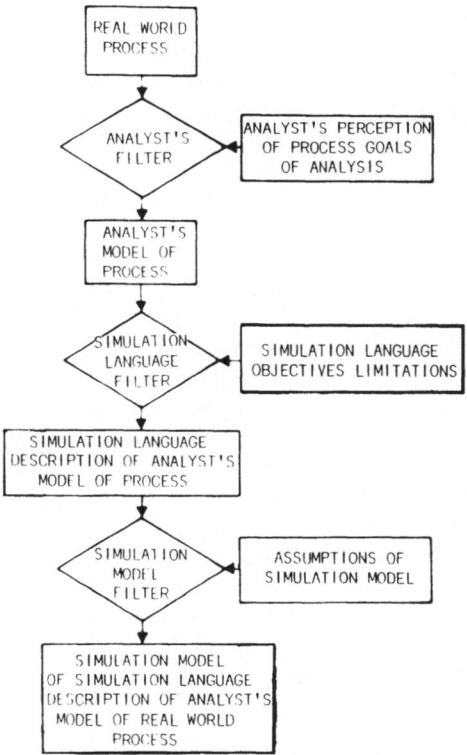

Fig. 3. Modeling Process Filters.

perform and the HOPROC language must be sufficiently capable of enabling operator tasks to be adequately described and the modeler
must correctly describe the decision rules and operator activities
that pertain in a given situation. Note that the operator model, the
language or the analyst's model do not have to be absolutely true
or complete - indeed, absolute truth and completeness are unrealis-
tic goals for a simulation model. Rather, the two models and the
language need only be sufficiently true or complete to adequately
represent the truth and thereby provide useful information.

These three issues also have implications with regard to the
generality of system simulation. There may, for example, be situations of interest that cannot be adequately represented in a simulation system because of limitations that exist either in the language
or in the simulation model. These limitations are the result of conscious choices made when the simulation system is designed to enable
the system to deal with the situations for which it was being designed. The modeling of operator error in HOS is an example. A conscious choice was made when HOS was designed to avoid introducing
certain types of random operator errors, although it is obvious that
such errors do occur in human performance. Because of this choice,
"truth" about human performance has obviously been compromised and

the argument can be made that the model is not "valid". But, HOS was not designed to evaluate the effects of such errors. Rather, it was designed to enable systems to be evaluated under a precise set of standardized conditions. Thus, it was considered inappropriate to routinely incorporate low probability events such as random error, and appropriate to model a trained operator whose performance is subject only to normal human performancè limitations.

The foregoing is not an apologia, but rather a statement of the ground rules that must be applied not just to an evaluation of HOS, but to any evaluation of any model's validity. Simply stated, a model cannot be judged as invalid because of either how it represents the processes that it purports to model or because it does not represent all the processes that pertain to the system that it is modeling. Such issues relate only to the usefulness of the model in solving specific problems. A model is valid to the extent that it predicts the performance of a system to an accuracy and level of detail adequate for solving the problems for which the model was designed. Implicit in this is the situational nature of validation. A model may produce verifiable results for all situations so far considered, but may fail when applied to a new situation because of limitations or unmodelled processes in the simulation model, the analyst's model, or the simulation language.

Approaches to HOS Validation

Four different types of validations, addressing progressively more realistic situations, have been used to evaluate the HOS system. As the situations involved come closer to real systems, the phenomena being simulated increase in complexity and become less repeatable in the real world, with a consequent decrease in direct verifiability of simulation outcomes. The conditions and objectives of these four verification approaches are as follows:

Type I: a) Is the HOPROC language sufficient as a mechanism for expressing the analyst's model of a system?

            b) Given the sufficiency of the language, is the analyst's model translated into reasonable sequences of operator activities?

Type II: a) Do the micromodels adequately reproduce human behavior on simple tasks?

            b) What are the appropriate parameters for the "standard" operator?

Type III: a) Does HOS reproduce operator behavior in part-task experimental and operational situations?

b) Is the assumption of micromodel additivity correct?

Type IV: Do the outcomes of HOS simulations correspond to known
            problems in full-scale operational systems?

The most basic evaluation level (Type I) was directed toward determining whether HOPROC could be used to describe conveniently and
precisely all desired aspects of operator and system performance and
whether HOS interprets and translates HOPROC code into sequences of
operator behaviors that are reasonable ways of accomplishing the simulated tasks. After dozens of HOS simulations, findings on these
questions have been positive. HOPROC has proven to be easy to learn
and surprisingly flexible in modeling operator and equipment procedures, including features of operator performance that were not intended for explicit modeling in HOS. An example is a series of simulations in which the operator was required to "learn" the response
characteristics of his system (gain, lag, etc.) before he could sa-
tisfactorily perform the required tasks (Glenn et al., 1977). As more
and more HOS simulations have been developed, it has been necessary
on occasion to modify aspects of HOPROC syntax or change the way an
instruction is processed to handle new task requirements, but the
original concepts behind HOPROC and HOS have remained virtually in-
tact.

The second type of evaluation has examined the validity of the
HOS micromodels. Since some of the micromodels were based directly
on experimental data and some were based on projections or concep-
tions of how the experimental micromodels should interact, this approach focused on:

1. Reproducing as simulations the baseline experiments from which
   the micromodels were derived, and

2. Simulating additional experiments from the human performance literature.

The objectives of simulating the baseline experiments were to
insure that micromodel interactions did not introduce any artifacts
that would prevent replication of baseline studies and to develop
appropriate parameters and equations to describe the "standard" operator. The objective of simulating a variety of additional experiments was to verify that the micromodels had predictive validity for
precisely controlled and controllable situations. The results of
these comparisons of simulations to experimental data indicated that
micromodel estimates were of sufficient accuracy for continued validations in more complex task situations. Reach times, for example,
were within 0.1 second of reported reach times under similar circumstances, and key literature on short-term memory could be reproduced
by varying parameters to compensate for different experimental procedures across comparison studies. Display reading times from the si-

mulations closely matched those reported for different types of displays. This extensive series of simulations is presented in a series of reports (Analytics, 1975 a, b, c).

The third type of evaluations has been a series of "part-task" simulations. Tasks in these studies closely resembled operator functions in systems but were of less complexity to allow more precise structuring of task conditions. The objectives were to verify the additivity of times generated by micromodels in interaction and to examine simulation performance in situations more closely approximating real systems than laboratory experiments. These simulations included:

1. A divided attention study in which the operator performed a manual tracking task with interference from a secondary task whose frequency and duration were varied (Lane et al., 1977).

2. A mail-sorting simulation which required the operator to use simple decision rules combined with keyboard entry (Analytics, 1975c).

3. A subset of the LAMPS helicopter Air Tactical Officer (ATO) functions, combining CRT tracking and control manipulation with key entry (Strieb et al., 1976).

Results from all three simulations gave additional support to the HOS model. The outcome of the LAMPS ATO subset simulation is particularly interesting. In this simulation, the operator must use a joystick to move a cursor (hook) on a CRT screen until the hook coincides with a symbol on the display. Two different keys are then depressed - one to alert the system and one to destroy the symbol. Fig. 4 shows the average time taken for each of three consecutive hook-and-destroy operations across four trials (an initial practice trial was eliminated). Each entry in the table represents the mean time for five operators performing the operations in a LAMPS simulator. Corresponding estimates from HOS are also shown. Note that HOS estimates vary slightly for the three symbols, since physical location of symbols on the display affects the HOS operator as they would a real operator, and that the HOS estimates are well within the range of variability for the real operators.

TRIAL \ ACTION	T_2	T_3	T_4	T_5	μ,σ	HOS
A_1	4.6	4.4	5.8	4.8	4.90,.62	4.61
A_2	5.2	4.8	6.0	5.8	5.45,.55	5.41
A_3	5.4	5.6	6.0	5.4	5.60,.28	5.27
μ,σ	5.07,.42	4.93,.61	5.93,.12	5.33,.50	5.32,.56	5.10,.43

Fig. 4. Comparison of HOS HOOK/DESTROY Simulation Results with Experimental Data.

   As the use of HOS has matured, simulations have been developed
for several large-scale weapon systems. Results of these applications are the fourth type of HOS validation. The major simulations
conducted with HOS to date are:

1. A simulation of the Air Tactical Officer (ATO) on-board the U.S.
   Navy's LAMPS helicopter during a generalized ASW mission (Strieb
   et al., 1976).

2. A simulation of the Sensor Station 3 (non-acoustic) operator (SS-3)
   on the U.S. Navy's P-3C ASW patrol aircraft during a reconaissance
   mission (Lane et al., 1979).

3. A simulation of the P-3C Sensor Station 1 (acoustic) operator
   (SS-1) during an open ocean convoy escort mission (currently under development) (Banowetz et al., 1980).

4. A simulation of the pilot on NASA's Terminal Configured Vehicle
   (TCV) during the approach and landing phases of both a curved
   and straight landing under both manual and automatic control
   (currently under development) (Glenn and Doane, 1980).

   The first of these (LAMPS ATO) served primarily to verify the
ability of HOS to simulate full-scale complex systems, since the simulation was developed after the major LAMPS design decisions already had been made. This study also served to provide the data for
the part-task studies previously described. The second simulation
(the P-3C SS-3) was designed to show that HOS could identify actual
operator/system problems and suggest solutions. This simulation was
actually a series of baseline and sensitivity studies with different system configurations, operator characteristics, and tactical
rules of engagement. The SS-3 operator station was selected because
there were known problems that had been reported by the fleet during
certain types of mission with certain of the P-3C aircraft. The HOS
simulations examined how operator performance varied under each condition and the extent to which overall mission goals were achievable
under each condition. The simulations clearly demonstrated that the
degradation in operator system effectiveness and mission goal achievement in certain configurations could have been predicted by the HOS
model. They also indicated how modifying the rules of engagement to
achieve mission goals would affect exposure of the aircrew in a hos-
tile environment.

   The third set of simulations (the P-3C SS-1) applies HOS to the
evaluation of a system currently under development. In this case,
the hardware system being modeled is the Navy's Advanced Signal Processor (ASP). The HOS simulation will be used to define how the operator should best use the system and suggest ways in which the operator's job could be aided in future modifications to the system. An
interesting feature of this simulation is the way in which it com-

bines simulated and real operator decision-making. The SS-1 operator's responsibilities with respect to the prosecution of the mission are limited to the recognition and classification of the acoustic signature data obtained through the ASP. Primary responsibility for tactical direction of the mission - e.g. determining where to lay sonobuoys and which buoys to monitor in order to optimize detection of the targets - rests with the Tactical Coordination Officer (TACCO). Because of the complexity of these tactical decisions, it was decided not to invest in modeling the TACCO for initial versions of the simulation. Thus current SS-1 simulations use an actual TACCO who can communicate with the simulated operator to control progress of the mission as in an actual exercise. If necessary, a procedure for executing the specific TACCO functions critical to ASP evaluation will be added to the simulation at a later time.

The difficulty with simulations such as the SS-1 and SS-3 as validation exercises is that the situations and scenarios involved are so complex that it is impossible to obtain controlled, reliable data against which to compare simulation outcomes. For both simulations it would be necessary to run actual fleet exercises with precisely the configuration of equipment, emitters and targets used in the simulation in order to compare numerical results and to do this with multiple operators, since HOS estimates "average" operator performance. This same problem is, of course, encountered with any model of any complex system. Predictions from models of complex systems can only be compared to known system problems for general agreement and reasonableness of outcomes, not for mathematical correspondence. For the SS-3 situation, extensive fleet studies and a limited data base on system problems were available for examination. In no case was there disagreement between HOS predictions and problems identified by operators in the fleet.

In contrast to the simulation of fleet systems, the TCV study represents a complex collection of tasks for which extremely detailed and well-controlled experimental data is available. NASA has been gathering data on pilot eye movements during final approach and landing in both ground-based simulators and actual flights, under a variety of conditions including curved and straight-in approaches, manual and auto-pilot control, varying traffic conditions, and with differing types of pilot instrumentation. The HOS TCV simulation models these conditions for comparison with experimental data.

ONGOING WORK AND FUTURE DIRECTIONS

Most of the development work and many of the validation studies for the basic version of HOS were completed in 1976. Although the model has undergone continuous refinement and considerable expansion in capability since that time, efforts during the last four years have focused principally on applications and on exploring a variety

of new uses for HOS that have arisen since its conception. Much of
the ongoing work using HOS is of a complexity and level of detail
considerably beyond the scope envisioned during initial formulation
of the model. Although the HOPROC/HOS architecture and the basic micromodel structure has been remarkably adaptable to these new requirements, the experience gained with the model during these applications,
together with the increasing maturity of human factors engineering
over the last 10 years, has pointed out many new directions for further HOS development.

Much of the impetus for detailed reexamination of HOS concepts
has come from use of the model in two areas - the evaluation of system cost effectiveness and the selection and verification of decision-
aiding algorithms. The HOS model is being used extensively in a program aimed at evaluating proposed avionics and software changes to
existing systems prior to actual hardware development. This program,
called Operator Interface Cost-Effectiveness Analysis (OICEA), has
been the catalyst for the SS-3 and SS-1 simulations (Lane et al.,
1979). A key aspect of OICEA is the development and maintenance of
baseline simulations and cost data for major Naval air systems to
allow quick evaluation of proposed system updates and modifications.
This work is continuing with the expansion of the SS-3 operator simulation to include a variety of proposed software alternatives for
the processing of electronic emission data. Algorithms will priori-
tize targets and provide varying degrees of automation in support of
the SS-3 operator in assessment of threats to the P-3 platform. The
HOS/OICEA simulation will be used during algorithm development to
evaluate and refine alternatives on the basis of predicted improvements in mission performance relative to development and support
costs.

The OICEA evaluation of the kind of automation needed by the
SS-3 is parallelled by a related but independently-derived requirement arising from generalized decision aid/decision automata development efforts. A network of programs devoted to augmenting decision-
making capabilities in Naval systems has demonstrated the need for
systematic techniques for allocating decision functions among system
components (Hopson et al., 1980; Zachary, 1979). Two major aspects
of this technique development are the use of operator models for
selecting among decision-augmentation approaches and the development of a standard language for describing the functions of proposed
decision algorithms. New uses of HOS are thus addressing applications
of the model as a detailed decision-analytic simulator and the re-
vision of HOPROC to serve as a standard decision-specification language.

The extraordinary diversity of these new demands has indicated
the possibility of major improvements in model performance by developing separate specialized versions of HOS for differing applications.
Experience gained in the large-scale ASW simulations suggests that

HOS may provide a level of sophistication and detail higher than necessary for large systems. The major problems affecting performance
in such systems almost invariably involve information - too much, un-
timely or improperly formatted. Relatively small amounts of the variance in performance among system alternatives is attributable to
variations in reach time, memory and other factors simulated in detail by the micromodels. A standard time charge could therefore be
used for many activities with little loss in predictive power. A new
version of HOS, still in the planning stage, will substitute macro-
models for selected micromodels, while retaining such desirable features as task resequencing and straightforward analysis of information management. This "HOS for Very Large Systems" (HOS/VLS) will
trade some precision in modeling for more rapid and less expensive
simulation development.

     In a similar trend toward specialization, another ongoing study
will determine whether more detail, rather than less, may be necessary for some applications. Evaluating what may be relatively subtle
performance differences among alternative decision algorithms requires both increased precision in decision modeling and a rich procedural language for specifying the decision-making process. The
specialized version of HOS proposed for decision allocation will
likely provide more primitive modeling of such functions as control
manipulation and anatomy movement in return for a HOPROC/HOS architecture materially enhanced in decision simulation detail. Another
feature of the "decision-making" HOS will allow the model to drive
actual system control/display devices, emulating a hardware simulator in real time for a real operator. This approach in a primitive
form is being employed on the SS-1 simulation.

     The basic micromodel structure of HOS was defined in the early
1970's, based on the literature existing at that time. Far more extensive work on human cognitive behavior has become available since
that period. Although no particular problems with the HOS micromodels have been encountered, it is probable that directions from recent research can be used to fine-tune or revise specific micromodels or to restructure the entire micromodel set. The sequence of experiments and specific data requirements to address these issues
have been outlined in a preliminary study by Glenn and Wherry (1979).

     There are two additional areas in which changes and improvements
to HOS will ultimately be necessary to insure its ability to simulate
future systems. It has been recognized for some time that the system
controller aspects of HOS are less well defined than other parts of
the operator's job. Although HOPROC has been used to simulate control tasks quite effectively in validation studies, it is less efficient for this purpose than classic control models. To date, no ap-
lication of HOS has required this capability. Several anticipated
systems, however, particularly those of the VSTOL class proposed for
antisubmarine warfare mission, require the controller (the pilot) to

perform extremely demanding vehicle control tasks while processing
large quantities of environmental and sensor data. For these situations, it would be desirable to integrate sophisticated control models into the cognitive modeling framework of HOS. The micromodel
architecture of HOS should allow this incorporation without major
changes and will add an element of "intelligence" to the controller,
a feature that is typically lacking in control models.

The dramatic growth in proposed system automation in recent
years points out the second area for potential HOS modification. The
trend toward increased reliance on automated processors will materially shift the role of future operators in many systems toward the
performance of monitor/supervisor duties. This shift is thoroughly
documented by papers in Sheridan and Johannsen (1976) and in George
(1979). Two kinds of additional capability will be required for HOS
to efficiently simulate highly automated systems. The first - expanded decision-making emphasis - is already underway. The second
involves the development of an additional micromodel to simulate operator monitoring behavior. Although the monitor role can be simulated
acceptably by combinations of micromodels within the current structure, the lack of an explicit monitoring model was specifically i-
dentified as an inefficiency in HOS during a series of workload studies (Glenn et al., 1977).

To date, HOS has been able to simulate with reasonable conve-
nience a wide variety of different systems. Applications over the
last four years have suggested the potential improvements noted above
to allow a similar capability for simulating the new kinds of systems expected in the next decade. Models like HOS, with the ability
to integrate a realistic operator component into a systems modeling
framework, will become more critical as systems depart from traditional roles for the operator. Human factors engineering technology
must keep pace with equipment technology, providing techniques which
allow credible, objective and detailed statements about probable
system (not just operator) performance. Without such techniques,
the inappropriate use of operator capabilities, already aggravated
by careless reliance on new electronics solutions, cannot be pre-
vented.

REFERENCES

Analytics, 1975a, Development of a Quantitative Display Reading Mod-
    el for HOS, Part I, Technical Report 1117-F, Analytics, Willow
    Grove, Pennsylvania.
Analytics, 19-5b, Development of a Quantitative display Reading Mod-
    el: Model Specifications. Part II, Technical Report 1117-G,
    Analytics, Willow Grove, Pennsylvania.
Analytics, 1975c, The Human Operator Simulator, Volume VI: Simula-
    tion Descriptions, Technical Report 1181-B, Analytics, Willow

Grove, Pennsylvania.

Anderson, A. and Miller, K., 1976, Timeline Analysis Program (TLA-1) User's Guide, Final Report Contract NAS1-13741, NASA Langley Research Center, Hampton, Virginia.

Asiala, C., 1975, Advanced Man/Machine Evaluation Techniques, American Defense Preparedness Association, Huntsville, Alabama.

Banowetz, V., Kelley, J. and Rea, M., 1980, A Human Operator Simulator Model of the P-3C Advanced Signal Processor Passive Acoustics, Draft Technical Report 1400.01-A, Analytics, Willow Grove, Pennsylvania.

Davies, B.T. and Watts, J.W., 1969, Preliminary Investigation of Movement Time Between Brake and Acceleration Pedals in Automobiles, Human Factors, 11 (4), 457-459.

Dodge, R. and Cline, T.S., 1901, The Angle Velocity of Eye Movements, Psychology Review, 8, 145-157.

Donders, F.C., 1869, Over de snelheid van Psychische Processen, Tweede Reeks, 11, 92-120. Translated by W.G. Koster, 1969, as: On the Speed of Mental Processes, Acta Psychologica, 30, 412-431.

Drury, C.G., 1975, Application of Fitts' Law to Foot-Pedal Design, Human Factors, 17 (4), 368-373.

Edwards, R., Curnow, R. and Ostrand, R., 1977, Workload Assessment Model: Users Manual, Report D180-20247-3, Boeing Airspace, Seattle, Washington.

Fitts, P.M., 1954, The Information Capacity of the Human Motor System in Controlling the Amplitude of Movement, Journal of Experimental Psychology, 47, 381-391.

Fitts, P.M. and Peterson, J.R., 1964, Information Capacity of Discrete Motor Responses, Journal of Experimental Psychology, 67, 103-112.

Fitts, P.M. and Posner, M.I., 1967, "Human Performance", Brooks/Cole Publishing Co., Belmont, California.

George, F. (ed.), 1979, Proceedings of the 15th Annual Conference on Manual Control, Technical Report AFFDL-TR-79-3134, Wright Patterson AFB, Ohio.

Glenn, F. and Doane, S., 1980, A Human Operator Simulator Model of the NASA Terminal Configured Vehicle (TCV), Draft Technical Report 1463, Analytics, Willow Grove, Pennsylvania.

Glenn, F., Strieb, M. and Wherry, R., 1977, The Human Operator Simulator, Vol. VIII: Applications to Assessment of Operator Loading, Technical Report 1233-A, Analytics, Willow Grove, Pennsylvania.

Glenn, F., and Wherry, R., 1979, Experimentation Requirements to Support the Human Operator Simulator, Draft Technical Report 1400.02-E, Analytics, Willow Grove, Pennsylvania.

Greening, C., 1978, Analysis of Crew/Cockpit Models for Advanced Airctaft, NWC TP-6020, Naval Weapons Center, China Lake, California.

Hick, W.E., 1948, The Discontinuous Functioning of the Human Operator in Pursuit Tasks, Quarterly Journal of Experimental Psychology, 1, 36-51.

Hopson, J., Stark, S., Detwiler, D., Zachary, W. and Fitzkee, S., 1980, A Generalized Decision Automation Algorithm for AEW Engagement/Intercept Planning, NADC-80104-60, Naval Air Development Center, Warminster, Pennsylvania.

Jahns, D.W., 1973, A Concept of Operator Workload on Manual Vehicle Operations, report no. 14, Forschungsinstitut für Anthropotechnik, Wachtberg, West-Germany.

Karger, D.W. and Bayha, F.H. 1966, "Engineered Work Management", Industrial Press, Inc., New York.

Klein, T. and Cassidy, W., 1972, Relating Operator Capabilities to System Demands, Proceedings of the Human Factors Society - 16th Annual Meeting.

Kristofferson, A.B., 1967, Attention and Psychophysical Time, Acta Psychologica, 26, 93-100.

Lane, N., Strieb, M.I. and Leyland, W.E., 1979, Modeling the Human Operator: Applications to System Cost Effectiveness, in: "Modeling and Simulation of Avionics Systems and Command Control and Communications Systems", NATO/AGARD Conference Proceedings CP-268.

Lane, N., Strieb, M.I. and Wherry, R.J., 1977, The Human Operator Simulator: Workload Estimation Using a Simulated Secondary Task, in: "Methods to Assess Workload", NATO/AGARD Conference Proceedings CP-216.

Linton, P., Jahns, D.W. and Chatelier, P., 1977, Operator Workload Assessment Model: An Evaluation of a VF/VA-V/STOL System, in: "Methods of Assess Workload", NATO/AGARD Conference Proceedings CP-216.

Pew, R., Baron, S., Feehrer, C.E. and Miller, D.C., 1977, Critical Review of Performance Models Applicable to Man-Machine Systems Evaluation, report no. 3446, Bolt, Beranek and Newman, Cambridge, Massachusetts.

Sheridan, T. and Johannsen, G. (eds.), 1976, "Monitoring Behavior and Supervisory Control", Plenum Publishing Corp., New York.

Siegel, A. and Wolf, J., 1961, A Technique for Evaluating Man-Machine Systems Design, Human Factors, 3, 18-27.

Strieb, M.I., Glenn, F.A., Fisher, C. and Fitts, L., 1976, The Human Operator Simulator, Vol. VII: LAMPS Air Tactical Officer Simulation, Technical report 1200, Analytics, Willow Grove, Pennsylvania.

Strieb, M.I., Glenn, F.A. and Wherry, R.J., 1977, Computer Aids for Operator Station Design, Technical report 1285-A, Analytics, Willow Grove, Pennsylvania.

Strieb, M.I. and Wherry, R.J., 1979, An Introduction to the Human Operator Simulator, Technical report 1400.020, Analytics, Willow Grove, Pennsylvania.

Stroud, J.M., 1955, The Fine Structure of Psychological Time, in: "Information Theory in Psychology", H. Quastler (ed.), Free Press, New York.

Topmiller, D.A. and Sharp, E.D., 1965, Effects of Visual Fixation and Uncertainty on Control Panel Layout, Technical report AMRL-TR-69-149, Aerospace Medical Research Laboratories, Wright Patterson Air Force Base, Ohio.

Wherry, R.J., 1969, The Development of Sophisticated Models of Man-Machine System Performance, in: Symposium on Applied Models of Man-Machine Systems Performance, report no. NR-69H-591, North American Aviation, Columbus, Ohio.

Wortman, D.B., Duket, S.D., Seifert, D.J., Hann, R.L. and Chubb, G.P., 1978, Simulation Using SAINT: A User Oriented Instruction Manual, AMRL-TR-77-61, Wright Patterson, AFB, Ohio.

Zachary, W., 1979, Decision Aids for Naval Air ASW, Technical report 1366-A, Analytics, Willow Grove, Pennsylvania.

SAINT, A DIGITAL SIMULATION LANGUAGE FOR THE STUDY OF MANNED SYSTEMS

Gerald P. Chubb

INTRODUCTION

Development   of SAINT evolved over a period of five years and was based upon experience gained in having used the Siegel-Wolf, two-man, operator simulation model (Siegel and Wolf, 1967). That particular model viewed system performance in terms of operator activities over time. It was written in FORTRAN but was not modularly constructed. Modifications of the program sometimes led to unforeseen difficulties which complicated debugging new routines.

If human factors were to be considered during design, there was a need to better describe man's impact on system performance, and conversely there was a need to improve the description of how the system impacted the operator. This needed to be accomplished in a form readily accepted by the engineering community.

Two languages were particularly popular at the time: GPSS (General Purpose Simulation System) (Schriber, 1974) and SIMSCRIPT. (Markowitz, et al., 1963). Both required special compilers, so neither could be easily modified. GASP II (Pritsker and Kiviat, 1969) was SIMSCRIPT-like in its modeling constructs or descriptive concepts, but executed under the FORTRAN compiler. The Siegel-Wolf model could be represented in GPSS. It was believed it could also be represented in GASP. However, it was subsequently discovered that P-GERT (Whitehouse, 1973), developed using the GASP subroutines, was a more suitable vehicle for capturing the basic character of the Siegel-Wolf model.

SAINT Evolution

Pritsker observed that the Siegel-Wolf model was basically an
activity-on-node representation of discrete event sequences (Elmaghraby, 1977). P-GERT had been developed to analyze this class of models. However, because of the dynamics of time stress formulated by
Siegel and Wolf, their networks were non-linear with time varying
parameters. Task times could vary as a function of defined levels of
workload "time-stress". P-GERT was only capable of portraying and
analyzing linear, time-invariant networks. Therefore, the first stage
of SAINT development was devoted to generalizing the formal representation of the Siegel-Wolf model (Pritsker et al., 1974) and extending
P-GERT to capture this model of man.

The second stage of SAINT development was a major advance and
exploited recent developments in GASP IV (Pritsker, 1974). This allowed combined modeling. Continuous processes could be represented
as differential or difference equations and solved by the Runge-
Kutta-England fourth-order integration algorithm. This permits one
to model the equations of motion for an aircraft that describe its
position and attitude over time. To illustrate a simple combined model, a very elementary representation of aerial refueling was developed (Wortman et al., 1974). No attempt was made to portray vehicle dynamics to any high degree of fidelity, and the operator's control actions were simplistically described by a strategy that mechanistical-
ly provided discrete regulation of vehicle acceleration. However, the
model did provide some interesting insights. The preliminary formulation led to excessive numbers of disconnects and breakaways. Systematic variations in scan rates and regulatory actions demonstrated that
with more frequent scanning and less dramatic regulatory changes,
better system performance was achieved. While there was no attempt
at exactness of representation, certain general conclusions were apparent from even this simplistic portrayal of man-machine interaction.

During the third stage of SAINT development, revisions were made
that completely divorced SAINT from the original Siegel-Wolf model
without precluding its implementation, should one choose to do so.
It was recognized that other formulations of behavioral dynamics
might be preferred, and a modeler should not be forced to use a prescribed formulation such as workload "time-stress" if that did not
appear relevant in the system context being addressed. SAINT III documentation was not separately prepared, but the technology was applied to modeling a remotely piloted vehicle drone control facility
(Wortman et al., 1976). SAINT permits a modeler to describe task
times (or any event duration) as either a sample from a prescribed
distribution, a derived value (as a function of prescribed parameters), or as some combination (as in the Siegel-Wolf model, where
the average and standard deviation of a normal deviate vary with
stress level).

During the fourth stage of SAINT development, additional attention was given to resource considerations and to simplifying what had
become a very complicated set of symbols for modeling. This effort
culminated in a redesign of the computer codes and preparation of a
thorough set of documentation (Wortman et al., 1977 a, b and c, and
Duket et al., 1977). A general overview of SAINT is presented in
Seifert and Chubb (1978), including some contemporary applications.

General Character of SAINT

SAINT is not itself a model. It is a network-oriented, combined
simulation language for modeling manned systems. Either the man, the
machine, or both can be represented as discrete events, continuous
processes, or both. Most system models have chosen one or the other
form of modeling man. The classical describing function approach
(McRuer et al., 1965) to modeling manual control emphasizes a frequency domain analysis of a continuous process model of man. The
more modern approach to this problem was formulated by Kleinman and
Baron (1971) and also describes manual control as a continuous process (with sampling) but does so in the time domain. In their original formulation of tracking behavior, Garner and Hake had cautioned
that man did not behave as a time-invariant, linear, differential
equation. The discrete task approach to modeling man is equally naive, because it generally suppresses the subtleties and sophistication of the formal approaches to modeling the manual tracking task
as a continuous process. Seifert (1979) illustrates how these two
approaches to modeling man can be combined in SAINT, if one chooses
to do so. They need not be mutually exclusive as prior work implicitly suggests. Unfortunately, few systems engineers are presently
equipped with the skills to exploit both aspects of this modeling
technology. However, SAINT is equipped to support either or a combination of approaches to modeling man in a system context.

SAINT constructs have been arbitrarily separated into three
groups (Table I): 1) elementary, 2) intermediate, and 3) advanced.
The elementary constructs provide a powerful set of modeling concepts
that require no programming experience of the user. Some of the more
useful features of SAINT become available in the intermediate set of
constructs. While none of these require any FORTRAN coding as such,
they are more difficult to employ correctly, and their inadvertent
misuse can create debugging problems more easily solved if one has
experience with software development. Although SAINT includes a substantial number of error checks on input data, it cannot be expected
to find all possible errors, and the ones it does not find may indeed
be subtle and difficult to isolate. Prior experience in this sort of
problem solving is generally acquired by anyone who designs, develops,
and successfully implements a computer program. So while use of intermediate SAINT constructs does not necessarily require such skill, it
may prove quite valuable when errors are made preparing the input data.

Table I. SAINT Concepts

Elementary Concepts (No Programming Required)

Node (Event/Task)	Resources	Duration
Precedents (Event)	Attributes	Branching
Different Precedents	Labeling	Statistics

Intermediate Concepts (Programming Experience Helpful)

Priority	Modification
Completion Precedence	Clearing (Pre-emption)
General Task Characteristics	Information Attributes
Information Packet Choice	Mechanisms

Advanced Concepts (Programming Necessary)

User Functions	Monitors
Moderators	Regulators
State Variables	Switching

Finally, the advanced constructs require FORTRAN programming or programming in some other language compatible with FORTRAN on the computer being used. Moreover, use of the continuous process modeling constructs requires the user be experienced at describing system dynamics mathematically in terms of algebraic, trigonometric, or differential (or difference) equations. This is not a trivial effort. While SAINT provides facilities for exercising such models, it is tacitly assumed the models themselves are readily available. This presumes some rather sophisticated engineering analysis and modeling not typically done by human factors specialists.

This report makes no attempt to extend SAINT applications or man models. Here we choose instead to give some elementary examples of SAINT that require no programming experience. These illustrate some generic modeling issues, but the models do not address any significant design problem. By contrast, a more complex model is also presented that has a variety of design implications. Although it is only rudimentary, the generic aspects of the problem context are presented, and the model could be profitably expanded in a variety of directions that are left unexplored.

FIVE SIMPLE TEST CASES

The SAINT diagrams will be presented without the accompanying program inputs or output listings. The latter will appear in a technical report still in preparation at this time.

Test Case One

This example (Fig. 1) has only two nodes: the source task that initiates a network and a sink task which ends the network. The duration of each is fixed, but in the first task this is specified directly by a scalar constant while in the second task the time specification for event duration is indirectly identified by reference to a distribution set. In general, the distribution set specifies the nature of the distribution and its parameter values. In this case, a constant time value is specified. If an event is <u>known</u> to be of fixed duration, it is simple enough to make the specification directly, but since it is convenient (if not necessary) to estimate the activity duration until data are obtained, use of the indirect specification may be preferred. The distribution type and parameter values can then be altered appropriately as model updates are made. The quality and quantity of data available often change as a system transitions through its development stages.

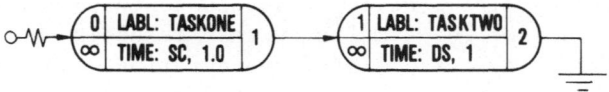

NOTE: DS, 1, CO, 1.0 — ACCOMPLISHES THE SAME THING
AS SC, 1.0 BUT IS MORE GENERAL AND EASIER TO
UPDATE (WITHOUT REDRAWING THE DIAGRAM).

Fig. 1. Test Case One.

In order to provide a more convenient identification of nodes (other than the assigned node number), SAINT permits the assignment of an eight character nmemonic label (LABL:) or name to any task. No processing is done on the label itself, but it serves as a tag on the task that is especially helpful when one examines output listings from a SAINT computer run.

Test Case Two

Here we label our tasks with more meaningful names (Fig. 2). In this case, we assume a navigation system capable of displaying heading error. Once the navigator has properly set up the system, he can flip a switch to send the signal to the pilot's station and tell the pilot to null the indicated error by centering the Flight Control Indicator (FCI). The total activity time is a joint function of the response time for switch setting and the time for the verbal communication. Hypothetical values are shown and the tasks are portrayed

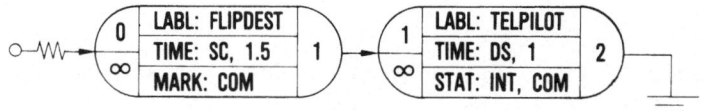

KEY: DS, 1, NO, 3.0, 0.0, 10.0, 0.25*
    DS   — DISTRIBUTION SET
    NO   — NORMAL (GAUSSIAN)
    INT   — INTERVAL
    COM   — COMPLETION

Fig. 2. Test Case Two.

as strictly sequential events. SAINT's ability to handle multiple
activity sequences permits modeling tasks as concurrent processes
if one chooses to do so. For simplicity, that issue is suppressed
here. The switch setting task is of fixed duration, 1.5 time units
long. The units themselves are arbitrary and may be used to represent seconds, minutes, hours, days, etc. The verbal notification
task is represented as a stochastic variable, normally distributed,
with a mean of 3.0 and a standard deviation of 0.25 (seconds). On
any one realization of this network, a sample will be drawn from
that distribution. Separate iterations will result in different sample values. The resultant variability over many iterations may be
interpreted as intra-operator variance. In this model, alternate
message formats are not considered explicitly, so alternate forms of
communication are implicitly subsumed in the chosen distribution set.

    Even this simple model has some interesting features. The com-
pletion of task one is marked and statistics are generated for the
time interval to completion of task two. This effectively measures
the duration of the second task. Several exercises might be attempted, some of which are suggested here. Different numbers of iterations
(5, 10, 15, 20, 25, 50, 75, 100) might be attempted to illustrate the
vagaries of sampling error as a function of sample size. The parameters of the normal distribution could be varied to illustrate how
well (or poorly) the default values for histogram generation capture
the output distribution. Alternate distributions could be specified
to discern how they compared with the normal. Fig. 3 portrays a set
of samples drawn from the same normal distribution. The mark time
could be set to the start of task one to demonstrate how the inclusion of a fixed additive constant affects overall activity time when
combined with a random variable (time duration for task two). While

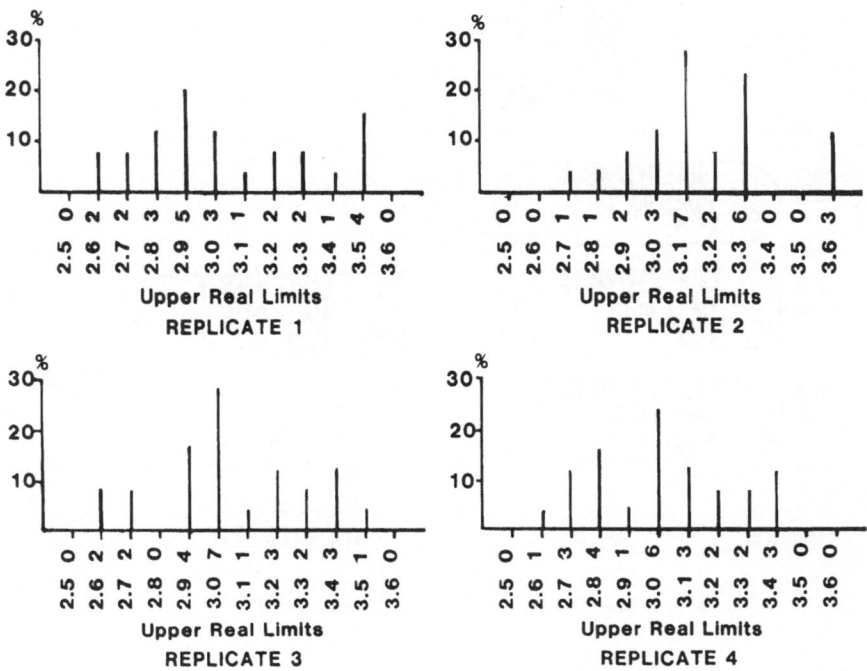

Fig. 3. Examples of sample distribution drawn from the same popula-
      tion.

these are somewhat academic issues with known results, they are con-
structive instructional uses of a simple model.

Test Case Three

Here (Fig. 4) we postulate three different forms the message
might take and explicitly consider how often each type is used:. 1)
message one "Center the FCI" is used 10% of the time, 2) message two
"FCI Good" is used 30% of the time, and 3) message three "FCI's Good;
Center Up" is used 60% of the time. Note that the duration of each
message is now portrayed as a constant task time. However, total time
for activity completion is variable because each message type is of
a different duration and occurs with a different frequency. The selection of modeling approach (lumped aggregate descriptions or detailed microprocess descriptions) will lead to different formulations
in SAINT. The language itself does not force one or the other philosophy but permits implementations of either approach. The choice of
an appropriate level of detail in activity description is left to
good judgment and depends upon the modeler's purpose and objective:
the nature of the problem to be solved or the decision to be made
from model results.

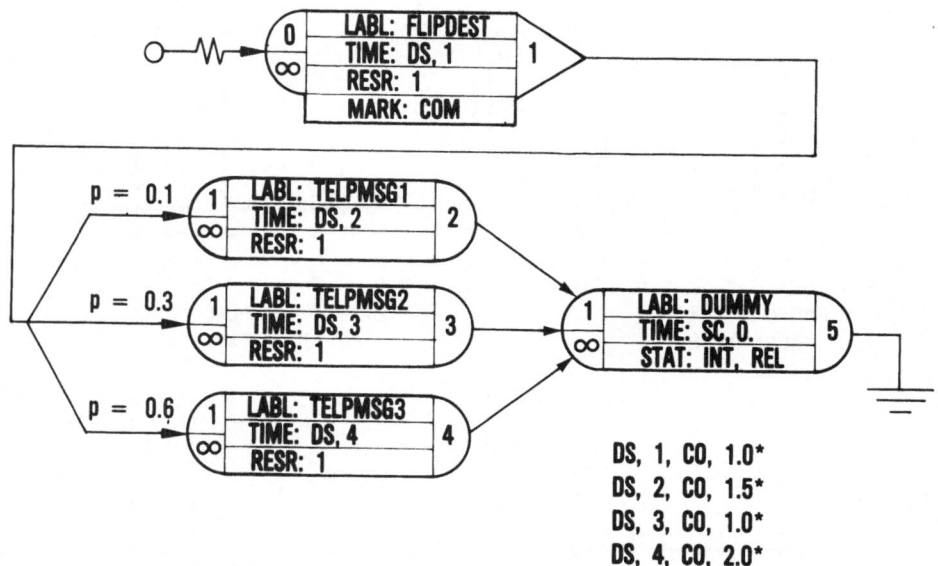

Fig. 4. Test Case Three.

One might also exercise this model in several ways. The branching probabilities might be altered. The number of message types might be expanded. The message durations might be made stochastic instead of fixed values. The statistics collected here only measure message duration. Other statistics could be collected. The whole process could be replicated to discern how stable the message frequencies are for a specified branching probability. Only "on the average" will the branching probabilities generate actual frequencies corresponding to the specifications. In any one set of iterations, the realized frequency of branching will vary from one replicate to the next.

Test Case Four

The model is now embellished (Fig. 5) to include a crude representation of possible interactions between navigator and pilot. The navigator is portrayed as first composing his thoughts before electing to speak, and on occasion he even forgets to coordinate with the pilot. Correspondingly, the pilot, presumably familiar with this ideosyncracy of his navigator, waits a period of time before asking whether he should center the FCI. This delay has been represented by a negative exponential distribution, a special case of the more general Erlang distribution. Hastings and Peacock (1975) provide a rather extensive description of a variety of statistical distributions.

If the navigator never forgets, the results of test case four should closely parallel those of test case three, but as the proba-

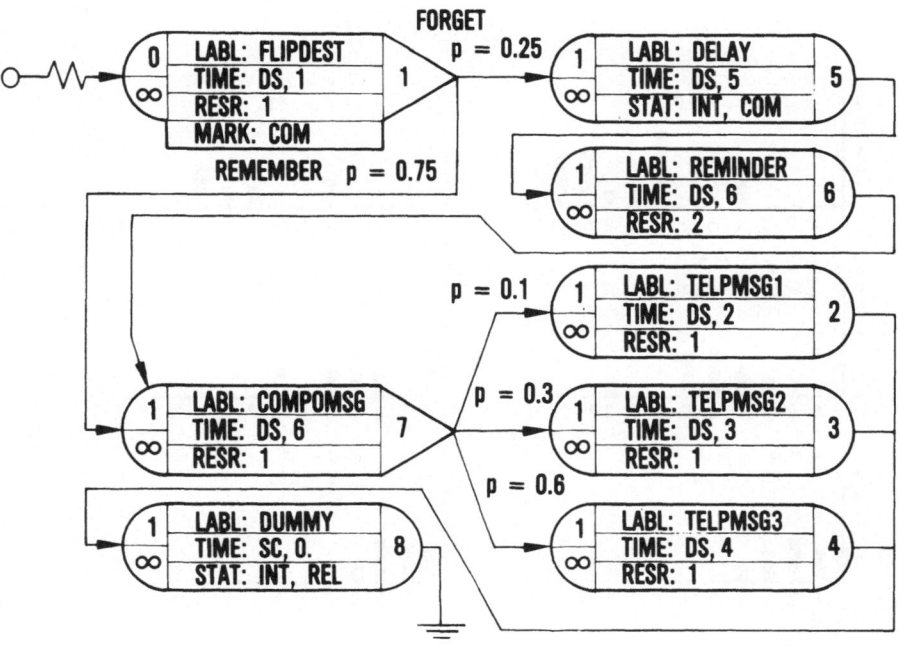

Fig. 5. Test Case Four.

bility of forgetting increases, we should observe more radical departures from prior results. Here the pilot always asks the same question: "FCI Good?". There may be cases where inter-operator variability is of interest, and the task duration for this message would require a separate distribution (DS, 6) to appropriately reflect the pilot's verbal response. However, if the difference is known or believed to be practically negligible, we could choose to use the same distribution (DS, 3) as used for the navigator's mandate: "FCI Good!".

Note also that the pilot and navigator are treated as resources necessary for task performance, but the delay event has no resource requirement. Each node in the network serves a purpose but need not be an operator task as such.

The number of delays encountered will be a function of the probability of forgetting to tell the pilot what to do. This is reflected in the histograms of Fig. 6, and is summarized in the linear decline in delays as a function of improved memory (Fig. 7). Total activity duration will be driven by the injection of delays. While the average duration is relatively sensitive to such effects (Fig. 8), the maximum in a series of iterations is even more sensitive (Fig. 9, as evidenced by the steeper slope on the regression line. Each of the six conditions was replicated ten times, twenty-five iterations of the model per replicate.

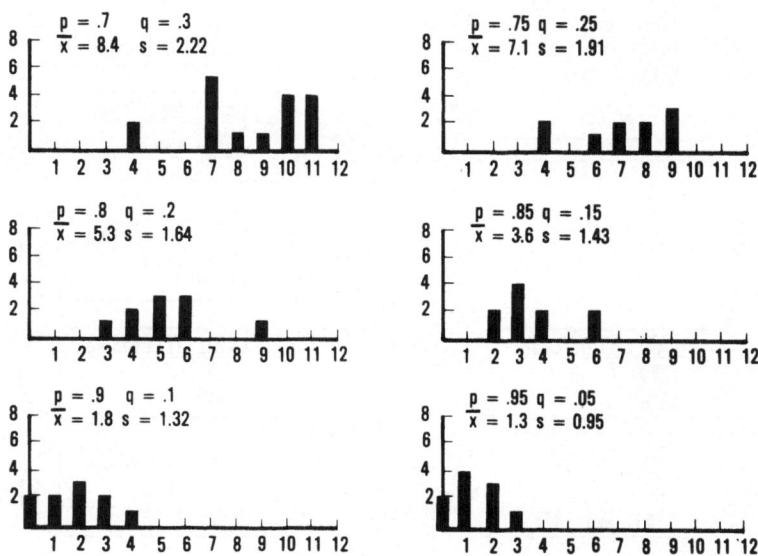

Fig. 6. Histograms for number of reminder messages as a function of branching probabilities (ten replicates in each of the six sets).

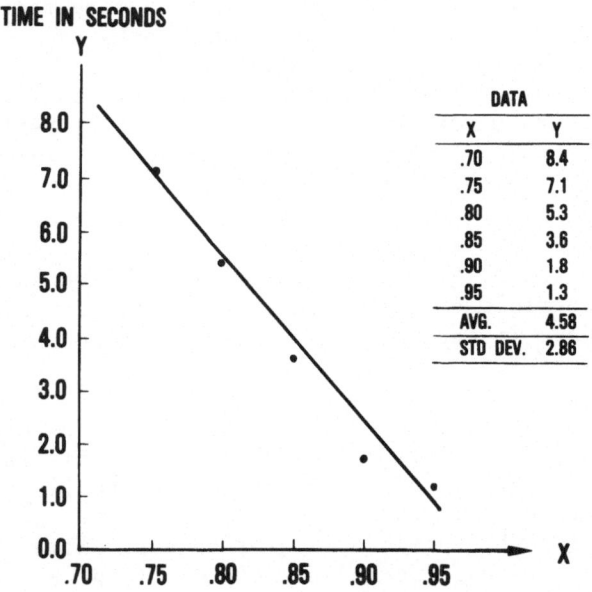

Fig. 7. Decline in reminders as a function of the probability of re-membering to issue the message.

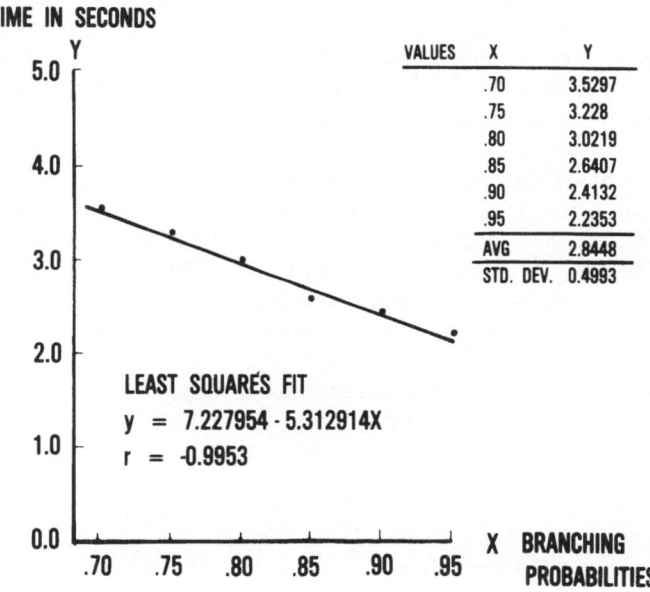

Fig. 8. Average activity duration as a function of remembering.

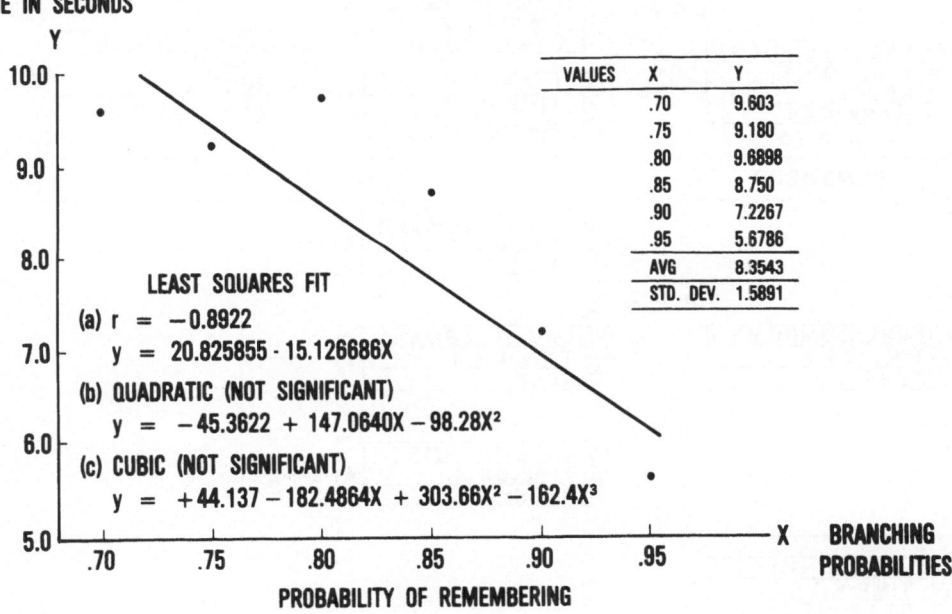

Fig. 9. Maximum activity duration as a function of remembering.

Some interesting exercises here would be the manipulation of the
delay distribution parameters, variation of the time for reminding
(pilot's message duration), and a repeat of the exercises proposed
for test case three. It is hypothesized that activity duration will
be more sensitive to the delay than the duration of the pilot's message under a fairly broad range of conditions.

It should be noted that this model implies a telepathic pilot.
One really ought to question the reasonableness of assuming a reminder occurs only after a delay induced by the unobservable act of
forgetting. This may be a small point, but it illust ates how various issues might be considered significant, ignored, or even be in-
tentionally surpressed depending on the modeler's intent or purpose.

An alternate formulation is presented in Fig. 10. Here the delay is initiated as a separate source node. If the navigator remembers, no reminder is initiated, but if he forgets, a reminder is only initiated after a delay has been incurred. This model still postulates forgetting as a necessary condition for the reminder, but it
is not a sufficient condition. A delay is necessary also, but since
it is not sufficient, the reminder is not automatically initiated.

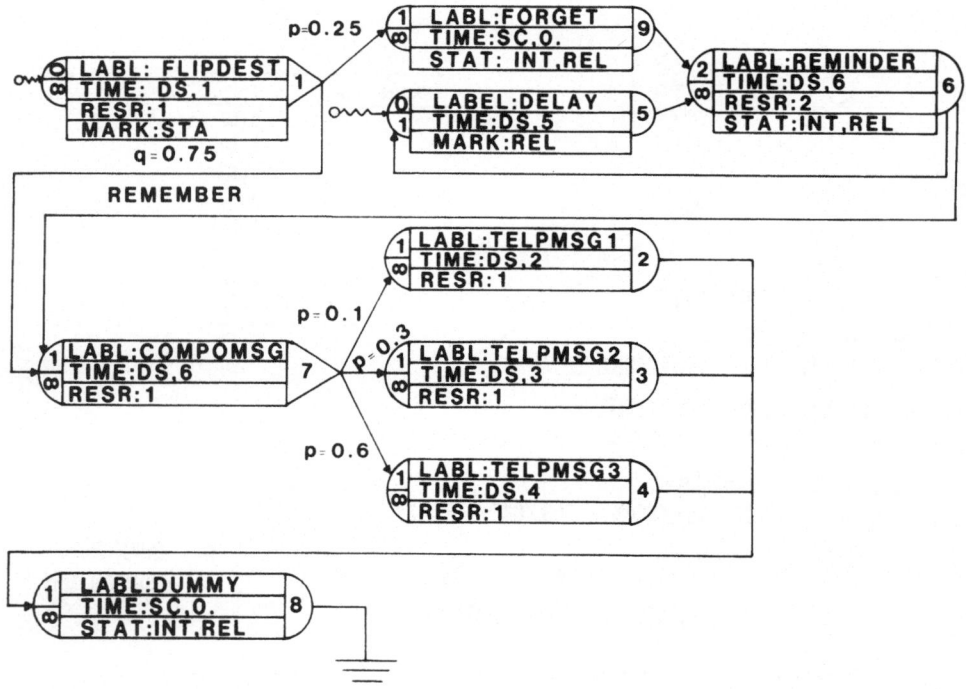

Fig. 10. An alternative formulation of test case four.

If it were, the reminder could be coincident with the navigator's attempt to notify the pilot, which is yet another formulation that
might be postulated. The formulation presented here should not provide results different from those of the original model. While their
behavior should be identical, the second formulation has more credibility in terms of face validity. This may be desirable even if not
necessary for predictive validity.

Test Case Five

Fig. 11 represents the pilot as a pest of sorts, because he will
attempt to periodically remind the navigator whether he needs to or
not, and not just once but repeatedly (due to the loop around the delay node that triggers the reminder). The navigator here does not so
much forget as perseverate: he may simply elect not to issue the message and will again go back to composing his thoughts. It is inter-.
esting to note our inability to discriminate (by observation) whether the operator forgets or chooses not to act. In a model, we can
postulate either or both cases and see whether it makes any difference as to which occurs. The net effect in either case is similar
here in that the activity does not complete until the message confirming system status is issued.

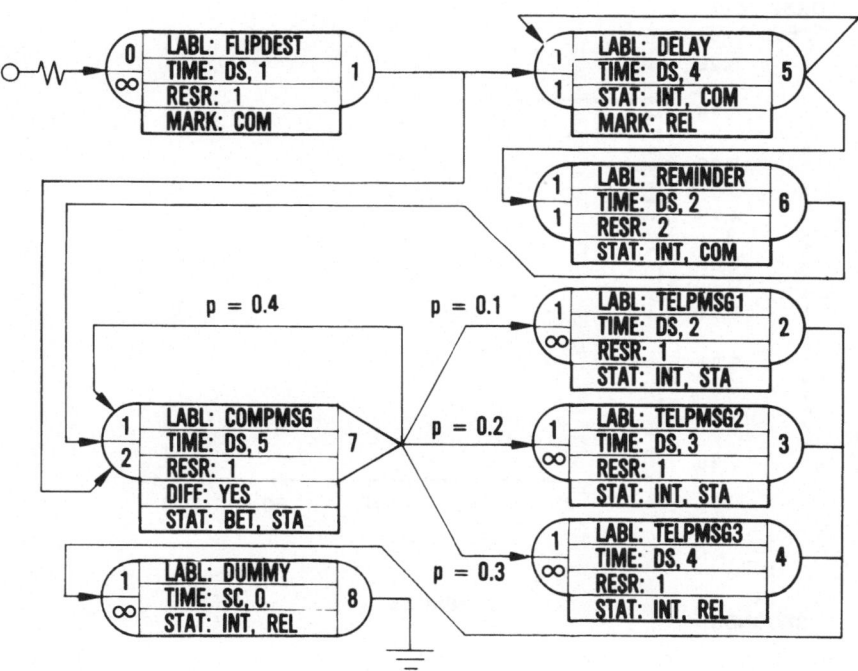

Fig. 11. Test Case Five.

Note that if the navigator chooses to perseverate, the message will not be issued until the pilot intervenes since two (different) precedents must be satisfied if node 2 is performed more than once. If the reminder is issued, but the navigator is not perseverating, the message is issued only once since the feedback loop out of node 7 is not taken when a message occurs, thereby precluding satisfaction of the precedence requirement for re-release of task 7. The problems induced by reminder/message coincidence are not explicitly treated. One might postulate several different consequences. The navigator might have to repeat his message. The pilot might instead repeat his interrogation. Both might be attempted with a possible repetition of the interference. All of these are possible embellishments of the model, none of which will be explored here.

Fig. 12 is an example of a histogram describing the delay distribution as it was realized in one set of 100 iterations where 87 delays were encountered (generated). This has a pronounced effect on the distribution for reminders, as shown in Fig. 13. While marked skewness is apparent, kurtosis is not as pronounced. Fig. 14 demon-

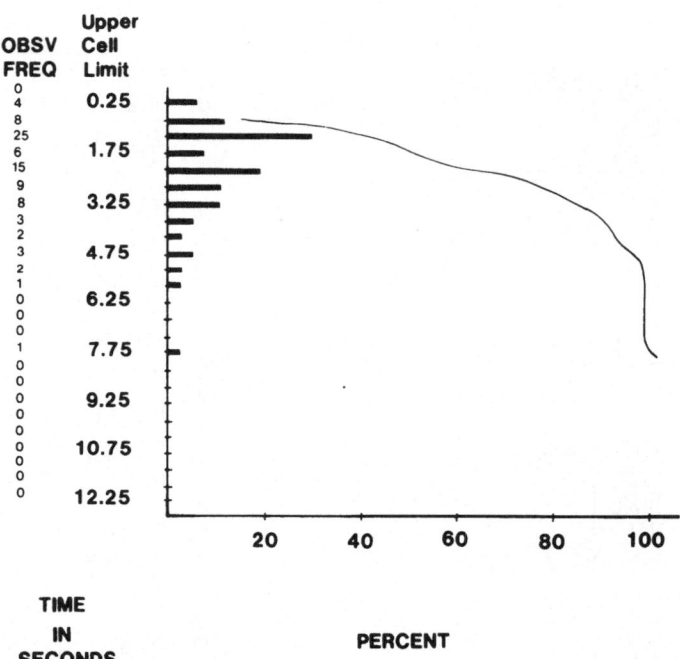

Fig. 12. Histogram of delays experienced from sampling the Erlang distribution.

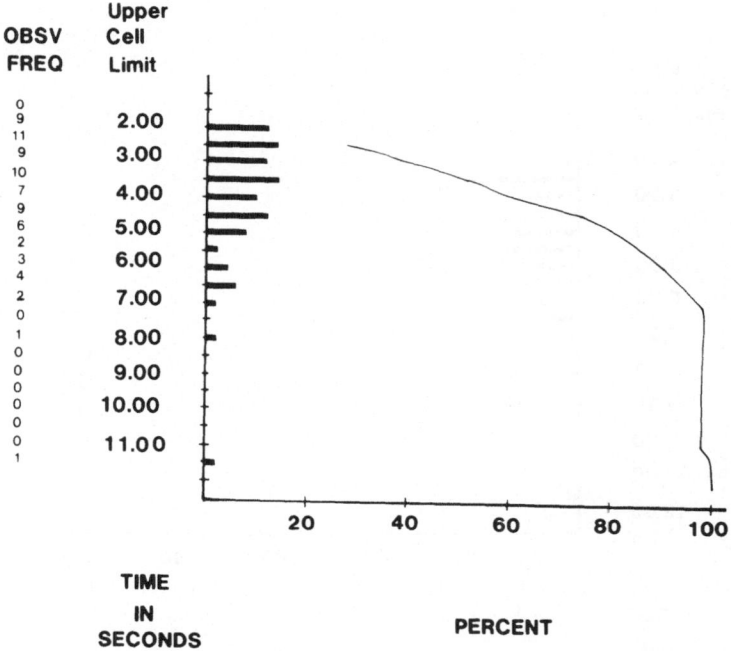

Fig. 13. Distribution for completion time of the reminder message.

strates that looping has a tendency to induce a similar effect:
marked skewness is evidenced in the distribution of time between message composition, despite the fact that the nominal time for this
task is normally distributed. In part, the observed skewness here
may be attributable to the effect of tasks 5 and 6, but further experimentation with the model is required to confirm or refute such
a hypothesis.

Fig. 15 evidences potential bimodality along with skewness,
but then, it includes the differential induced by message duration.

It is apparent from these demonstrations that the nature of the
distribution describing activity duration will change its character
markedly depending upon the structure of the tasks which are subelements of the activity and one's definition of the beginning and end-
ing events that define that activity whose duration is to be measured. While some component elements may themselves be normally distributed, there is no basis for assuming that the overall activity
will be. Skewness, kurtosis, and bimodality may all be induced by
the structural integration and interaction among the subelements of
the activity.

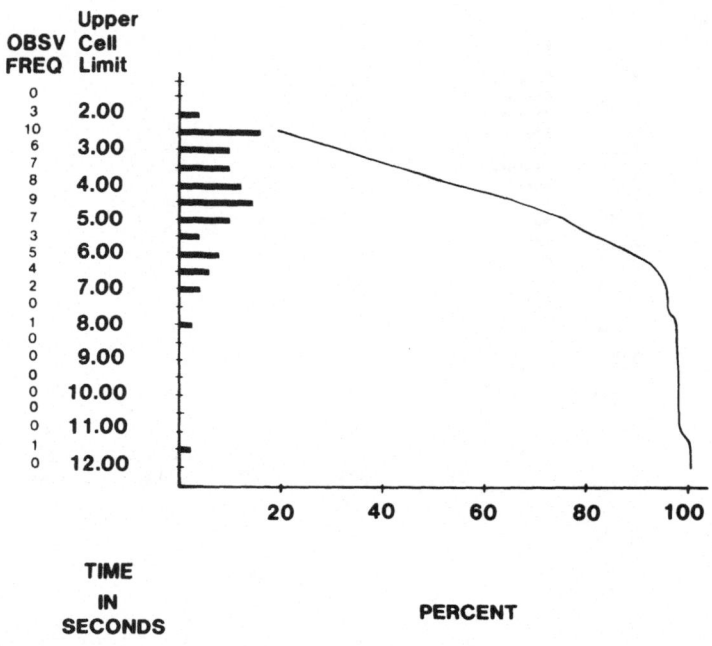

Fig. 14. Time between message composition by the navigator.

Summary

None of these examples bears on any critical aspects of system performance, but the test cases, simple as they are, illustrate a surprisingly diverse set of modeling issues. Further, none of these examples even exploited the more sophisticated features of SAINT. A model need not be complex or even wholly valid to be useful. Insights into subtle aspects of a problem can often be gained from a superficial and highly abstract representation. While no one should recommend that poor models are acceptable for problem solution, neither should one fail to capitalize on any opportunity to gain understanding from experience with less than the ultimate, final product. In any case, modeling is most successful when it begins modestly and builds on a base of proven success in capturing aspects of the problem of special interest to the modeler.

A MORE COMPLEX MODEL

None of the preceding examples illustrated how man impacts machine or vice versa. Now we will formulate a model that draws on some of the advanced SAINT concepts. It is assumed that the hard-

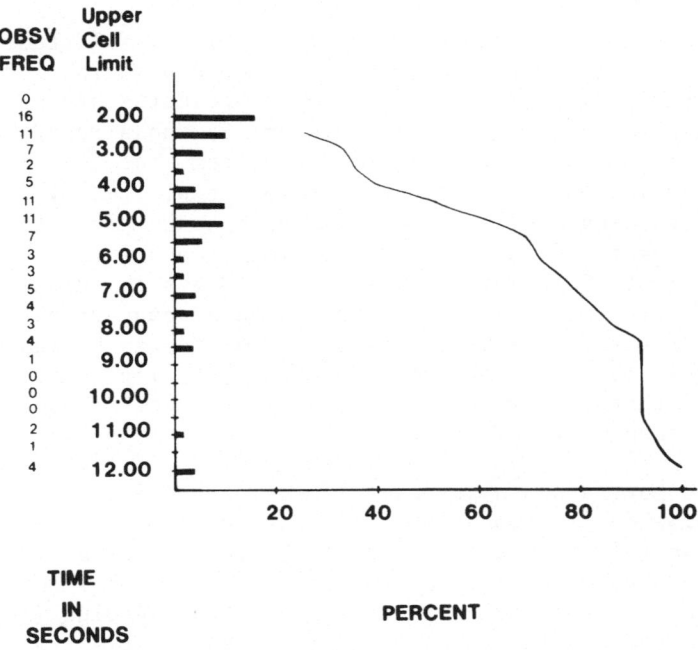

Fig. 15. Total activity duration.

ware predicts aircraft present position by extrapolation. Velocity
and heading measures are used to project present position forward in
time. The operator attempts to correct system prediction error by
performing position fix updates using radar imagery. Three measures
of aircraft position need to be represented: 1) planned position,
2) system position estimates, and 3) true position. If the goal is
to stay oncourse and on-time, then differences between the first and
third measures is the real error. Differences between the second and
third are the system or instrument error. More modern (inertial) navigation systems tend to minimize the impact of operator error (by
Kalman filtering). The system architecture portrayed here is of an
older era, one in which dead-reckoning was based on simple integration of sensor inputs. There were several opportunities for operator
error to impact system performance. This form of navigation still
persists as a backup in case of failure in the modern aids. Consequently, it is still a worthy problem to address and one often ignored with present emphasis on pilotage as a manual control task.

The Sighting Problem

In theory, for a known constant flight level (altitude) a set of crosshairs will accurately overlay a ground position indication (GPI) on a radar display if: 1) the GPI coordinates are known exactly and presented accurately to the system, 2) the aircraft present position is accurately portrayed in the system, and 3) all velocities are known, accurate, and appropriately exploited by the system to permit "tracking" the point as time passes and aircraft position changes. If the first and third assumptions are presumed valid, then any misalignment of the crosshairs is due to system error in predicting aircraft present position. If the operator can accurately reposition the crosshairs, appropriate compensation can be made to the system representation of present position. For simplicity, it is also assumed that the system appropriately corrects sensor image scaling for changes in aircraft attitude.

Fig. 16 portrays a hypothetical mission segment. The waypoints are successive destinations in a flight path or mission plan. The angle between this path and true polar North is typically designated as the true course. The aircraft heading is in general not known exactly. Its estimated value is designated true heading (in distinction to magnetic heading which may vary substantially from true heading).

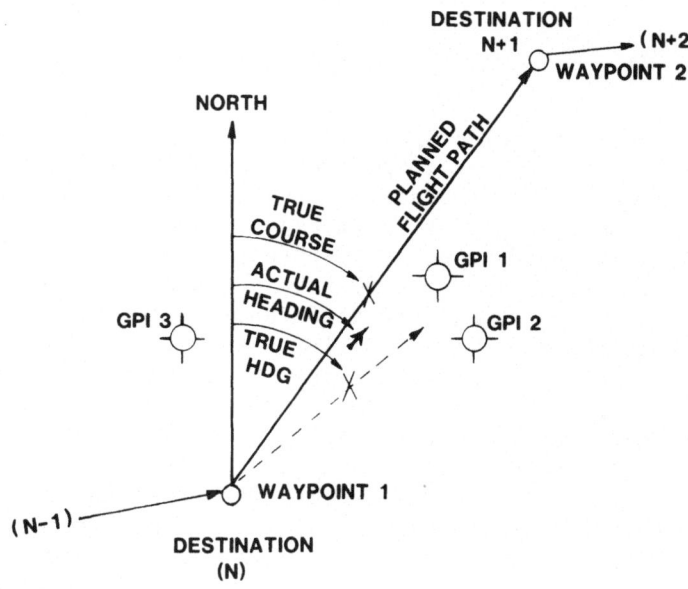

Fig. 16. A hypothetical mission segment.

Three factors affect system error: 1) present position error (initially), 2) true heading error, and 3) aircraft velocity errors. The basic formulation is: RT = D (Rate times Time equals Distance). Consequently, if position is known at some time (t) and velocity flown since then is known, present position can be predicted. Since the problem is two dimensional (assuming constant altitude and a flat earth), velocities North and East need to be separately considered (South being the negative of North and West being the negative of East). Position excursions North are reflected in latitude changes while excursions East are reflected in changes in longitude. Any course other than a cardinal heading ($0°$, $90°$, $180°$, or $270°$) will result in changes to both latitude and longitude coordinates of present position. If present position errors are compensated by repositioning the crosshairs, then the crosshairs will stay on a point if the velocity and heading are correct. Otherwise, the crosshair will begin to drift off the point at a rate proportional to those errors. If the crosshairs are later placed back on this point. adjustments can be made to compensate for assumed (constant) velocity errors. If the crosshairs are now moved to a new point, position and velocity errors having been corrected, positioning error may be attributed to system error in heading estimates and crosshair repositioning will again provide a compensatory input. Neither of these latter two compensations will be treated here, only the first: position error update. Velocity updates and heading fixes should be treated, along with altitude calibration, for a more complex model. The flat earth model becomes unacceptable as distance traveled increases (to several hundred miles).

Sources of Error

While Strieb and Wherry (1979) have argued against the concept of "human error" for evaluating systems during development, the concept is still appealing and may in fact be necessary in describing extant systems. Here we envision three potential "human" related errors: 1) reading the GPI coordinates from a map, 2) entering these into the system, and 3) physically positioning crosshairs on the radar image of the GPI. It is apparent that the GPI may itself be poorly mensurated, the chart or map may have been printed with an error, or aids prepared to preclude the need to interpret the map may themselves be in error (e.g., transcription or typographical error).

In reading the GPI coordinates, the operator may most easily transpose digits. This might occur either in degrees or minutes (decimal minutes or seconds were ignored here and therefore implicitly assumed to be correct).

The operator is assumed to communicate these values to the system by setting latitude and longitude counters. It is presumed that hardware mechanization will intrinsically limit the accuracy achiev-

able by the operator, but he will himself exhibit some variability
in the value entered.

Finally, the operator is presumed to be incapable of positioning
the crosshairs over the exact (true) position of the GPI because of
a variety of handicaps. These include azimuth resolution of the radar, reflectivity of the point itself, size and extent of the feature
of interest, the workload, the stress, and possibly the anxiety of
the operating environment, as well as the nature of the operator's
attempts to tune his radar set properly.

From this discussion, it should be apparent that at this level
of description, what we have chosen to call "human error" in fact con-
founds the system mechanization and human performance. This confounding may or may not be circumvented by more detailed analysis. While one
does not wish to accept such confounding, many practical applications,
especially for an existing system, may present cases where it is obviously impractical to insist on analyses which eliminate confounding. Time and cost constraints may preclude attempt(s) to do so anyway.

The other "errors" are worth considering. The first is the selection of the GPI to be used for a position fix update when more than
one is within radar viewing range. The second is selecting which point
to move the crosshairs onto when they do not fall where they should
(due to system and/or operator error(s) in positioning the crosshairs
previously).

By edict or experience the operator should know that the closer
the point is when the fix is taken, the better the resulting update,
all factors considered. However, the closer point (in fact) may not
be the one chosen because of errors in radar scope interpretation,
changes in aircraft heading after the choice was made, or system error in the displayed value for present position. No attempt will be
made to treat these problems. It will be naively assumed the choice
is based instead on the minimum range between the system estimate of
aircraft position and the operator entered counter values for the GPI
position.

Considering the accumulated system error, the crosshairs will
probably fall off (or away from) the designated GPI. In fact, they
may lie closer to some other GPI (known or unknown) to the operator.
This can occur whether or not the operator made a careful study of
the area and correlated the displayed imagery with his maps and
charts. This entire process may be of interest but will be suppressed here. For simplicity, we will assume a stochastic process governs
his decision. First, the range from the crosshairs to adjacent points
will be taken. Second, the normalized range will be used as the argument of a negative exponential function. This will generate a num-

ber between 0 and 1 that will be used as a criterion for the decision. Third, a random number will be drawn and compared with the criterion for the first such point considered. If the criterion is larger, the range is sufficiently small that the point will be chosen.
Otherwise another point will be considered and compared to a new random number. If all the points are reviewed without a selection being
made, the process is repeated twice and then an arbitrary choice is
made. This rather elaborate scheme is meant to mimic the indecision
of the operator in using range as a singular criterion in the face
of the other uncertainties mentioned. One might refute the reasonableness of the stated model, and certainly other credible formulations
might be proposed.

Any such arbitrary formulation may be employed in the absence
of additional empirical insight, further study, analysis, and/or empirical research. If sensitivity analyses demonstrate that system
performance is affected by the model as formulated one should challenge the model's validity. If alternate formulations have little or
no effect on system consequences, one might question the wisdom of
extensive analysis or empirical research, however objectionable the
present formulation may appear. Why waste the effort in model validation if the system is insensitive to any differences between models?

Implementation

The discrete network for this model is presented in Fig. 17. In
task 1 the operator reads the GPI coordinates from the map. If no
error was made, USERF sets information attribute 1 to zero, otherwise
it is set to one and either a transposition error is made in the degrees (information attribute 2 is set to zero) or the minutes (information attribute 2 is set to 1) position. The start of the task is
marked to permit the collection of interval statistics. Task duration is governed by distribution set 1, which is a normal distribution with a mean of 3.0 (secs), a standard deviation of 1.0 (secs),
and a minimum and maximum of 1.0 and 10.0 (secs) respectively.

At the second node, the latitude and longitude counters are
loaded with values as read from the map, plus or minus a random error representing operator entry inaccuracy. The values entered are
carried as system attributes (1 and 2 respectively). Task duration
is governed by distribution set 2, designated as a log normal distribution, with a mean of 5.0 (secs), standard deviation of 1.5 (secs),
and a maximum and minimum value of 1.0 (secs) and 30.0 (secs) respectively. Statistics are collected on the interval to release, effectively measuring the duration of task 1.

In task 3, the Automatic Crosshair Laying (ACL) button is depressed, and the crosshairs are positioned at the location determined
by: 1) system coordinates for aircraft present position, and 2) the

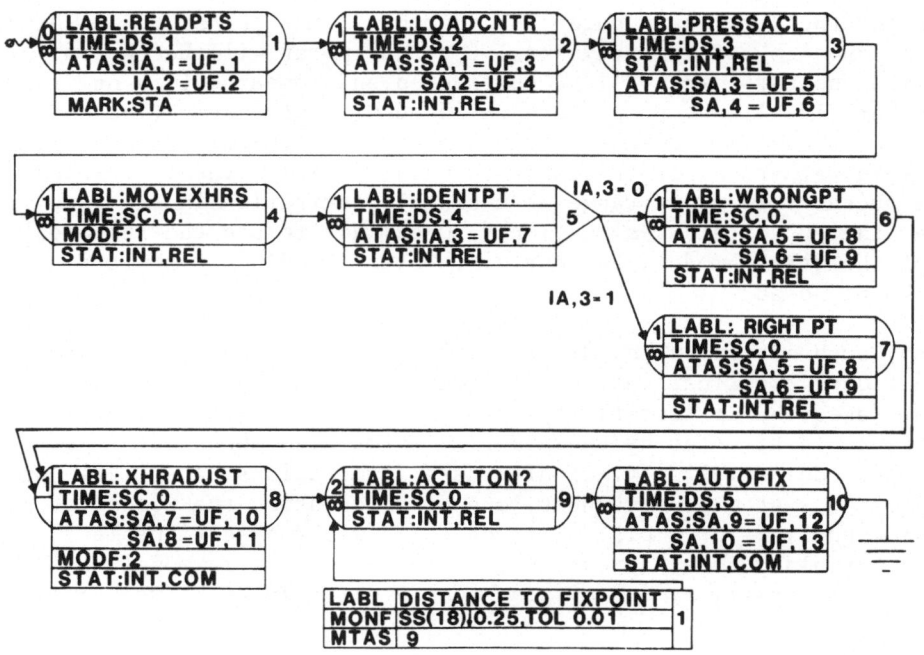

Fig. 17.   Generic model of present position update via radar fix
           taking.

counter values for GPI coordinates. The resultant crosshair coordinates are stored in system attributes 3 and 4. The time to press the
ACL button is determined by a normal distribution with a mean of 0.5
(secs), standard deviation of 0.25 (secs), and a minimum and maximum
values of 0.1 (secs) and 3.0 (sec) respectively. The interval statistics collected at task release can be used to determine the duration of task 2.

     This initiates the displacement of the crosshairs from their pre-
vious position to the newly designated position. The time required is
a function of distance moved and equipment drive rates. While task
duration is specified as zero, moderator function entry 1 adjusts
this to the appropriate value. Time (in secs) is set to twice the
angular separation of the last and next points for crosshair position.
The assumed drive rates will depend on system mechanization and would
probably differ from those used here. Statistics on the interval to
release can be used to derive the elapsed time to completion of task
3.

     Once the crosshairs have been set by the system, the operator
needs to identify what feature on his radar scope is the designated

GPI. He can err in this assessment, and his choice governs where he intends to move the crosshairs. The time for this decision is described by distribution set 4, a normal distribution with a mean of 0.75 (secs), standard deviation of 0.25 (secs), and a minimum and maximum of 0.1 (secs) and 2.0 (secs) respectively. Information attribute 3 is set to zero if the wrong choice is made or to one if the right choice is made. Branching out of the task depends on the nature of the choice made. Statistics on the interval to release actually determine the time to completion of task 4.

Task 6 and 7 accomplish the same objective. They set system attributes 5 and 6 to the latitude and longitude of the point the operator plans to put the crosshairs on, as determined in task 5. Since this essentially sets up the next task, these are dummy nodes that consume no time. The statistics for the interval to release effectively describe the time to completion of task 5.

The crosshair adjustment in task 8 sets system attributes 7 and 8 to reflect operator positioning error (whatever the multiplicity of causes), and while the time is defined as zero, this is adjusted by moderator function entry 2 which sets the time based on the difference between the crosshair coordinates and the selected point coordinates. A random time is added to reflect terminal adjustment time that is not a function of distance moved. The statistics here reflect the interval to task 8 completion, since task release is coincident with task 5 completion.

Task 9 is **a gating** function that precludes fixtaking until the crosshairs are within a prescribed distance from assumed aircraft present position. When this occurs (as evidenced by signalling from monitor 1), the ACL light "comes on", and the operator may depress the autofix button. The statistics to release of task 9 effectively determine the delay incurred waiting for the system prescribed condition to be met.

Task 10 ends the network simulation by updating present position, as reflected in the values of system attributes 9 and 10. The time to press the button is reflected in distribution set 5, a normally distributed time with a mean of 0.5 (secs), standard deviation of 0.25 (secs), and a minimum and maximum of 0.1 (secs) and 1.5 (secs) respectively. The statistics for the interval to completion of this task actually describe the total activity duration and may be used to derive the duration of this last task in the network.

While this is still a relatively simplistic description of both man and machine, it does convey some of the complexity of the potential interactions between human and hardware errors and has the potential for describing how these can affect navigation accuracy. At any point in time, a comparison can be made between the planned position and the actual position. This vector can be separated into two

components: 1) along track error, and 2) cross track error as diagrammed in Fig. 18. If the aircraft is "on-track" and "on-time",
these errors would be zero. Along track error reflects timing or mission pacing, largely accomplished through commanded airspeed changes.
Such changes are often necessary to compensate for wind conditions
(among other factors). Cross track error will typically be tolerable
if kept within some specified limits. If excessive, corrective action
is warranted and is typically accomplished by a commanded change in
aircraft heading. These corrective actions have not been incorporated
into the present model presented here.

If one postulates a system with more than one navigator, the
fixtaking could be allocated to one operator while monitoring and compensating for along and cross track errors could be allocated to the
other. The model should then become concerned with the need to coordinate and synchronize the parallel activities of the two operators.
While SAINT permits such consideration, no attempt will be made to
demonstrate that capability here. If there is only one operator, he
must elect the strategy for time sharing between these activities.
This too would be an interesting extension to the present formulation but will not be pursued here.

In addition to the histograms and tabular data routinely provided by SAINT, state variables may be plotted over time in a fashion
analogous to a strip chart recording. Variables can be scaled either
of two ways. First, the scale can be governed by the minimum and maximum values observed in the simulation or by a user specified range
of values. The danger of the first approach is that if the minimum

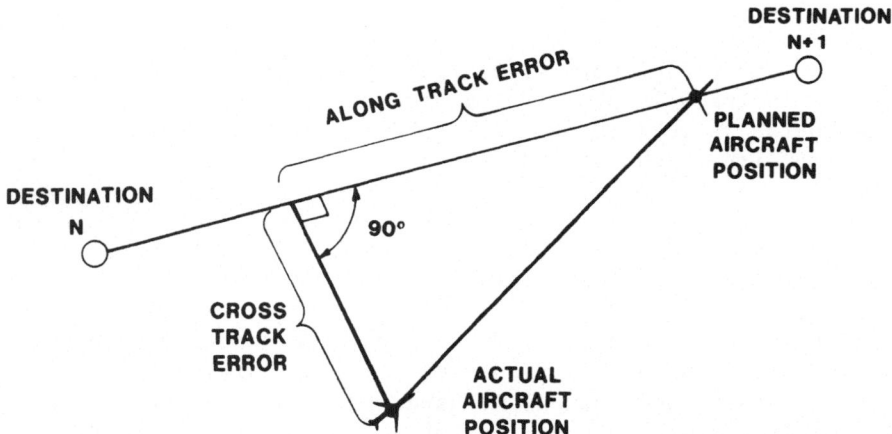

Fig. 18. Schematic showing along and cross track error components of
         the difference between planned and actual position.

turns out to be the same value as the maximum, nothing will be plotted. In the second approach, the user specifies the prescribed range. Then if an identical minimum/maximum is within the range, the plot will show a constant value over time. If the variable does not hold to a constant value, a prespecified range of values for the plot may not be as sensitive to small variations as would be a range dictated by the observed maximum and minimum. Also, a specified range can inadvertently truncate the plot if the value goes beyond the range of that scale. With a dynamic system model that is not itself well understood, it may take several trial runs to get the plot set up as you want it to appear.

CONCLUSION

There are a number of concepts available in SAINT which have not been illustrated here, but the basic features of SAINT have been demonstrated at least in part.

Two deficiencies have been noted in SAINT applications to date. There is no built-in capability to define and manage queues. This is often a useful construct to have and presently requires extensive user programming to accomplish. Second, there is no way to "hold" resources and mask them from being reassigned upon task completion. Revisions to SAINT have been accomplished to achieve this objective, but a better implementation needs to be developed.

SAINT could also be improved by developing a tutorial, interactive "front-end" to assist users prepare their input deck. A graphics "back-end" would be helpful to draw the SAINT network diagrams so the model documentation used to communicate among the modeling team members can be kept up to date with the computer code.

Since SAINT runs under the FORTRAN compiler, all memory allocation needs to be defined in the FORTRAN source program's COMMON statements. These have pre-defined values which may be too large for efficient computer use with small models and too small for exercising large models. The REDIMEN program (Seifert et al., 1980) was designed to help make such updates to the SAINT source code less painful and more efficient.

REFERENCES

Duket, S.D., Wortman, D.B., Seifert, D.J., Hann, R.L. and Chubb, G.P., 1977, Documentation for the SAINT Simulation Program, AMRL-TR-77-63, Air Force Aerospace Medical Research Laboratory, Wright-Patterson Air Force Base, Ohio.

Elmaghraby, S.E., 1977, "Activity Networks: Project Planning and Control by Network Models", John Wiley and Sons, New York.

Hastings, N.A.J. and Peacock, J.B., 1975, "Statistical Distributions", Butterworths, London.

Kleinman, D.L. and Baron, S., 1971, Manned Vehicle Systems Analysis by Means of Modern Control Theory, NASA-CR-1972.

Markowitz, H.M., Hausner, B. and Karr, H.W., 1963, "SIMSCRIPT: A Simulation Programming Language", Prentice-Hall, Inc., Englewood Cliffs, NJ.

McRuer, D., Graham, D., Krendal, E. and Reisner, W.J.D., 1965, Human Pilot Dynamics in Compensatory Systems: Theory, Models and Experiments with Controlled Element and Forcing Function Variations, AFFDL-TR-65-15, Air Force Flight Dynamics Laboratory, Wright-Patterson, Air Force Base, Ohio.

Pritsker, A.A.B., 1974, "The GASP IV Simulation Language", John Wiley and Sons, Inc., New York.

Pritsker, A.A.B. and Kiviat, P.J., 1969, "Simulation with GASP II, A FORTRAN Based Simulation Language", Prentice-Hall, Inc., Englewood Cliffs, NJ.

Pritsker, A.A.B., Wortman, D.B., Seum, Ch.S., Chubb, G.P. and Seifert, D.J., 1974, SAINT: Volume I, Systems Analysis of Integrated Networks of Tasks, AMRL-TR-78-126, Air Force Aerospace Medical Research Laboratory, Wright-Patterson Air Force Base, Ohio.

Schriber, Th.J., 1974, "Simulation Using GPSS", John Wiley and Sons, Inc., New York.

Seifert, D.J., 1979, Combined Discrete Network – Continuous Control Model of Man-Machine Systems, AMRL-TR-79-34, Air Force Aerospace Medical Research Laboratory, Wright-Patterson Air Force Base, Ohio.

Seifert, D.J. and Chubb, G.P., 1978, SAINT, A Combined Simulation Language for Modeling Large Complex Systems, AMRL-TR-78-48, Air Force Aerospace Medical Research Laboratory, Wright-Patterson Air Force Base, Ohio.

Seifert, D.J., Koeplinger, G. and Hoyland, C.M., 1980, REDIMEN: SAINT Redimensioning Program, AFAMRL-TR-80-5, Air Force Aerospace Medical Research Laboratory, Wright-Patterson Air Force Base, Ohio.

Siegel, A.I. and Wolf, J.J., 1967, "Man-Machine Simulation Models: Performance and Psychosocial Interaction", John Wiley and Sons, Inc., New York.

Strieb, M. and Robert, J., Jr., 1979, An Introduction to the Human Operator Simulator, Technical Report 1400.02D, Analytics, 2500 Maryland Rd., Willow Grove, PA.

Whitehouse, G., 1973, "Systems Analysis and Design Using Network Techniques", Prentice-Hall, Inc., Englewood Cliffs, NJ.

Wortman, D.B., Duket, S.D. and Seifert, D.J., 1976, SAINT Simulation of a Remotely Piloted Vehicle/Drone Control Facility: Model Development and Analysis, AMRL-TR-75-118, Air Force Aerospace Medical Research Laboratory, Wright-Patterson Air Force Base, Ohio.

Wortman, D.B., Duket, S.D., Seifert, D.J., Hann, R.L. and Chubb, G.P., 1977, Simulation Using SAINT: A User-Oriented Instruction Manual,

AMRL-TR-77-61, Aerospace Medical Research Laboratory, Wright-Patterson Air Force Base, Ohio.

Wortman, D.B., Duket, S.D., Seifert, D.J., Hann, R.L. and Chubb, G.P., 1977, The SAINT User's Manual, AMRL-TR-77-62, Aerospace Medical Research Laboratory, Wright-Patterson Air Force Base, Ohio.

Wortman, D.B., Duket, S.D., Seifert, D.J., Hann, R.L. and Chubb, G.P., 1977, An Example to Illustrate the Use of SPSS for the Analysis of a SAINT Model, AMRL-TR-77-64, Aerospace Medical Research Laboratory, Wright-Patterson Air Force Base, Ohio.

Wortman, D.B., Sigal, C.E., Pritsker, A.A.B. and Seifert, D.J., 1974, New SAINT Concepts and the SAINT II Simulation Program, AMRL-TR-74-119, Aerospace Medical Research Laboratory, Wright-Patterson Air Force Base, Ohio.

ANALYTICAL EVALUATION OF MANNED SYSTEMS WITH TASK NETWORK MODELS

Karl-Friedrich Kraiss

INTRODUCTION

Numerous efforts have been made to simulate operator behaviour. A wide variety of models for manual control, display scanning, monitoring and decision making are described in the literature (Pew et al., 1977, Sheridan and Johannsen, 1976). Those part task models usually address only a specific part of the operator's job and not the whole system in a mission context. Task analytic models on the other hand (Greening, 1978) address a larger portion of the operator job. Here human behaviour is looked at as a time sequential decision and control process. Modelling of the process starts with a detailed task analysis down to elementary functions. As soon as those elementary functions have been isolated, a desirable, i.e., normative sequence of actions has to be worked out. During this step, usually extensive discussions and contributions by experts are needed. Subsequently, a task network is synthesized from these elements. Linkages and transition probabilities between tasks are selected in such a way that the model behaves realistically under dynamic operational conditions.

Time sequential network models yield results not only about the system performance but also about the frequency and duration where attention is attributed to single tasks. It is thus revealed to which degree the attention of the operator is requested and this may be taken as a strong indicator for the actual workload.

From this short description, it appears that a combination of relatively simple, well established part task models with the flexible task-analytic approach should be most promising (Veldhuyzen and Stassen, 1977; Cavalli, 1978; Card and Moran, 1980; and Kraiss, 1980).

181

For the simulation of manned systems usually the following components are needed:

Operator: Discrete functions (push buttons, switches)
            Continuous functions (manual control)
Machine :   Discrete functions (mode of operation)
            Continuous functions (system dynamics)

These functions may be simulated by the system analyst to the required level of detail. It turns however out that the interactions between discrete and continuous functions are very difficult to handle. Therefore the implementation of the simulation software often becomes too complicated for practical ergonomic applications.

This situation has been markedly improved since special purpose simulation languages for the analyses of manned systems have become available as e.g., SAINT (Chubb, 1980). With this instrument network simulation becomes a reasonable tool for ergonomic analysis.

MODELLING PRINCIPLES

The synthesis of a network model unavoidably has to be heuristic to a certain degree. It depends on the input the analyst gets from experts and on the kind of model that is desired.

Modelling may be descriptive, i.e., operator behaviour is simulated as observed in comparable situations. In contrast to that, a normative procedure simulates a desirable or optimal strategy. In addition network models may differ considerably in the degree of detail.

In order to keep the heuristic part in the model as small as possible, basic knowledge on human behaviour in monitoring and manual control should be taken into account.

a) Manual Control:
- For the simulation of manual control tasks McRuer's crossover model has been applied with success (Sheridan and Ferrell, 1974). Since in the network model human control strategy is assumed to be time sequential, a transition from simultaneous multi loop control to sequential single loop control is necessary. This means that any loop can be closed only in particular time intervals, the sequence of which is controlled subjectively by the operator.

b) Monitoring and Attention Allocation:
- The simulation may be based on the assumption that the trained and motivated operator has an inner model of the process structure and dynamics.
- The operator is able to make, within certain limits, predictions about the future system status.
- Various tasks are handled in sequence, not in parallel.

- Attention allocation between various tasks is a critical factor in the network synthesis. It is assumed that priorities are determined based on a mental comparison between observed situation and inner model expectation.
- Bookkeeping of priorities and task sequencing is assumed to take place in the operators short term memory.
- Time estimates about body and eye movements, choice reactions etc. should be drawn from reliable data bases as far as available (see Figs. 1 and 2.

NETWORK MODELLING PARADIGMS

I. <u>Evaluation of a Submarine Instrument Panel</u>

As a paradigm for the network approach to the analytic design of manned systems, the instrumentation of a submarine is addressed. A submarine is a high inertia system. The dynamics are rather compli- cated with a variety of nonlinearities and crosscouplings between longitudinal and lateral axis. Diving is performed with coupled bow and stern depth rudders while heading is controlled with a side rud- der. In addition water may be pumped back and forth in the boat in order to change pitch moment. It is also possible to increase or de- crease the weight by pumping water in and out of the boat. In former boat generations (which are still operational) each of these control actions was worked upon by a dedicated operator. The actions of this personnel were coordinated by verbal direction of an additional of-

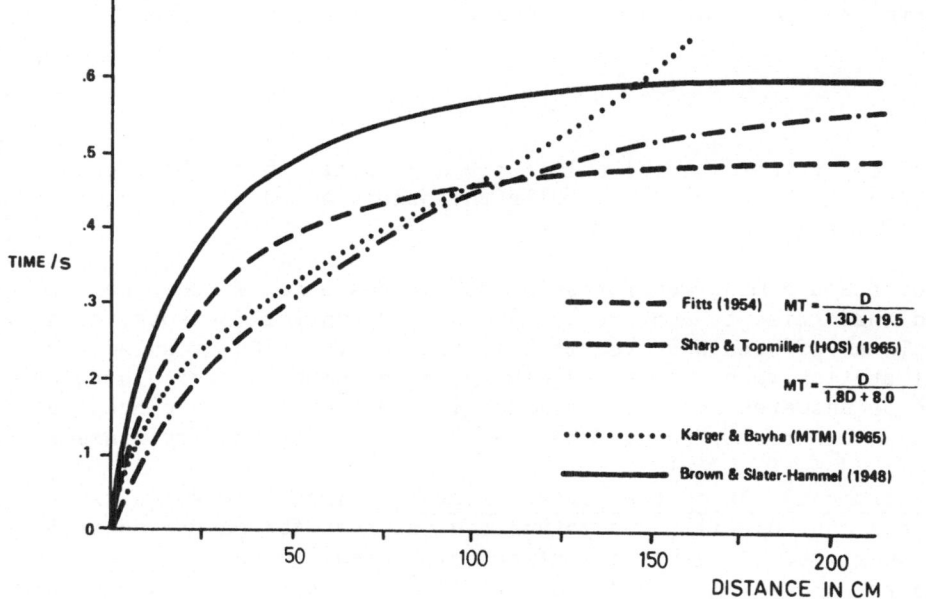

Fig. 1. Hand movement times as a function of distance.

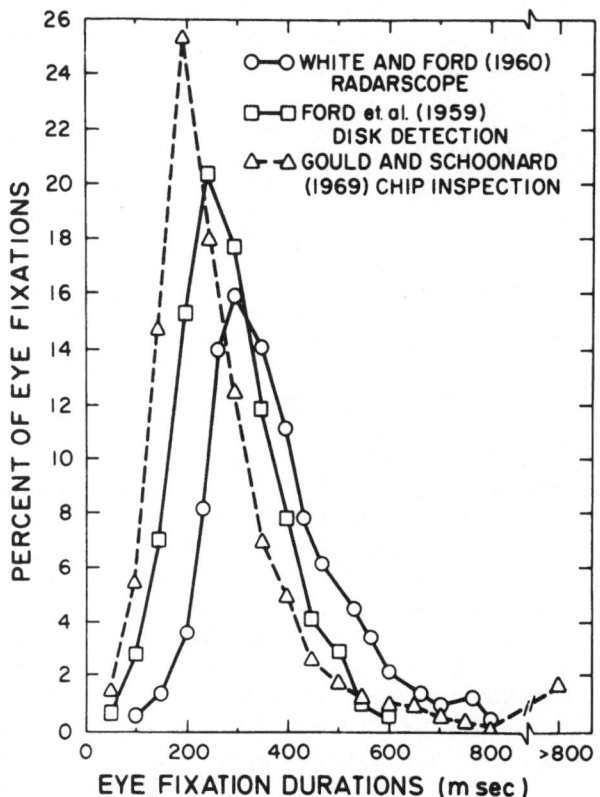

Fig. 2. Eye fixation durations for various visual tasks.

ficer. This officer was the only trained capacity aboard having insight into the boat dynamics, while the others acted as genuine servo amplifiers.

   For the upcoming generation of submarines a new concept was developed that tries to achieve a one man boat control (Bernotat et al.,
1975). These efforts have resulted in the construction of an advanced
control station with a combined diving and heading control. The question to be answered by analytic means is, whether in this concept a
single man can really handle the boat in an operational environment.

   Instrumentation of the control station. Since it was realized
that depth control will be a rather tedious task for manual control
it was suggested, to study the effect of a predictor display on operator performance (Smith and Kennedy. 1975). This display as depicted
by Fig. 3 was worked out in cooperation with navy officers and contains on a CRT-screen a boat symbol with an analog representation of

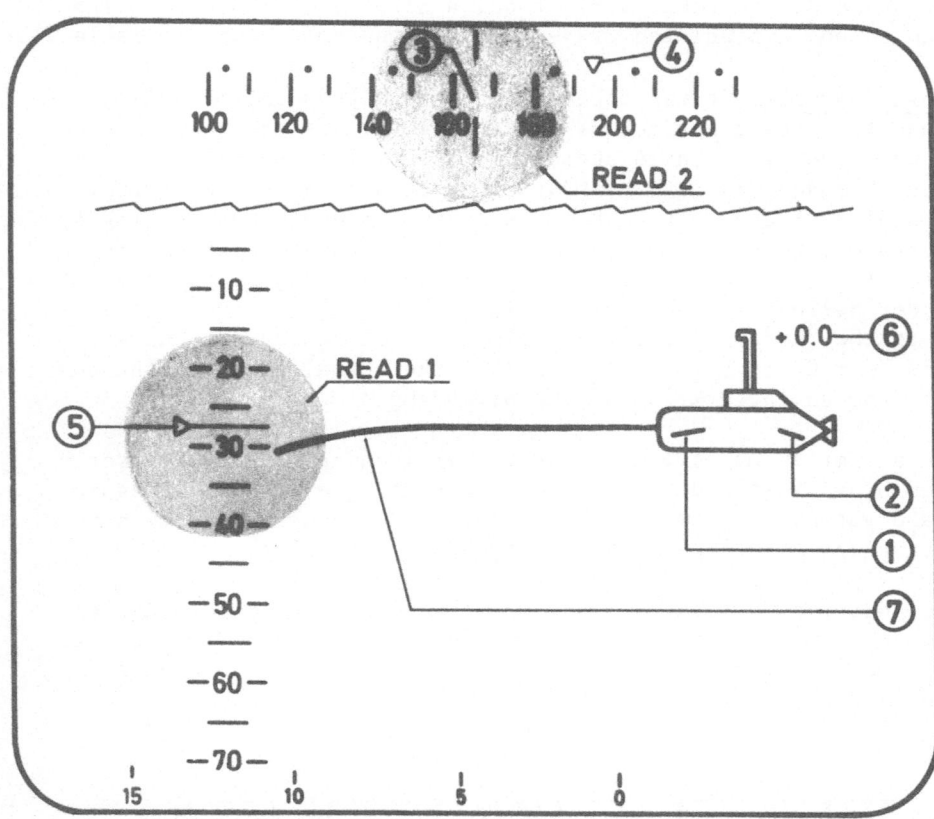

Fig. 3. Panel layout for the Predictor Display. For explanation of elements see Table I (READ 1, 2 are assumed fixation areas of the operator, see text).

Table I. Description of display elements.

DISPLAY ITEM No.	INDICATION	SYMBOL	RANGE	RESOLUTION
1	Bow depth rudder		$\pm 25\,(°)$	$1\,(°)$
2	Stern depth rudder	δ	$\pm 25\,(°)$	$1\,(°)$
3	Heading	ψ	$360\,(°)$	$1\,(°)$
4	Turning rate	$\dot{\psi}$	$\pm 6\,(°/s)$	$1\,(°/s)$
5	Depth	D		$1\,(m)$
6	Pitch angle (digital)	Θ	$\pm 30\,(°)$	$0,1\,(°)$
7	Predictor			

bow and stern depth rudder deflection, a digital indication of the
pitch angle and a predicted trajectory for the boat (compare Table I).

Boat dynamics. It was decided to perform this instrumentation
study and the subsequent network modelling for a simplified version
of the boat and at a single speed (10 Kn). At this speed, only the
stern depth rudder is active and the bow rudder can be neglected.
Fig. 4 depicts the resulting block diagram for depth and heading dynamics. As may be seen from this scheme, the depth rudder machine
consists of a first order lag ($T_\delta = 1/a$) where manual control input
and rudder deflection are limited. The pitch angle of the boat is
linked to the rudder deflection angle by a second order transfer
function ($\omega = 0.05$; $\zeta = .707$). Finally, the actual depth of the boat
results from an integration of the pitch angle.

In a similar manner, the side rudder machine is rather exactly
modelled by a first order lag (time constant $T_\eta = 1/b$) where again
input and output are limited. The actual side rudder deflection is
integrated and this finally yields the boat's heading.

In addition, a rather strong coupling from side rudder deflection to depth must be taken into consideration as indicated in Fig.
4. Any side rudder activity results in a tendency of the boat to
pitch up and subsequently climb.

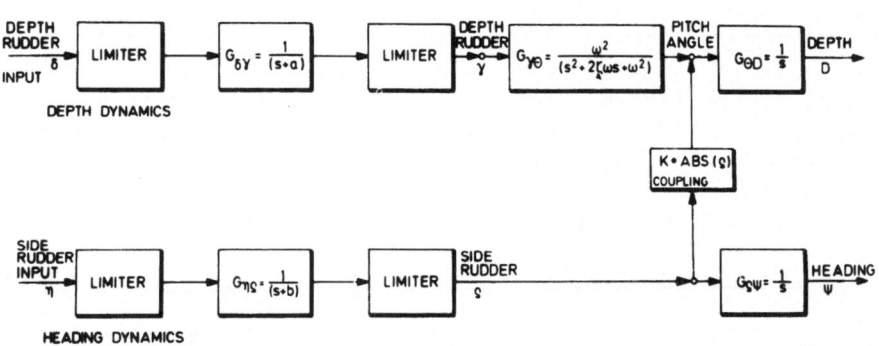

$$\omega = .05; \quad \zeta = .707; \quad T_\delta = 1/a = 4,2; \quad T_\eta = 1/b = 3.5$$

Fig. 4. Simplified boat dynamics for a constant speed of 10 (Kn).

In view of these simplified dynamics the assumption is justified
that depth manoeuvres will be extremely difficult to cope with by manual control while heading changes should be manageable without difficulties (Kraiss, 1981).

Human operator modelling for the submarine control tasks. Modeling of manual diving and heading control makes use of the simplified

boat dynamics described earlier. In addition a normative strategy for the execution of manual control actions has to be established.

From the simplified boat dynamics shown in Fig. 4 it appears that diving involves a forth order transfer function ($G_{\delta D}$):

$$G_{\delta D} = \frac{\omega^2}{(s+a)\ (s^2+2\zeta\omega s+\omega^2)\ s} \tag{1}$$

In order to reduce the apparent high order of the submarines longitudinal transfer function a predictor display is designed, that shows the depth trajectory of the boat extended into the future (see Fig. 3).

Using the Taylor expansion, the predicted depth at time $(t+\alpha)$ may be calculated from the actual depth at time t by the equation:

$$D_P = D(t+\alpha) = D(t) + \alpha\dot{D}(t) + \frac{\alpha^2}{2}\ddot{D}(t) + \text{remnant} \tag{2}$$

After application of the Laplace transform the predictor transfer function results in:

$$G_P = \frac{D_P}{D} = 1+\alpha s + \frac{\alpha^2}{2}s^2 \tag{3}$$

and the longitudinal boat dynamics, including the predictor display, are described by:

$$G_{\delta D_P} = \frac{\omega^2}{(s+a)\ (s^2+2\zeta\omega s+\omega^2)\ s} \cdot \frac{(s^2+ \frac{2}{\alpha}s + \frac{2}{\alpha^2})}{\frac{2}{\alpha^2}} \tag{4}$$

As may be seen from this equation, the second order dynamics in the submarine transfer function may be compensated by the predictor transfer function if prediction time α is selected in a suitable way. With $\alpha = \frac{\sqrt{2}}{\omega} = \frac{\sqrt{2}}{.05} = 28,3$ (s) equation (4) reduces to:

$$G_{\delta D_P} = \frac{1}{(s+a)\ s} \tag{5}$$

According to McRuer's crossover model the human operator behaves such, that in the vicinity of the crossover frequency the open loop transfer function is (Sheridan and Ferrell, 1974, p. 236):

$$G_o = G_{H.O.} \cdot G_{\delta\Theta} = \frac{\omega_{co}}{s} \cdot e^{-s\tau} \tag{6}$$

and the human operator transfer function calculated according to McRuer's crossover model results in (see Fig. 5):

$$G_{H.O.} = \omega_{co}(s+a) \ e^{-s\tau} \tag{7}$$

With this transfer function a direct control of depth appears to be possible. The feedback loops to be closed by manual control thus involve predicted depth and the first derivative (vertical speed). It depends on the selected instrument layout, how accurate these readings can be made from the available instruments (compare Table I).

Attention allocation strategy. Permanent depth control can not be permitted, since attention of the operator has to be shared with heading control. Therefore a heuristic approach to the definition of a normative attention allocation strategy must be performed. The selected strategy makes use of the following basic principles: (1) availability of an "inner model" representing a well trained operator, (2) capability to perform a mental short term prediction of process status, and (3) availability of a short term memory to store task priorities and sequencing (see Fig. 5):

- At a voluntary moment (t_{now}) the operator reads the actual predicted depth $D_p(t_{now})$ from the endpoint of the predictor display and the commanded depth D_c from the depth meter. The observed depth error is $\Delta D_p(t_{now})$. He then decides whether this reading at this time is in accordance with his expectation of an optimal boat trajectory. This expectation is based on an inner process model $D_{opt} = f(t)$.

- If depth error ΔD_p is "critical" at a time t_{now}, immediate corrective action is required (switch in Fig. 5 closed). "Critical" values for the depth error affect diving performance to a degree that can not be tolerated. These values must be determined by a sensitivity analysis for this parameter. As long as ΔD_p remains below critical values no immediate action is requested.

- If predicted depth error ΔD_p is not critical at the instance t_{now} of observation a mental prediction of the error is performed. This yields an estimate of the time t_c where the predicted depth error $\Delta D_p(t_c)$ will have reached a degree that requires immediate depth control. Time t_c is stored in the short term memory to make sure, that attention is returned to depth control at this time. The computation of error prediction is performed in analogy to Eq. (2).

- The time interval $\Delta t_c = t_c - t_{now}$ is free for other actions as heading control or simply idle (no workload). However, for practical considerations an upper limit is imposed on the maximum permitted idle time ($\Delta t_c \leq MIT$) to make sure that depth is controlled not only at high depth errors but also at least after MIT seconds.

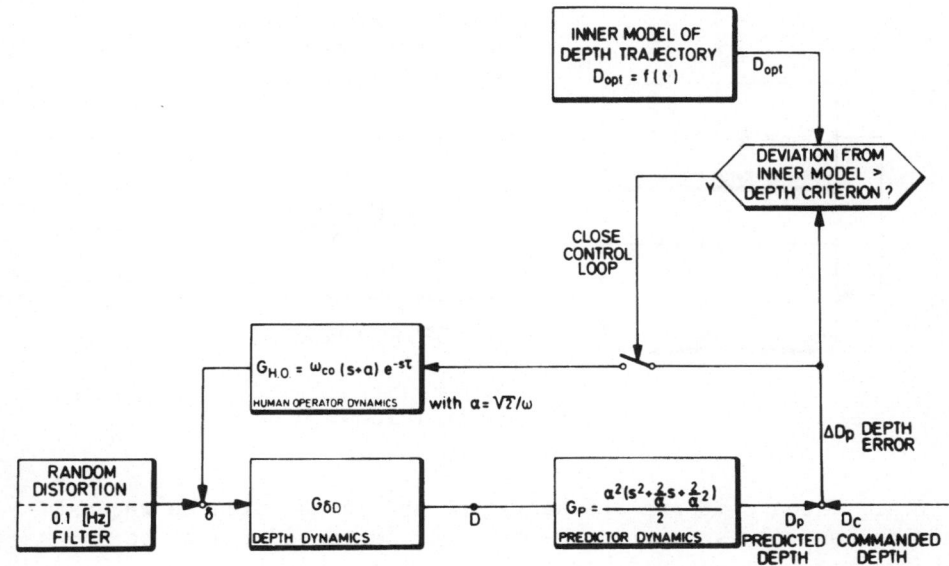

Fig. 5. Depth control model for the predictor display instrumentation.

Heading control. The simplified heading transfer function is depicted in Fig. 4 as:

$$G_{\eta\Psi} = \frac{1}{s(s+b)} \tag{8}$$

Human operator control performance may again be described by the crossover frequency model and thus results in (see Fig. 6):

$$G_{H.O.} = \omega_{co}(s+b) \, e^{-s\tau} \tag{9}$$

From operational considerations it is derived that heading control has lower priority than diving control and may be considered as a secondary task. Heading control thus should only be performed in intervals where no diving control is needed. Consequently three criteria have been established for the attention allocation to heading control:
- As soon as $\Delta\Psi$ is lower than $0.5°$ no further control is required (switch in Fig. 6 open) since it is not necessary to maintain heading more accurate than 0.5 degrees.
- If $\Delta\Psi$ is larger than $0.5°$ control is only requested if the error is increasing that is if ABS $(\Delta\Psi(t+\Delta t)) > $ ABS$(\Delta\Psi(t))$.
- Independently from the two criteria stated above heading control should be looked at any five seconds at least.

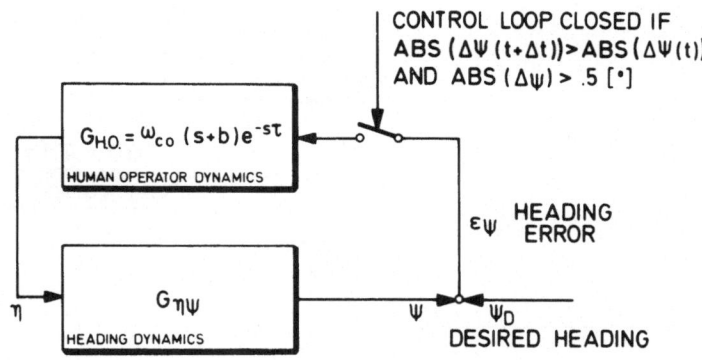

Fig. 6. Heading control loop.

If all three criteria are satisfied and in addition no request for diving control is waiting the operator is considered to be momentarily free from any workload.

Task network synthesis. In the sections above the instrumentation and the selected normative control strategies have been explained. Now, the problem remains to synthesize a network of elementary tasks in such a way, that it shows the desired control behaviour.

As a tool for network synthesis the above mentioned simulation language SAINT is used. SAINT offers a symbology to represent the interconnections between tasks in a network graphically. The network is transposed later into control cards and read into a computer. Simultaneously, various subroutines have to be written in FORTRAN (which are not represented in the network graphic). In this case subroutine STATE, e.g., contains the submarine dynamics and the feedback loops shown in Figs. 4, 5 and 6.

The resulting network, as depicted in Fig. 7, consists of 7 tasks with the following labels: START, STOP, READ1, READ2, PRIOR, IDLE, CLOSLOOP. Main functions of each of these tasks are described below in detail:

START: TASK 1 starts the simulation run for a predetermined time (e.g., for 300 s); calls up a random number every 0.25 seconds to generate the disturbance of the boats pitch moment; collects data for plotting.

STOP: TASK 2 stops the simulation run after completion.

READ1:   TASK 3 simulates the operator absorbing information from the panel area labelled READ1 in Fig. 3. Read values are rounded off corresponding to the display layout (see Table I). These calculations are performed in user function UF,3.

PRIOR: In TASK 4 the attention allocation computations are performed. The information gained by observation of the READ1

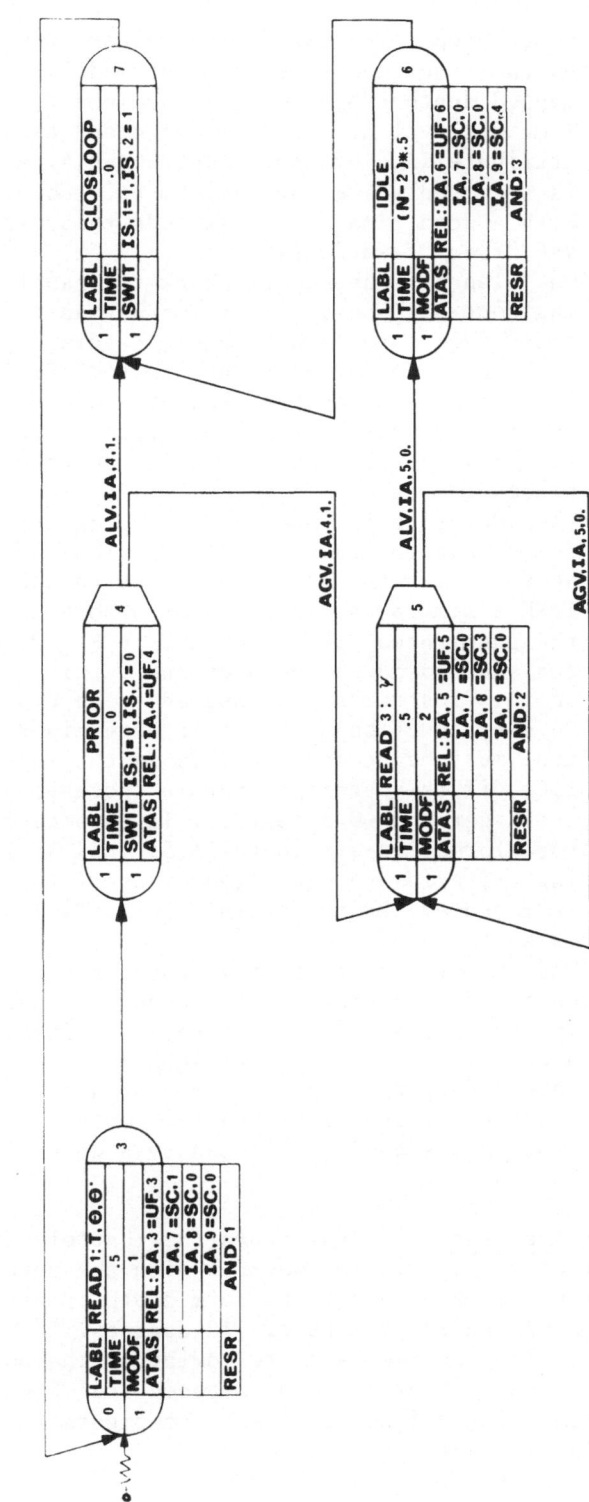

Fig. 7. SAINT network model for the predictor display instrumentation.

panel area is compared with values commanded and predicted
by the inner model. This inner model, representing a commanded depth trajectory, is implemented in subroutine STATE.
Time interval Δt_c is determined and assigned to information
attribute IA,4 by user function UF,4. A number of 3 seconds
is selected to be the maximum tolerable idle time (MIT).
With a scan time of 0.5 seconds this yields a maximum of
N=6 "free" fixations.
One glance is needed to check the panel area READ1 for diving control. Thus, only if more than one fixation is "free"
branching is to TASK 5 (check heading), otherwise to TASK
7. However, a forced branching to TASK 5 takes place every
five seconds to ensure regular heading observation.
Before branching to TASK 5, depth control is switched off
(IS,1=0) and heading control is switched on (IS,2=1) as depicted in Fig. 7.
Task duration is indicated as being zero, because it is assumed that the required computations are performed simultaneously with the instrument observation.

READ2:   TASK 5 simulates the operator absorbing information from
the panel area labelled READ2 in Fig. 3. Readings can be actual and commanded heading and turning rate ($\dot{\psi}$). Dwell time
on this instrument is assumed to be 0.5 s.
In user function UF,5, it is determined whether heading control must be performed. This is the case, as described earlier, if the heading error is increasing and if this error
is larger than 0.5 degrees. In this case a "1" is assigned
to information attribute IA,5. TASK 5 is iterated until heading criteria are satisfied or until the available idle time
is consumed. Branching is then to TASK 6 and heading control
is switched off.

IDLE: TASK 6 collects statistics about the operator time requested
neither for depth, nor for heading control, thus representing the operator's free capacity. After completion there is
a deterministic branch to TASK 7.

CLOSLOOP: TASK 7 performs the switch-on of depth control (IS,1=1)
and the switch-off of heading control (IS,2=0). Task duration is 0 seconds, followed by a deterministic branch to
TASK 3.

Model Analysis. To study submarine control with the designed instrumentation a test mission was run with the network model. During
this mission the model had to perform depth transitions of 50 m with
a simultaneous heading change of 110 degrees. It was assumed that during the mission crew members are moving in the boat and hence a random pitch moment disturbance was imposed (see Fig. 5). The mission
was performed with a fixed parameter configuration:

Dwell time per panel area   DT : 0.5 s
Maximum permitted idle time   MIT: 3.0 s
Depth error requiring immediate control ΔD_c: 2.0 m

Results are shown for a 300 s interval in Fig. 8. With the selected predictor display depth transition is quick and smooth. Scanning behaviour, as indicated in the lower part of the figure appears
to be regular over the entire depth transition interval and a considerable amount of the operator scan time remains idle. In general
the simulator studies lead to the conclusion that the submarine can
be easily handled by one operator.

Sensitivity analysis of network parameters. The task networks
synthesized to simulate the operator performance during the submarine
control contain a series of parameters that may, at the first glance,
appear to be arbitrary values. Obviously, the selection of correct
parameter values is a crucial point in network modelling. Even if
these values are backed up by experimental or psychophysiological
data like, e.g., instrument dwell times, it is essential to know,
how sensitive the model reacts to changes in these parameters.

A series of 300 s simulation runs for 50 m depth transition has
been performed while parameter values were modified as indicated in
Table II. The indicated average depth errors may be related to the
performance with standard parameters.

If we look to the results in Table II it turns out that the
largest observed change in performance amounts to an average of 0.26
m. Operationally this is only relevant in case of periscope diving.

Table II. Sensitivity analysis for various parameters. Indicated is
the average depth error in meters for 50 m depth transitions in a 300 s interval.

$\Delta\Theta_c$ (°)	MIT (s)	DT (s)	AV. DEPTH ERROR (m) **	DEVIATION (m)
2	3	.3	5.88	-.24
		.5	6.12*	.0
		.7	6.21	.09
2	2	.5	5.98	-.14
	3		6.12*	.0
	4		5.97	-.15
	5		5.87	-.25
1	3	.5	6.05	-.07
2			6.12*	.0
4			5.86	-.26

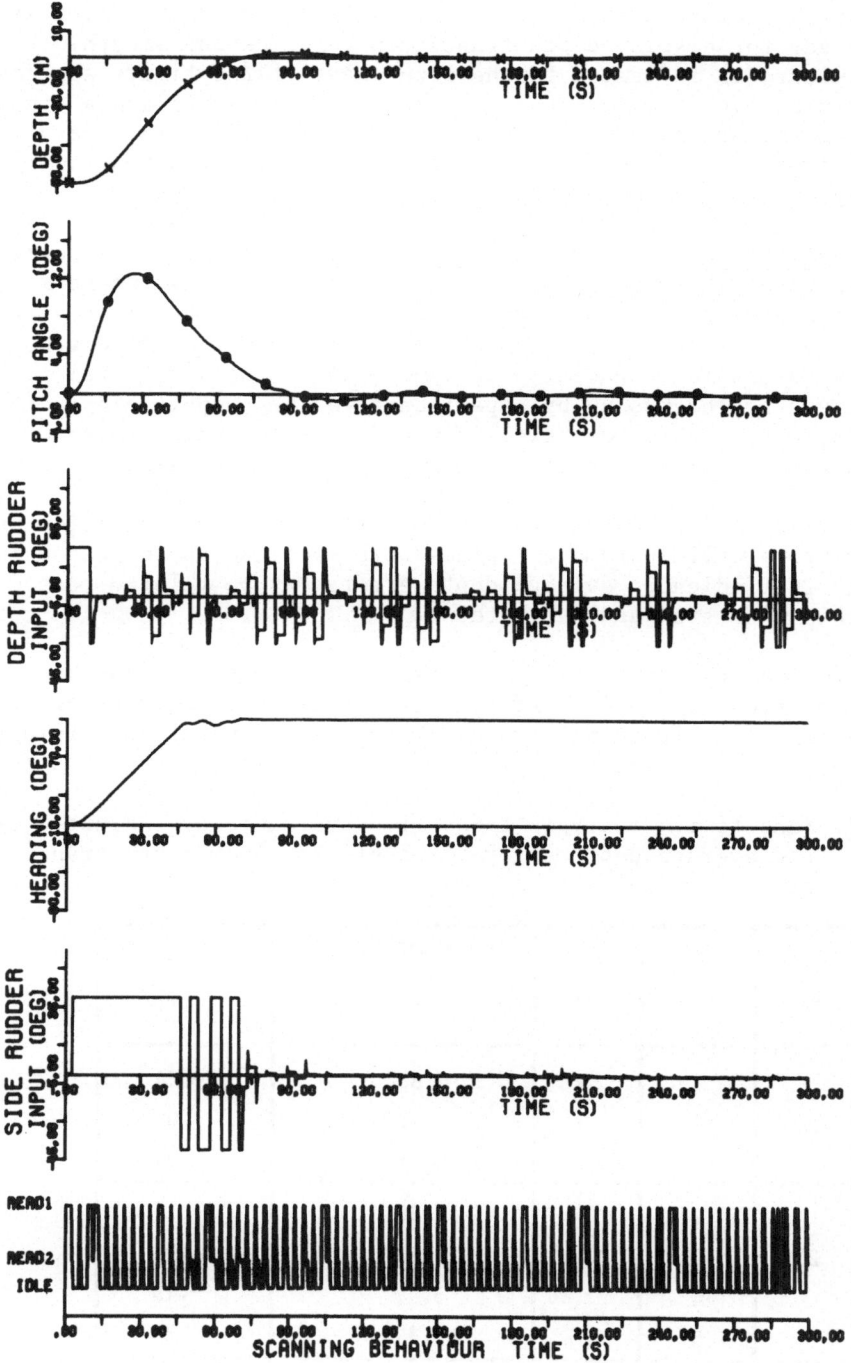

Fig. 8. Model results for a depth transition (50 m) with simultaneous heading change (110°) and random distortion of pitch moment.

In general is may be stated that the predictor display model is rather insensitive to changes in the analysed parameters. The selection of parameter values is hence not critical in this paradigm.

Model validation by comparison with experimental data. In the course of the submarine instrumentation study a part task simulator for manual control of the boat was built up. This gave the opportunity to test the model validity by comparison with experimental data.

For testing purposes, a mission was worked out in cooperation with navy personnel which combines various simultaneous dive and heading control manoeuvres. The total mission of 920 s duration includes 5 segments. For the experiments a matched group of 4 subjects was selected. This group was given an extensive explanation about the instrumentation of the boat, the mission profile and the accuracy requirements and limitations in each segment. It was pointed to the fact that diving is considered to be the main task, while heading control should only be performed as a secondary task.

Each subject performed 16 mission runs. The last 8 runs were used to calculate the experimental data, while the first 8 runs were considered as training phase. Thus, for any mission segment 32 runs are registered.

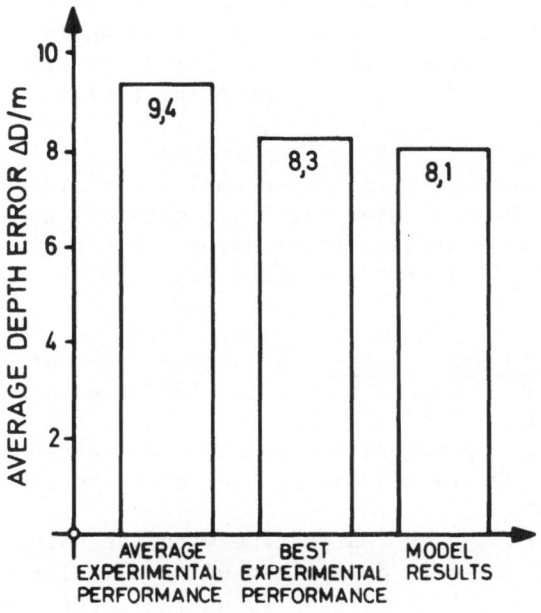

Fig. 9. Comparison of experimental and analytical results for depth control performance.

Experimental results are presented in Fig. 9. It may be seen that the average depth error determined in the experiments amounts to 9.4 meters and the best performance observed during the experiments was 8.3 meters.

Model results which are also shown in Fig. 9 correspond only poorly with the average experimental scores. A much better matching in qualitative as well as quantitative respect can be stated if reference is made not to the average but to the best performance during the experiments.

This result comes not as a surprise, since an optimal strategy has been selected as a basis for modelling. Obviously the subjects can achieve this optimal behaviour only by exception and not to the average. If desired, a less than optimal control strategy could be assumed for the model, resulting in a better matching of the model to average experimental scores.

II. Prediction of Monitoring Behaviour in an Instrument Reading Task

A second paradigm for the task network modelling approach addresses the simultaneous supervision of four vertical instruments. The experimental setup is described in detail by Stein together with comprehensive experimental data (Stein, 1980).

On each instrument an upper and lower limit for the pointer position is indicated. Pointers are moved by filtered white noise (ω = 0.25; 0.5; 1.0; 1.5). The probability that an instrument is outside the limits is varied (p(H1) = 0.125; 0.25; 0.5). Correspondingly p(H0) is the probability that the pointer remains within the tolerance area.

As soon as at least one instrument pointer violates the limit a push button must be kept pressed until all pointers are again within tolerance. In the experiments a total error P has been determined which consists of false alarms and missed signals (P = P(H0,D1) + + P(H1,Do)).

Normative monitoring strategy. To prepare the network simulation of the described monitoring situation the following normative strategy has been worked out (Fig. 10):
- At the beginning all pointers are observed and attention is concentrated to the one who is nearest to the tolerance limit (highest priority).
- For this instrument the estimated time when the pointer goes outside the tolerance area is determined using a Taylor extrapolation. The time interval until transition is an indicator for the importance of the instrument.
- If no immediate violation of the tolerance area is impending, other instruments may be observed. In Fig. 10, e.g., there is time for

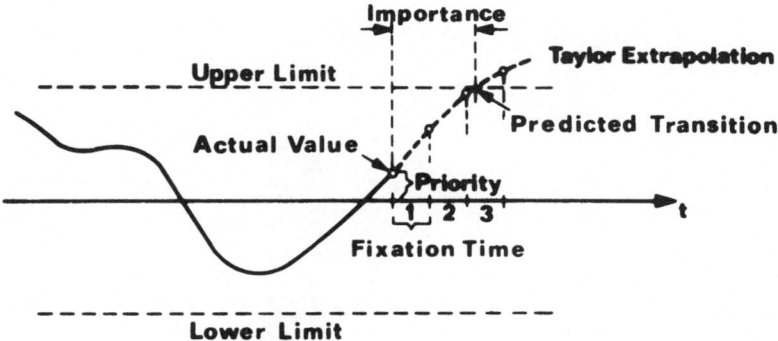

Fig. 10. Normative strategy for instrument monitoring.

two changes of fixation point.
- The sequence in which spare time is devoted to particular instruments is determined using the instrument priorities as remembered from the last observation.
- Permanent observation in any case is attached to the most important instrument.
- If one instrument is neglected more than 1 second, a forced observation takes place.
- After any observation the operator has to check which reaction on the push button is appropriate at this instance.

Network synthesis. The normative strategy outlined above is represented by the task network shown in Fig. 11. In this network task 1 (RANDOM) and task 2 (SINK 1) generate statistically independent and filtered random disturbance functions which are needed to move the instrument pointers. Task 3 (ZEITSTOP) registers the exact times, when pointers leave the tolerance region. Tasks 11 to 14 (CHECK 1 to 4) symbolize the observation of four instruments. Each fixation is assumed to take .3 seconds to the average. In task 21 priorities for further observations according to the described rules are calculated. Finally task 22 (REACTION) simulates pushing or releasing of the push button as convenient.

Model validation by comparison with experimental data. The left diagram of Fig. 12 shows experimentally determined data for the total error as determined by Stein (1980). It may be seen that the total error increases if the pointers move quicker due to an increased ω and if the tolerance area is narrowed (p(H1) is increased). For comparison the right diagram in Fig. 12 shows analytical data as determined by computer simulation using the task network of Fig. 11. As may be seen, a reasonably good coincidence can be stated. The model results are in most cases within one standard deviation of the experimental data.

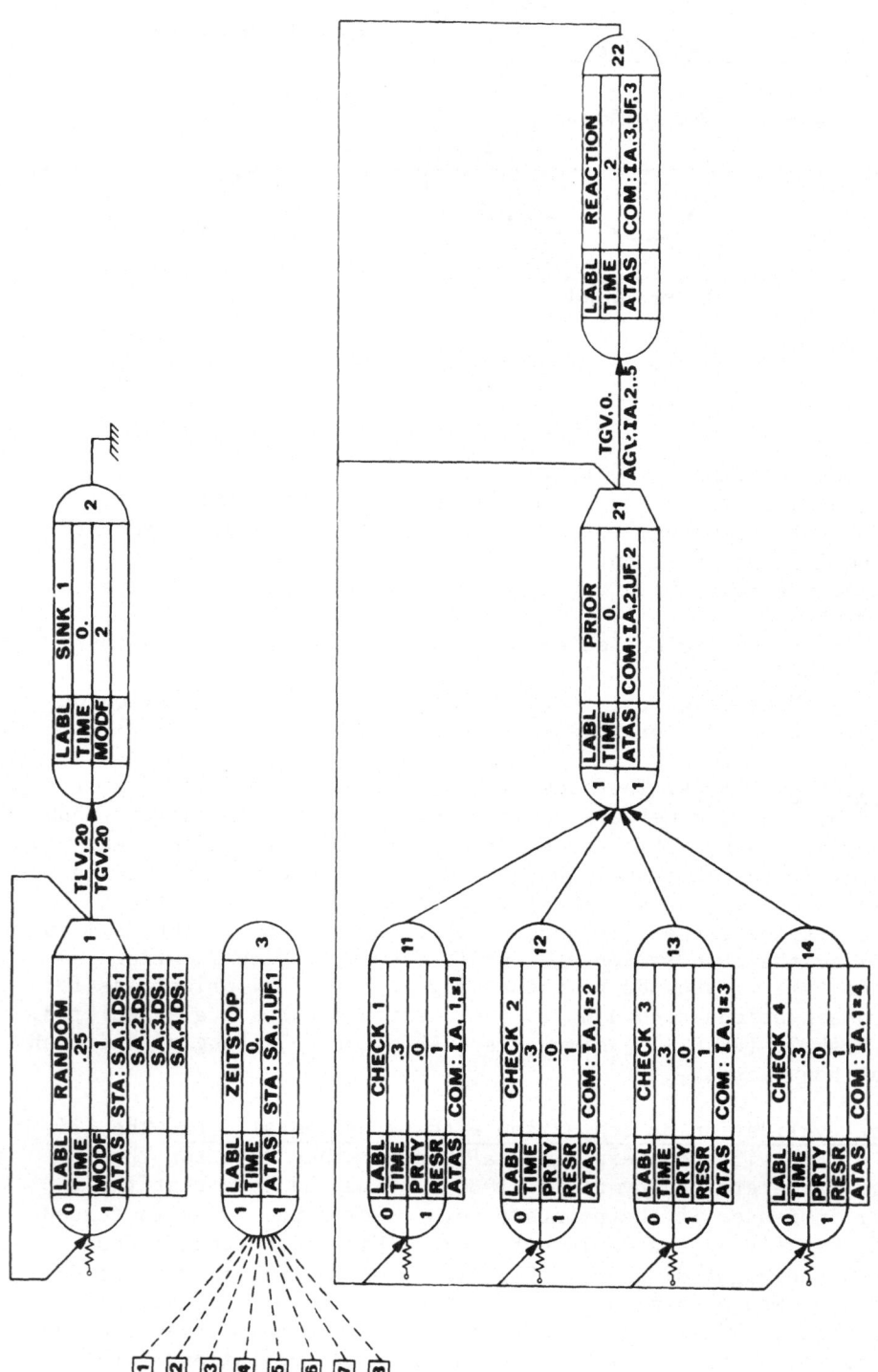

Fig. 11. Task network model for multi-instrument monitoring.

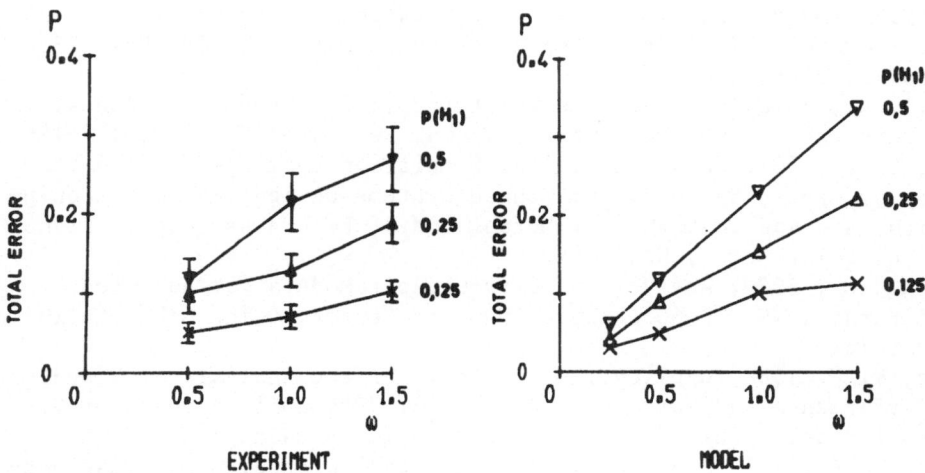

Fig. 12. Experimental and analytical data for simultaneous four instrument monitoring. The total error combines false reactions and missed signals. p(H1) is the probability that one
instrument is out of tolerance (experimental data from
Stein, 1980).

CONCLUSION

The described paradigms show, that operator behaviour may be
simulated in the context of rather complicated tasks using network
models. Implementation of the model structure is considerably facilitated by the application of SAINT as a special purpose simulation
language. The described methodology of simulating monitoring/deci-
sion making by task network models in combination with well established models for manual control generally appears to be a promising tool for the analytical evaluation of manned systems.

ACKNOWLEDGEMENT

The author gratefully acknowledges the assistance of A Knäuper
in implementing the SAINT models and of W. Stein in providing experimental data. Thanks are also due to E. Schubert and K. Steinheuer for
building up the part task simulator and for running the experiments.

REFERENCES

Bernotat, R., 1975, Anthropotechnische Optimierung eines typfreien
Uboot-Lenkstandes, Forschungsinstitut für Anthropotechnik,
Wachtberg, West-Germany, Bericht Nr. 25.

Card, S.K. and Moran, T.P., 1980, The Keystroke Level Model for User Performance Time with Interactive Systems, Comm. ACM, Vol. 23, No. 7, 396-410.

Cavalli, D., 1978, Discret-Time Pilot Model, in: Proc. 14th Annual Conference on Manual Control, Techn. Rep. NASA CP-2060, 177-186.

Chubb, G.P., 1981, SAINT, a Digital Simulation Language for the Study of Manned Systems, in:"Manned Systems Design; Methods, Equipment and Applications", J. Moraal and K.-F. Kraiss (eds.), Plenum Publishing Corp., New York.

Greening, C., 1978, Analysis of Crew-Cockpit Models for Advanced Aircraft, NWC TP 6020, Naval Weapons Center, China Lake, California.

Kraiss, K.-F., 1980, Application of SAINT for the Analysis of Visual Performance and Workplace Layout, in: "Ergonomic Aspects of Visual Display Units", Taylor and Francis, London, 153-160.

Kraiss, K.-F., 1981, A Submarine Instrumentation Study Using Task Network Models, IEEE Transactions on Systems, Man and Cybernetics, to be published.

Pew, R., Baron, S., Feehrer, C.E. and Miller, D.C., 1977, Critical Review of Performance Models Applicable to Man-Machine Systems Evaluation, Report No. 3446, Bolt, Beranek and Newman, Cambridge, Massachusetts.

Sheridan, T.B. and Ferrell, W.R., 1974, "Man-Machine Systems", MIT, Cambridge, Massachusetts.

Sheridan, T.B. and Johannsen, G. (eds.), 1976, "Monitoring Behavior and Supervisory Control", Plenum Publishing Corp., New York.

Smith, R. and Kennedy, R., 1975, Predictor Displays: History, Research and Applications, Dunlap and Associate Inc., Inglewood, California.

Stein, W., 1980, Leistungsbestimmende Faktoren von Anzeigefeldern und Oberwachungsaufgaben bei der Fahrzeugführung, FA Anthropotechnik der DGLR, Berlin.

Veldhuyzen, W. and Stassen, H.G., 1977, The Internal Model Concept: An Application to Modeling Human Control of Large Ships, Human Factors, 19 (4), 367-380.

STATISTICAL TECHNIQUES FOR INSTRUMENT PANEL ARRANGEMENT

Walter W. Wierwille

INTRODUCTION

The proper arrangement of instruments on a panel, to be used by
a human operator in performing a specified task, has been the subject of several technical papers and reports. However, a unified presentation involving the most important topics associated with instrument arrangement has not yet been made. Furthermore, specific important details appear not to have been addressed previously. This paper
represents an attempt to provide a unified but brief presentation capable of direct application to arrangement problems.

Four important aspects of the arrangement problem will be presented. The analysis of a given instrument arrangement usually involves all four. The first of these problems deals with the important
relationships that exist between link values and the associated first
and second order probabilities. It will be shown that, if the link
values are defined properly, they become derivable from the second
order probabilities. Furthermore, it is shown that under the assumptions of "independent scanning", the link values are derivable from
the first order probabilities. Additional relationships are also
presented which involve the transition method of data gathering.

The second problem described is that of record length estimation. If eye movement data are gathered as a means on which to base
a rearrangement, the question arises as to how much data to take.
This question is answerable using double sampling plans which predict the confidence limits on the probability estimates as a function of record length. The third problem is associated with the criterion to be used for the optimization process. It is shown that
four subclasses of criteria exist for possible optimization: first

and second order importance optimizations and first and second order
probability optimizations. Optimization may involve all or any weighted grouping of the four subclasses. The fourth and final problem involves the execution of the optimization process itself. It is shown
that all but the most complex of problems can be handled by exhaustive search methods to obtain a global optimum. Those for which exhaustive search is not feasible, because of computational limitations,
require the use of procedures resulting in local optima and subsequent selection from among them. The paper concludes with a brief
description of two important remaining problems in instrument panel
arrangement.

STATISTICAL RELATIONSHIPS

It is not generally recognized that mathematical relationships
exist among the various statistical entities associated with instrument panel arrangement. This section will present the most important
of these relationships, in an explanatory style.

For purposes of explanation, consider that a panel redesign is
to be based on eye-movement data in an existing system or a simulation. In particular, eye fixation position data are obtained at e-
qually-spaced time samples. For most applications a sampling rate
between five and ten samples per second would prove adequate. At each
sample point, the instrument on which the eyes are fixated is determined. If the eyes are in transition on a given sample, the sample
can be deleted.

From the data it is possible to obtain many different kinds of
statistical parameters. Among the most important are the first order
statistics, that is the frequency of fixation parameters. Frequency
of fixation for instrument i is simply the number of samples indicating fixation on instrument i, divided by any convenient constant. If,
however, the constant is taken as the sum of all fixations, then the
frequency of fixation can be considered as a probability estimate.
In other words

$$p_i = \frac{\text{number of fixations on instrument i}}{\text{total number of fixations}} \qquad (1)$$

where p_i is sample probability of fixating on instrument i. Assuming
that there are N instruments

$$\sum_{i=1}^{N} p_i = 1 \qquad (2)$$

The p_i's contain information about the relative frequency of use of
each instrument, and therefore become the basic ingredient for op-

timization schemes based on first-order statistical data (Huebner and Ryack, 1961; McCormick, 1976).

There are two important second-order statistics in instrument panel optimization. One of them is an estimate of joint probability of fixation and the other is an estimate of the link value. The joint-probability estimate can be obtained by determining the number of eye-transitions from instrument i to instrument j on consecutive sample pairs, and then normalizing the results:

$$p_{ij} = \frac{\text{number of transition pairs from i to j}}{\text{total number of transition pairs}} \tag{3}$$

In (3) is there is no change in eye-position from one sample to the next, this is considered a transition from i to i. Again, if there are N instruments

$$\sum_{i=1}^{N} \sum_{j=1}^{N} p_{ij} = 1 \tag{4}$$

It should be noted here that p_{ij} does not necessarily equal p_{ji}, since there are many different routes by which the eyes may traverse the instruments. Also, provided that the total number of transition pairs is large, the following relationships can be demonstrated:

$$\sum_{i=1}^{N} p_{ij} = p_{j}$$

and   (5)

$$\sum_{j=1}^{N} p_{ij} = p_{i}$$

In other words, the joint probabilities contain the fixation probabilities. These relationships become exact as the sample size approaches infinity.

The other second order statistic of importance is the link value. For many years link values have been used in human factors work for instrument panel arrangement (Jones et al., 1949; Chapanis, 1959; Haygood et al., 1964; and Woodson and Conover, 1970). However, precise definitions of link values have been lacking, and corresponding confusion has resulted. In this paper, a single definition is proposed which makes it possible to relate the link values to both the joint probabilities and the fixation probabilities.

Consider that the essential element desired in the definition of a link value is an assessment of the strength of the association

between two instruments, i and j. In other words, if there are many
direct eye transitions from i to j and from j to i, then the link
value L_{ij} should be relatively large. Therefore, if the link value
is defined to be proportional to the sum of the joint probabilities,
p_{ij} and p_{ji}, it will have such a property:

$$
L_{ij} = \begin{cases} K_o(p_{ij} + p_{ji}); & i < j \\ \text{undefined} & ; i \geq j \end{cases} \tag{6}
$$

In this definition, K_o is an arbitrary positive constant. K_o may be
selected in several different ways; for example, it may be selected
so that the maximum link value is 10.

   The link value is undefined for $i > j$ to avoid confusion with regard to direction. The value L_{ij} assesses the "traffic" between i
and j, but it does not assess the direction of the traffic.

   In general, the number of link values for any given problem will
be $\frac{1}{2}N(N-1)$. In a three-instrument problem there are three link values; L_{12}, L_{13}, L_{23}; in a four instrument problem, there are six, and
so on. The number of joint probabilities is of course N^2.

   It is advantageous to take the link value definition one step
further by normalizing the values. Suppose that each link value is
divided by the sum of all link values. The result can be defined as
a link value probability,

$$
p_{Lij} = \frac{L_{ij}}{\displaystyle\sum_{i=1}^{N-1} \sum_{j=i+1}^{N} L_{ij}} \tag{7}
$$

and, correspondingly,

$$
\sum_{i=1}^{N-1} \sum_{j=i+1}^{N} p_{Lij} = 1 \tag{8}
$$

Note that the denominator in (7) represents summation over all link
values. In other words, the link value probability is the link value with K_o so specified that the sum of the link values is unity.

   Working now with the link value probability, several important
relationships will be derived. First, it is desirable to relate the
link value probability directly to the joint probabilities. To accomplish this, note that

$$
p_{ij} + p_{ji} = K_1 \, p_{Lij}; \quad i < j \tag{9}
$$

where K_1 is to be determined. Summing this expression over all link value probabilities yields

$$K_1 = \sum_{i=1}^{N-1} \sum_{j=i+1}^{N} (p_{ij} + p_{ji})$$

$$= \sum_{i=1}^{N} \sum_{j=1}^{N} p_{ij} - \sum_{i=1}^{N} p_{ii} \qquad (10)$$

$$= 1 - \sum_{i=1}^{N} p_{ii}$$

Therefore, the link value probability can be expressed in terms of the joint probabilities in two important ways:

$$p_{Lij} = \frac{p_{ij} + p_{ji}}{\displaystyle\sum_{i=1}^{N-1} \sum_{j=i+1}^{N} (p_{ij} + p_{ji})} \qquad i<j \qquad (11)$$

or, equivalently,

$$p_{Lij} = \frac{p_{ij} + p_{ji}}{1 - \displaystyle\sum_{i=1}^{N} p_{ii}} \qquad i<j \qquad (12)$$

In (11), only the "off-diagonal" joint probabilities are required to calculate link value probabilities. In (12), all joint probabilities are required, but the computation is simpler.

It should be noted that the link value probabilities contain the necessary information for so-called sequence-of-use optimization of panels. However, the link value probabilities are less general than the joint probabilities. Whereas the link value probabilities can be derived from the joint probabilities, as seen in (11) and (12), the reverse is not true. Additional information must be provided before the joint probabilities can be reconstructed from the link value probabilities.

There are other equations that are important in the instrument panel arrangement problem. They involve the relationship between the link value probabilities and the fixation probabilities. Intuitively one would surmise that there must be such a relationship, since large numbers of fixations on one instrument would tend to incur large numbers of transitions to and from that instrument. To find the relationship, it is necessary to define "independent scanning". This type

Table I. Hypothetical numerical example showing (a) the joint probabilities and corresponding fixation probabilities, (b) the
link value probabilities derived from the joint probabilities, and (c) the link value probabilities derived from the
fixation probabilities under the assumption of independent
scanning.

	P_{ij}	j 1	2	3	
(a)	i 1	.067	.083	.090	.240
	2	.132	.110	.152	.394 P_i
	3	.041	.201	.124	.366
		.240	.394	.366	$\sum_{i=1}^{N} P_{ii} = .301$
			P_j		$\sum_{i=1}^{N} \sum_{j=1}^{N} P_{ij} = 1$

	P_{Lij}	j 1	2	3
(b)	i 1	----	.308	.187
	2	----	----	.505
	3	----	----	----

$$\sum_{i=1}^{N-1} \sum_{j=i+1}^{N} P_{Lij} = 1$$

	P_{Lij}	j 1	2	3
(c)	i 1	----	.289	.269
	2	----	----	.442
	3	----	----	----

$$\sum_{i=1}^{N-1} \sum_{j=i+1}^{N} P_{Lij} = 1$$

of instrument scanning is said to exist when

$$P_{ij} = P_i \, P_j; \quad \begin{array}{l} i=1,2,\ldots,N \\ j=1,2,\ldots,N, \end{array} \tag{13}$$

or in other words, when independence exists in all joint probabilities. Under these conditions, the link value probabilities are obtained by substituting (13) into (12):

$$P_{Lij} = \frac{2p_i p_j}{1 - \sum\limits_{i=1}^{N} p_i^2} \qquad i<j \tag{14}$$

Therefore, under the conditions of independent scanning, the link value probabilities are derivable from the fixation probabilities.

In practice, it would be difficult to sustain the assumptions imbodied in (13). However, Senders (1964, 1966) showed, using an expression similar to (14) (though not derived), that experimental link values were similar to those obtained by using only the fixation probabilities.

Table I shows a hypothetical numerical example of the main relationships derived to this point. The matrix of joint probabilities is given in (a). Note that the fixation probabilities are obtainable by summing either the rows or the columns (Eq. (5)). The link value probabilities (Eq.(12)) are shown in (b). As expected they sum to unity. They are defined only for i<j. If the fixation probabilities in (a) are used to estimate the link value probabilities under the assumption of independent scanning (Eqs. (13) and (14)), the values shown in (c) are obtained. Note that the link value probabilities of (b) and (c), although different, have the same ordinal relationship.

All of the results thus far in this section have been based on the equally-spaced-sample method of data gathering. However, portions are equally applicable to the transition method as well. In the transition method, data are gathered by noting only when eye-fixation transitions take place from one instrument to another. The numbers of transitions between instrument pairs are then recorded. From them the link values are computed using the definition:

$$L_{ij\tau} = K_2 \begin{array}{l} \text{(number of transition pairs from i to j} \\ \text{+ number of transition pairs from j to i)} \end{array} \tag{15}$$

where the subscript τ is used to designate the transition method of data gathering, and K_2 is an arbitrary positive constant. To show that the transition method is equivalent to the equally-spaced-sample method for obtaining link values, it is only necessary to substitute (3) into (6) and observe that

$$L_{ij} = K_3 L_{ij\tau}; \quad i<j \qquad (16)$$

where

$$K_3 = \frac{K_o}{K_2 \text{ (total number of transition pairs)}} \qquad (17)$$

Consequently, it does not matter which of the two methods of data gathering is used. The link value relative magnitudes are the same, and eqs. (7) through (14) apply to either method of link-value data gathering.

RECORD LENGTH ESTIMATION

It is necessary to recognize that estimation of probabilities described in the previous section is a problem of statistical estimation. If insufficient record length is used, the resulting estimates may contain large errors resulting from instability. To illustrate how this occurs, consider taking several different sample means of n points from a random process. For small values of n, say 3 or 4, the sample means fluctuate widely. But, as n becomes larger, the variation among the sample means decreases, and they tend to converge about the population mean of the process. In using the definitions provided in the last section, sample means of random processes (or functions of random processes) are being computed. Consequently, they are subject to the same statistical fluctuations.

This section will present two methods for estimating the amount of eye-fixation data required to attain a specified accuracy with known confidence level. Both methods have been presented previously, so that details of the derivations need not be presented here.

The first method was developed by Seeberger and Wierwille (1976), and was directed specifically at data gathering for instrument arrangement. The technique is based on a parametrically derived double sampling plan, in which an initial sample is taken for record length estimation purposes. The procedure is as follows:

1. Break the same record into n equal segments of T seconds each.

2. Select a fixation or joint probability or link value probability to be estimated, and calculate the sample probability for each segment of T seconds.

3. Calculate the variance, S^2, of the n sample probabilities.

4. Choose the total length of the confidence interval, 1 (the distance between the upper and lower desired limits).

5. Using the equation

$$N_o \geq n_o = \frac{4t^2_{\frac{\alpha}{2}, n-1} \, s^2}{\underline{1}^2} \qquad (18)$$

where N_o is an integer and where

$$0 \leq N_o - n_o < 1, \qquad (19)$$

calculate the number of segments, N_o, required to obtain the total tolerance $\underline{1}$ with confidence $1-\alpha$. In eq. (18), $t_{\frac{\alpha}{2}, n-1}$ is the t-distribution cumulative value (t-table value) with $n-1$ degrees of freedom.

6. If $N_o \leq n$, the present short record is adequate (an unlikely possibility). If $N_o > n$, extra record length must be obtained. The total record required is $N_o T$ seconds.

As an example of the procedure, assume that a three-minute interval of data is available for analysis and that it is desired to estimate the population fixation probability p_3 within an interval $\pm$ .05 (e.i., $\underline{1}$ = .1), with 90% confidence. The interval is arbitrarily broken into 6 equal-length segments, yielding the following estimates for p_3: .22, .18, .06, .24, .37, and .27. The variance of the estimates of p_3 is .0105. Using eq. (18),

$$n_o = \frac{4(1.895)^2 \, (.0105)}{.01} = 15.1 \qquad (20)$$

Therefore N_o = 16, and the total length of record required is 8 minutes.

The second procedure available for record length estimation is based on the industrial engineering technique of work sampling (Barnes, 1979; Konz, 1979). It was originally evolved to obtain accurate estimates of the percentages of time a worker spends on various activities. Such estimates could then be used to overcome time-bottlenecks in the work process. The work sampling approach is applicable to instrument panel arrangement problems if the approach is properly interpreted. The procedure when applied to estimation of eye-movement probabilities would be as follows:

1. Sample a short record of eye movement data at random intervals and determine the ratio, p, of the number of occurrences of the event of interest (e.g., eyes fixated on instrument 2) to the

total number of samples. (This ratio, p, is actually a probability estimate.)

2. Specify the desired confidence level and convert this to the number of standard deviations, Z, from the mean. (For a confidence
level of 90%, Z = 1.64; for a level of 95%, Z = 1.96.)

3. Specify the desired relative accuracy (tolerance). If, for example, a tolerance of $\pm$ 10% of p is desired, then S_T, the relative
accuracy, is 0.1p.

4. Using the equation

$$N_{01} \geq n_{01} = \left(\frac{Z}{S_T}\right)^2 \frac{1-p}{p} \tag{21}$$

calculate the number of samples required. In this equation N_{01} is
an integer representing the total number of random samples required and

$$0 \leq N_{01} - n_{01} < 1 \tag{22}$$

As an example of this technique, assume that an initial random
sample provides an estimate of p_2 of 0.29. It is desired to estimate
p_2 with relative accuracy of $\pm$ 20% and with 90% confidence. Using
eq. (21)

$$n_{01} = \left[\frac{1.64}{(0.20)\,(.29)}\right]^2 \frac{1 - 0.29}{0.29} = 1957.5 \tag{23}$$

Therefore the number of random samples required is N_{01} = 1958. It
should be observed that while this number is large, it represents
individual sample points, not segments of data as in the Seeberger
and Wierwille approach.

Both techniques presented in this section require that the data
satisfy the usual parametric assumptions. In particular, this requires that the sample values be normally distributed. Seeberger and
Wierwille tested this assumption and found that for various segment
lengths the assumption of normality was reasonable (could not be re-
jected). However, when dealing with joint or link value probabilities,
it is very important to choose segments of sufficient length so that
the probability being estimated is not zero (a no-occurrence situation) in any appreciable number of segments. The zero-occurrence situation must be avoided if parametric assumptions are to be sustained.

CRITERION SELECTION

In general, any optimization problem requires selection of a criterion prior to performing the optimization. The criterion is essentially the set of ground rules under which the optimum is to be judged.

Most human factors publications deal with instrument arrangement criterion selection only in a general way (Applied Ergonomics, 1970; Van Cott and Kinkade, 1972; and McCormick, 1976). Among these general criteria are instrument importance, frequency of use, sequence of use, and functional grouping. In this section attention will be directed to the first three criteria, with importance being divided into two sub-categories. Functional grouping is a macroscopic criterion in the sense that it can be used to decide which instruments are to be included in each given instrument panel or panel area. However, arrangement of groups is subject to optimization techniques similar to those described in this paper.

There are two distinct types of criteria that can be used to optimize an instrument arrangement: first order criteria and second order criteria. First order criteria describe the relationship between each given instrument and its location on the panel. The assumption is usually made that there is a panel center or prime location and other less desirable locations in descending order. The usual procedure is to specify the positions (holes) where instruments are to be located prior to performing the optimization. The optimization process then is one of ordering the weightings of the instruments and placing them in locations based on their weightings. Second order criteria, on the other hand, describe the relationships that exist among instrument pairs and their distances from one another. The optimization process in this case involves arranging the instruments in locations which allow the strongest relationships to have the least distance between corresponding instrument pairs. It is possible of course, to specify overall criteria which contain both first- and second order sub-criteria. In the remainder of this section first order, second order, and combined criteria will be described in detail.

First order criteria may involve two sub-criteria: instrument importance and frequency of use. Importance is determined primarily by engineering judgment. It must be based on the system's intended use and the criticality of each instrument to that use. Frequency of use on the other hand can be assumed to be equivalent to fixation probability as discussed previously. Therefore, it can be determined experimentally. The steps in developing a first order criterion are as follows:

1. From an experimental situation determine the fixation probability, p_i, for each instrument in the arrangement.

2. From background data on the system, determine the relative importance, M_i, of each instrument. Normalize the importance ratings
   using the equation

$$m_i = \frac{M_i}{\sum\limits_{i=1}^{N} M_i}. \tag{24}$$

3. Select a panel center and determine the positions where the N instruments may be located. It is assumed that for N instruments
   there will be N available locations (or holes) in the panel.

4. Develop a criterion for each instrument that includes both a
   weighted importance and weighted frequency of use component, multiplied by the distance to the panel center for each arrangement:

$$Q_{1ai} = \left[\beta p_i + (1-\beta)M_i\right] d_{ai}. \tag{25}$$

In this equation β is a relative weighting constant $(0 \le \beta \le 1)$,
which must be selected prior to optimization, d_{ai} is the distance
of instrument i from panel center for arrangement a, and Q_{1ai} is
the instrument criterion value for instrument i with arrangement
a.

5. Sum over the instruments to obtain the overall first order criterion:

$$Q_{1a} = \sum_{i=1}^{N} Q_{1ai}. \tag{26}$$

Note that each arrangement, a, may result in a different value of
Q_{1a}. The best arrangement is the one having the minimum Q_{1a}.

   Second order criteria may be developed in a similar way, using
two sub-criteria: importance of the relationship between each pair
of instruments, and relative number of transitions between each pair
(link values). Again the importance criteria are a matter of judgment and must be based on design information involving criticality.
The steps involved in development of a second order criterion are as
follows:

1. From an experimental situation, determine the link value probabilities, p_{Lij}, for each pair of instruments in the arrangement.

2. From background data on the system, determine the relative importance, M_{ij}, of each "link". Normalize the importance values using
   the equation

$$m_{ij} = \frac{M_{ij}}{\displaystyle\sum_{i=1}^{N-1} \sum_{j=i+1}^{N} M_{ij}}. \tag{27}$$

3. Determine the available locations (holes) which the instruments may occupy and the distances between them.

4. Develop a criterion for each instrument pair that includes both a weighted link importance and a weighted link use component, multiplied by the distance between the pair for each arrangement:

$$Q_{2aij} = \left[\gamma P_{Lij} + (1-\gamma) m_{ij} \right] d_{aij}. \tag{28}$$

In this equation α is a relative weighting constant ($0 \le \alpha \le 1$), d_{aij} is the distance between instruments i and j for arrangement a, and Q_{2aij} is the criterion value for the i, j pair.

5. Sum over the instrument pairs to obtain the overall second order criterion:

$$Q_{2a} = \sum_{i=1}^{N-1} \sum_{j=i+1}^{N} Q_{2aij}. \tag{29}$$

For each arrangement, a, the criterion Q_{2a} may assume a different value. The best arrangement is the one having the minimum value of Q_{2a}.

Finally, in terms of criteria, it is possible to combine first and second order criteria to form a single overall criterion for optimization. In this case, the optimization contains a total of four components, two first order and two second order. The procedure is as follows:

1. Perform all five steps in developing a first order criterion and all five steps in developing a second order criterion.

2. Combine the first order and second order overall criterion values to produce a single weighted criterion:

$$Q_a = \lambda Q_{1a} + (1-\lambda) Q_{2a} \tag{30}$$

In this equation λ is a weighting constant such that $0 \le \lambda \le 1$. Again, the arrangement producing the minimum Q_a is the optimum.

It should be noted that eq. (30) actually includes all previous criteria as specific cases. Suppose for example that a designer wishes to include only first order importance information in the op-

timization process. This is easily accomodated by setting $\lambda=1$ and
$\beta=0$. All other combinations with arbitrary relative weightings are
also possible.

Before leaving the problem of criterion selection, one additional matter should be addressed. In eq. (30), if $\lambda=0.5$, it cannot be
assumed that first and second order criteria are equally weighted.
This is a result of the fact that the average distance between instrument locations (holes) is not necessarily equal to the average
distance of locations from panel center. To overcome this difficulty,
an alternative formulation can be used:

$$Q_{ao} = \lambda Q_{1a} + (1-\lambda)k_d \, Q_{2a} \tag{31}$$

where

$$k_d = \frac{(N-1) \sum\limits_{i=1}^{N} d_{ai}}{2 \sum\limits_{i=1}^{N-1} \sum\limits_{j=i+1}^{N} d_{aij}}. \tag{32}$$

In eq. (32) k_d has the same value regardless of the arrangement, so
it need only be computed for one of them. Actually, k_d is the average distance from location to center over the average distance between locations.

OPTIMIZATION PROCEDURES

Optimization is the process of selecting the minimum value of
the overall criterion, as given in eqs. (26), (29), (30), or (31).
It must be recognized that that process of optimization is not trivial, once second order data are introduced. Substantial efforts have
been directed at attempting to solve optimization problems similar to
those described here. Most of them have arisen in facilities location
problems of industrial engineering and operations research (Buffa,
1955; Armour and Buffa, 1963; Lee and Moore, 1967; Heider, 1973; and
Parker, 1976). A number of studies involving instrument panel optimization have also been made (Haygood et al., 1964; Freund and Sadosky,
1967; Bartlett and Smith, 1973; Bonney and Williams, 1977; Pulat and
Ayoub, 1979; and Rabideau and Luk, 1975). The difficulty is that the
number of possible ways that N instruments can be fitted into N locations is N!

It would be desirable to have a technique that would optimize
arrangements of up to ten instruments. Usually, beyond ten, preliminary grouping by function is required anyway. The difficulty is com-
pounded by the fact that at present there is no general procedure,
other than exhaustive search, which guarantees convergence at the

global optimum for the second order case. In this section, two procedures will be presented, one which is strictly limited to first
order optimization and is easily implemented, and the other for the
remaining cases (second order alone, or first and second order combined).

First order optimization is easily accomplished, regardless of
the number of instruments, by a procedure known as the product technique (Huebner and Ryack, 1961; Freund and Sadosky, 1967). In this
procedure, the instrument having the maximum combined importance/
frequency of use value (the bracketed quantity in eq. (25)) is placed
in a location having the minimum distance from the panel center. The
instrument with the second highest value is placed in a location
having the second shortest distance from panel center. This process
is continued until all instruments are located. The procedure results
in an arrangement producing the minimum value of Q_{1a}.

Unfortunately, no similar general procedure exists for optimizations involving second order criteria. Therefore, it becomes necessary to choose one of two alternatives: exhaustivs search or limited
search. Exhaustive search will result in determination of the global
optimum, but may be expensive or computationally not feasible. Limited
search may result in either a local optimum or possibly a global optimum, but a global optimum cannot be guaranteed. Therefore, in using
limited search, a designer never knows whether the global optimum has
been achieved. If it has not been achieved, the amount by which the
arrangement could be improved is unknown.

This paper advocates the use of exhaustive search which takes
advantage of techniques for reducing the number of iterations. A program was recently developed at Virginia Polytechnic Institute and
State University, both for application and to determine running times
for problems of various sizes. A number of solutions have been obtained. However, before presenting the results obtained with the program, certain aspects of exhaustive search will be described.

As stated, exhaustive search requires N! trial solutions for
the general case. Trouble is encountered in the practical range of
$7 \leq N \leq 10$. Suppose, however, that the locations for the instruments
possess one axis of symmetry which passes through the panel center.
Then for every potential solution there is a mirror-image solution
having exactly the same criterion value. (It is assumed that at least
one instrument location is not on the axis of symmetry.) Symmetry in
one axis can therefore reduce the size of the exhaustive search by a
factor of 2. The program must be designed to recognize this symmetry
and to preclude the mirror image computations.

Assume there are two axes of symmetry in the locations, both
passing through the designated panel center and at right angles to
one-another. The reduction in the exhaustive search in this case is

by a factor of 4, again assuming that at least one instrument is not located on either axis.

Yet another way of reducing an exhaustive search is to specify that a given instrument must be positioned in a given location. Actually, it should be recognized that such a procedure may result in other than a global optimum. However, from a practical standpoint this may not matter. Very often a designer wishes to have one instrument constrained to a given location. The effect of constraining one instrument is to reduce the number of iterations required by N. In other words (N-1)! iterations are then required. Again, however, it must be recognized that reduction in computation time is somewhat less than by a factor of N, because the arrays that must be manipulated on each iteration still have dimensions of N, not N-1.

Results obtained with the digital program thus far indicate that exhaustive search is substantially more feasible than had been previously believed. The program is written in PL-1 language and has been run on an IBM-370-3032. It was accessed through an HP 2621-A terminal for the results shown in Table II. The non-starred values are measured system response times, whereas the starred values are estimates. CPU time can be assumed to be approximately 40% of the system response time. No results as yet have been obtained for the case in which one instrument is constrained. However, it is estimated that response time would be double that for N-1. For important problems, therefore, up to ten instruments can be handled, assuming one is constrained. Computational costs under such circumstances, while high, would not be prohibitive.

Table II. Total system response time in seconds.

		Axes of Symmetry		
		0	1	2
Number of Instruments	5	1.5		
	6	5.8	6.1	3.1
	7	48	19	12
	8	440*	240*	108*
	9	4400*	2400*	1080*

*Indicates estimate

FUTURE WORK

While there are many remaining problems in instrument panel arrangement, two problems in particular should be addressed. First, the results of this paper are based on eye movement and instrument importance information. Consequently, the results are limited to visual display arrangements. Many panels contain controls as well as displays. Additional work therefore is needed in extending the approaches described in this paper to cover the rigorous optimization of control and control/display arrangements. Much of the necessary background information to accomplish this already exists (Topmiller and Sharp, 1965; Sharp and Hornseth, 1965; Fowler et al., 1968; Bartlett and Smith, 1973; Udolf and Gilbert, 1973; and Dow and Reiss, 1980).

The second problem may require new avenues of approach. Work is desperately needed on arrangement and formatting with CRT displays. Panel arrangement procedures do not readily apply to this problem. Because many panels of instruments are being replaced by multipurpose CRT's, design guidelines and optimization procedures must be developed. The problem is severe because information can be presented sequentially in time and because there are no constrained "locations" in such a display. The amount of design freedom that can be employed is large, but this freedom also makes problem definition and optimization difficult.

REFERENCES

Armour, G.C. and Buffa, E.S., 1963, A heuristic algorithm and simulation approach to relative location of facilities, Mathematical Science, 9, 294-309.

Barnes, R.M., 1979, "Work sampling", 2nd ed., Krieger Publishing Co., Huntington, N.Y.

Bartlett, M.W. and Smith, L.A., 1973, Design of control and display panels using computer algorithms, Human Factors, 15, 1-7.

Bonney, M.C. and Williams, R.W., 1977, CAPABLE; A computer program to layout controls and panels, Ergonomics, 20, 297-316.

Buffa, E.S., 1955, Sequence analysis for functional layout, Journal of Industrial Engineering, 6, 12-25.

Chapanis, A., 1959, "Research techniques in human engineering", Johns Hopkins Press, Baltimore, Maryland, 51-62.

Dow, L. and Reiss, A., 1980, CUBITS, a control panel area allocation algorithm, Analytics, Willow Grove, PA, Technical Report 1400. 06A.

Fowler, R.L., Williams, W.E., Fowler, M.G. and Young, D.D., 1968, An investigation of the relationship between operator performance and operator panel layout for continuous tasks, Wright-Patterson Air Force Base, Ohio, AMRL-TR-68-170.

Freund, L.E. and Sadosky, T.L., 1967, Linear programming applied to optimization of instrument panel and workplace layout, Human Factors, 9, 295-300.

Jones, R.E., Milton, J.L. and Fitts, P.M., 1949, Eye fixations of aircraft pilots, IV, in: Frequency, duration and sequence of fixations during routine instrument flight, Wright-Patterson Air Force Base, Ohio, AF-TR-5975.

Konz, S., 1979, "Work design", Grid Publishing, Inc., Columbus, Ohio, 145-161.

Haygood, R.C., Teel, K.S. and Greening, C.P., 1964, Link analysis by computer, Human Factors, 6, 63-70.

Heider, C.H., 1973, An N-step, 2 variable search algorithm for the component placement problem, Naval Research Logistics Quarterly, 20, 699-724.

Huebner, W.J., Jr. and Ryack, B.C., 1961, Linear programming and workplace arrangement: solution of assignment problems by the product technique, Wright-Patterson Air Force Base, Ohio, WADC Techn. Report 61-143.

Layout of panels and machines, 1970, Applied Ergonomics, 1, 107-112.

Lee, R. C. and Moore, J.M., 1967, CORELAP-Computerized relative layout planning, Journal of Industrial Engineering, 18, 195-200.

McCormick, E.J., 1976, "Human factors in engineering and design", McGraw-Hill Book Co., New York, 290-299.

Parker, C.S., 1976, An experimental investigation of some heuristic strategies for component placement, Operational Research Quarterly, 27, 71-81.

Pulat, B.M. and Ayoub, M.M., 1979, A computer aided instrument panel design procedure-LAYGEN, Proceedings of the 23rd Annual Meeting of the Human Factors Society, Boston, Massachusetts, 191-192.

Rabideau, G.F. and Luk, R.H., 1975, A monte-Carlo algorithm for workplace optimization and layout planning-WOLAP, Proceedings of the 19th Annual Meeting of the Human Factors Society, Dallas, Texas, 187-192.

Seeberger, J.J. and Wierwille, W.W., 1976, Estimating the amount of eye movement data required for panel design and instrument placement, Human Factors, 18, 281-292.

Senders, J.W., 1964, The human operator as a monitor and controller of multidegree of freedom systems, IEEE Transactions on Human Factors in Electronics, HFE-5, 2-5.

Senders, J.W., 1966, A re-analysis of the pilot eye-movement data, IEEE Transactions on Human Factors in Electronics, HFE-7, 103-106.

Sharp, E.D. and Hornseth, J.P., 1965, The effect of control location upon performance time for knob, toggle switch, and push button, Wright-Patterson Air Force Base, Ohio, AMRL-TR-65-41.

Topmiller, D.A. and Sharp, E.D., 1965, Effects of visual fixation and uncertainty on control panel layout, Wright-Patterson Air Force Base, Ohio, AMRL-TR-65-149.

Udolf, R. and Gilbert, I., 1973, Behind the front panel, IEEE Spectrum, 10, 28-31.

Van Cott, H.P. and Kinkade, R.G. (eds.), 1972, "Human engineering guide to equipment design", rev. ed., U.S. Government Printing Office, Washington D.C.

Woodson, W.E. and Conover, D.W., 1970, Human engineering guide for equipment designers, 2nd ed., University of California Press, Los Angeles.

PART III
MEASUREMENT OF PERFORMANCE

A METHOD FOR SEMI-AUTOMATIC ANALYSIS OF EYE MOVEMENTS

Fred V. Schick and Hans Radke

INTRODUCTION

The assessment and analysis of the visual behaviour of the human operator in a manned system provides a useful tool for the evaluation of man-machine systems, in terms of system performance and operator workload.

Since more than 100 years, scientific efforts have been made to develop reliable techniques and procedures for the record and analysis of eye movements. The start of these efforts might be dated around 1880, when Landolt observed the eyes of probands through a glass plate, while they were reading inscriptions mounted on the rear of the plate.

In the course of time, numerous principles and variants of eye movement recording were created. To treat them all, would go far beyond the scope of the present paper. A very good survey of the methods which are in use today is given by Young and Sheena (1975).

From the viewpoint of manned systems research, it may be sufficient to discriminate roughly between two main groups of procedures:

- the assessment of the processes, when the eye is actually in motion. This is referred to as "eye-movements research" by Mackworth (1976).

- the assessment of the visual targets fixated by an observer, which is generally of much more importance for system design (Mackworth: "line-of-sight research").

The first group of methods deals with the eye when it is moving.

221

Here, the pauses or rest periods during the fixation of stationary
objects are only concerned as far as they are again associated with
movements, in particular with small, generally involuntary movements,
such as microsaccades or convergences, which are necessary to maintain a sharp image on the retina during fixation. The methods applied
in this field of research are able to detect and record eye movements
relatively to the skull - some of them with high accuracies of less
than one tenth of a degree of visual angle - but a relation between
the movements of the eyeball and the objects in the observer's environment is not necessarily established. Eye-movement research is
more suitable to provide fundamental knowledge of the general functions of the visual apparatus, such as speed, frequency, range, accuracy, types of eye movements and limits of visual capacities.

However, for many problems in system design, information on the
pauses between movements is of more value than the movements themselves. To become aware of the particular targets in the visual field,
where the subject is looking upon during periods of fixation, is a
most helpful tool for investigations on cognitive behaviour, which
comprises perception and information processing. Here, the task is
to determine the point of eye fixation in the observer's field of
view, often referred to as the EPR (eye-point of regard).

The problem can easily be solved, when the subject's head is
tightly fixed, and the spatial locations of the particular stimuli in
the visual field are known.

However, for the many situations, where the requirement of head
fixation cannot be met, additional effort has to be made for the re-
gistration of head position or field of view, in order to relate the
recorded eye positions to fixated targets.

Among the methods for continuous record and analysis of visual
targets are oculometry (which is the subject of an own contribution
to this session) and the techniques of eye-mark recording, which are
based on the principle of corneal reflection.

Apart from the highlights of oculometry, such as digital data
output and the lack of head-mounted devices, that might cause discomfort or interference with normal vision, there are certain drawbacks of this method. Particularly in field studies, the relatively
spacious equipment might be disadvantageous, but still more detrimental could be the limited range of head movements (the system permits
variations of head position within the limits of one cubic foot only).

Eye-mark recorders, with their head-mounted optical devices allow nearly completely free head movements. This is achieved at the
expense of accuracy. However, for the purpose of simply identifying
where the subject is looking, in terms of the item or region being

inspected, this does not have to be expressed in fractions of degrees of the visual angle. An accuracy of ± 2 degrees in both, vertical and horizontal axis, as it is given by the eye-mark recording
system described below, has proved to be totally sufficient for the
safe identification of visual targets, such as individual instruments on an instrument panel, etc.

FUNCTIONAL PRINCIPLE OF EYE-MARK RECORDING

    The recording technique described here and illustrated in Fig.
1 is basically a variant of the corneal reflex camera system proposed by Mackworth in 1958.

    The subject is wearing some kind of a face mask (see Fig. 2),
at the right side of which a small source of light is fixed. This
lamp produces a small, V-shaped spot of visible light, which is projected onto the subject's eye. The front surface of the eye, the
cornea, has a smaller radius of curvature than the eyeball itself.
So, due to variation of eyeball position, the V-mark is reflected
off the cornea in various angles. Then it is projected into the optical system, which is contained in the body of the face mask, by a
half-transparent mirror mounted in front of the eye.

    Simultaneously, the field of view that corresponds to the sub-
ject's actual head position is taken by another lens in the center
of the face mask. The V-mark is then superimposed to the field of
view image, and the combined image is transferred to a camera adap-

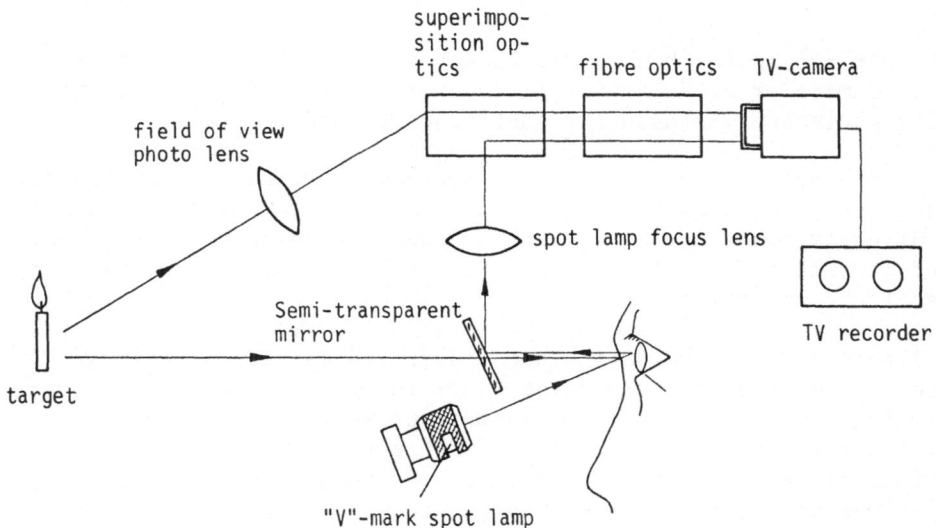

Fig. 1. Schematic diagram of eye-mark recording.

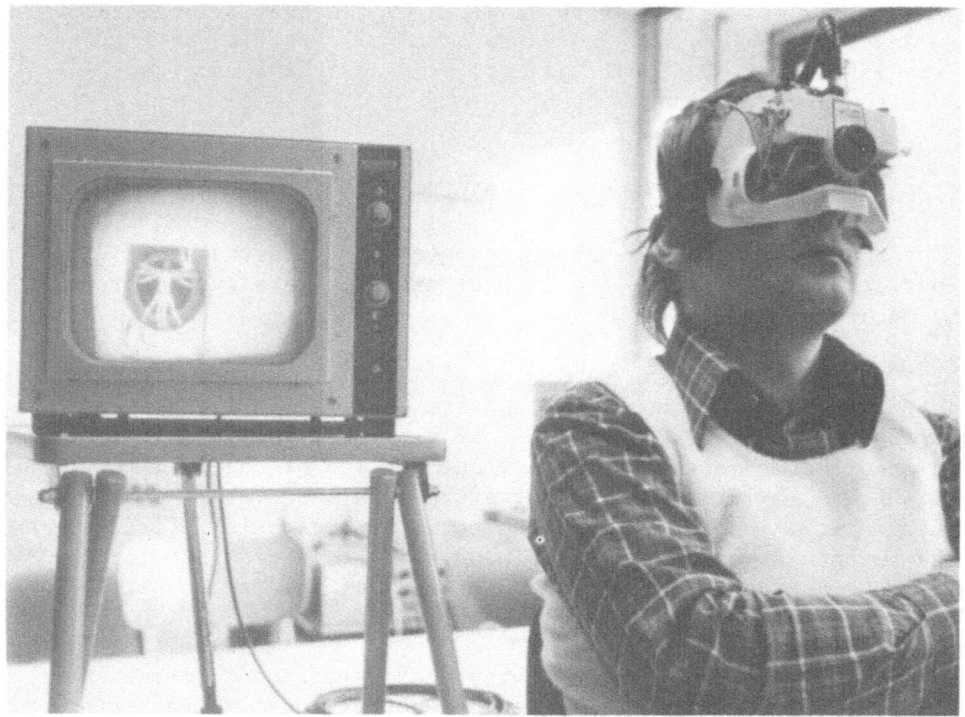

Fig. 2. Subject wearing NAC face mask; subject's eye-point of regard can be seen as 'V'-mark on the control monitor.

ter by a fibre optics bundle. The adapter fits with a 16 mm motion
picture camera or a TV camera. In the version treated here, TV recording equipment was used for continuous records of EPRs.

In the original version of the Mackworth corneal reflex camera,
the subjects had to sustain the weight of the camera too, which had
been directly fixed to the proband's head. The advanced version described here is a Model IV Eye Mark Recorder manufactured by the Ja-
panese NAC corporation.

Although a very versatile and widespread method, a serious
shortcoming of eye mark recording is given by its form of data output, which consists of analog informations stored on film or TV
tape. Analyses of eye mark records, with regard to statistical descriptions of EPR characteristics (such as average look frequencies,
dwell times on distinct targets, etc.) require as an intermediate
step the transformation of the optical information into a data format which is amenable to further processing. Generally, this transformation has turned out to be rather complicated and time-consuming.

METHODS CONSIDERED FOR THE ANALYSIS OF EYE MARK RECORDS

Manual Procedures

The most simple, but also the most labouring method is the inspection of the film or TV records frame by frame. The manual notification of the specified visual targets, which are hit by the EPR light signal in every single frame, can be done by means of frequency lists. But, as a TV record contains 50 frames per second, and an evaluator's working speed for the inspection of the records amounts about 0.5 frames per second (which is a realistic value observed in several evaluation trials), it can easily be seen that this procedure is far too time-consuming as to become applicable for the evaluation of records even of moderate length.

An advancement of this paper and pencil method are time-interval sampling techniques. For instance, a method proposed by Heinze (1977). Here, the TV records are inspected when they are replayed at the original speed. A time pulse generator is used which produces acoustic signals in one-second intervals. The evaluator again has a list of specified visual targets that may occur in the record. He continuously marks in his list the appropriate targets which are actually fixated during coincidence with the acoustic signals. Thereafter, the frequency list data have to be punched on cards for the computation of the per cent fixation frequency and the per cent fixation time of every single target.

To overcome the paper and pencil methods described above, a procedure for use on TV records has been theoretically considered, which might be named "manual eye mark tracking". A map of the visual field under investigation has to be programmed and stored in a processor. This can be done in terms of the coordinates of individual elements (i.e. possible visual targets) in the visual field, in relation to a reference mark in the center of the field. During the replay of the eye mark records, two electronically generated symbols, which can be remotely steered by control sticks, are superposed to the TV screen. An evaluator steers the first symbol as to follow the reference mark, which changes its position on the screen due to head movements of the experimental subject. Another evaluator makes the second symbol to follow the V-shaped EPR mark. The digital processor interfaced with the control monitor, computes the relative position of both symbols, determines the targets actually fixated by comparing the position data with the internally stored map, and stores the target data for further processing.

Fully Automatic Analysis

An extension of the above-mentioned procedure could be the fully automatic tracking of both, a fixed reference mark and the EPR mark contained in an eye mark record, by means of an electronic tracking system, which makes use of the brightness of both marks in contrast to the adjacent environment. However, the tracking systems which are in use at present are still unable to track more than one of those marks at a time. Recent inquiries of the manufacturers of these systems showed that a system which would reliably fulfil the requirements of this special application cannot be realized at justifiable costs at the present time.

There is still another disadvantage of the automatic as well as the manual tracking method. Both would not be applicable in cases where individual visual targets are changing their position relatively to the fixed reference mark. For instance, in an aircraft during approach, the runway borderlines will appear under different visual angles when the pilot is looking through the windscreen at different phases of the approach. To give another example, in car driving experiments too, a classification of the visual targets (other vehicles, road signs etc.) would be required rather than information on the position of the fixation point in a coordinate system.

Summing up, the crucial point is that the programming and computer storing of a fixed map for the identification of visual targets is inappropriate for a number of applications of eye mark recording.

The semi-automatic evaluation procedure presented below overcomes this difficulty, as it works with classification categories for visual targets, not with their exact locations in terms of visual angle coordinates.

SEMI-AUTOMATIC EVALUATION PROCEDURE

For the description of the semi-automatic evaluation procedure, an example of its recent application is thought to be helpful. During the field tests of a new central data entry system for transport aircraft (Schick et al., 1979), the visual behaviour of airline pilots has been recorded while they were performing flight tasks in a Boeing 747 aircraft simulator cockpit. The records have been taken, using TV eye mark recording equipment (see Fig. 1), under adverse conditions for the most other recording techniques, such as limited space, bad lighting and large head movements of the pilots.
For analysis of the eye mark records, the cockpit environment displayed in Fig. 3 was subdivided into eighteen distinct visual target areas, as shown in Fig. 4. Areas 7-9, 11, and 14-17 represent the alternative installation areas for the hardware units of the central data entry system under investigation.
To anticipate in short the basic working principle, the transformation of the eye mark signals from the TV tape records into digitally processable data works as follows. An encoding board is constructed, on which each of the visual target areas is represented by an own pushbutton switch. The switches are arranged according to

Fig. 3. Boeing 747 cockpit.

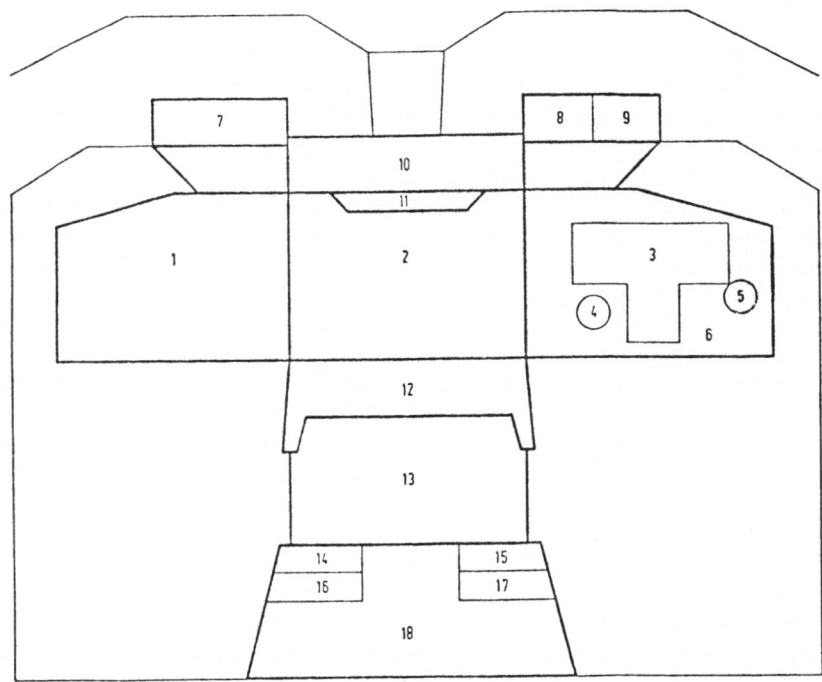

Fig. 4. Subdivision of the pilot's field of view into eighteen target areas considered for evaluation.

their correct spatial locations, in an outline drawing of the cockpit, which is mounted at the front of the encoding board. An observer watching the replay of the eye mark records on a control monitor has to classify the actual eyepoint of regard by pressing the appropriate switch on the encoding board, and to keep the switch depressed as long as the EPR 'V'-mark dwells in the corresponding area. TV recorder and encoding board are both linked with a digital processor, which calculates the duration of the dwells for the intervals in which the keys are actuated. The time intervals are computed by counting the TV frame frequency. For the entire record length, the sequence of switch codes plus the related dwell times are registered in this way, and stored on digital magnetic tape. In order to achieve high accuracy of encoding, the TV records are presented to the observer in slow motion, at about a quarter of the original speed. Next, the digitally stored data are processed for working out essential features of an EPR analysis: duration of looks, look frequencies and frequencies of transitions between the specified target areas.

Hardware

More in detail, the hardware of the evaluation system, which is schematically displayed in Fig. 5, has as its main element the digital processor with a magnetic tape unit. The interface consists of an analog/digital converter with twenty multiplex input channels and digital input and output registers.

Fig. 5 shows two encoding boards, which were introduced because the large number of 18 target areas was felt to be too difficult to supervise for one evaluator only. Thus, two observers were involved in the particular evaluation given here as an example. The first one operated an encoding board with the switches for those six target areas which were mostly fixated by the pilots. The second one supervised the remaining twelve areas that were less often fixated, by simultaneously operating the second encoding board. Fig. 6 gives an illustration of this procedure.

During operation, the actuation of a particular switch causes a voltage to be applied to the belonging input channel of the analog/digital converter, which is used to identify the switch (or target area, respectively) and the duration of its actuation.

The TV recorder had been modified to allow the counting of the number of frames during operation in slow motion.

Software

The software designed for the semi-automatic evaluation procedure consists of two major parts. The first part manages the basic data acquisition during the encoding procedure, the second part contains arithmetical and statistical algorithms for the computation of the main EPR characteristics.

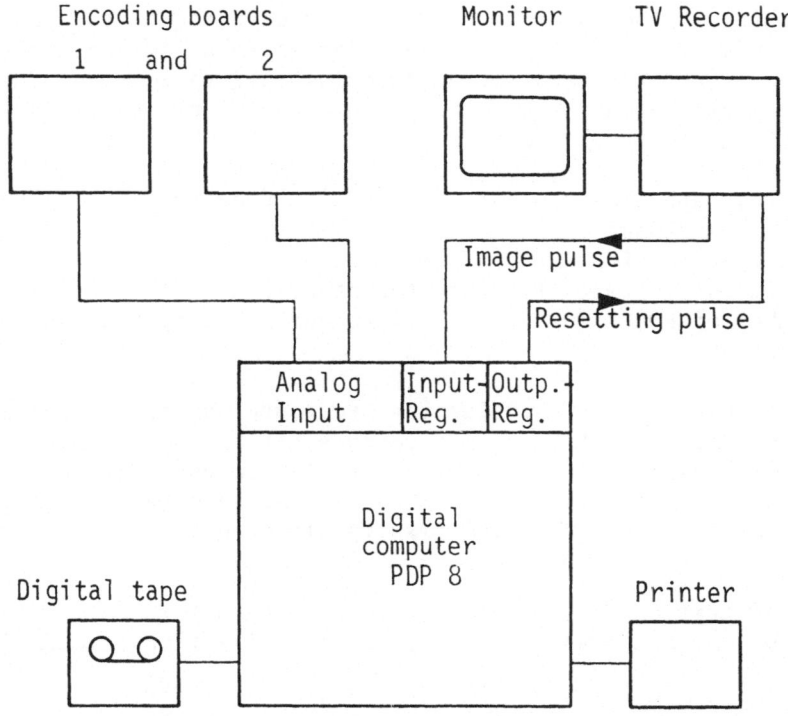

Fig. 5. Block diagram of the evaluation hardware.

Fig. 6. Illustration of the data acquisition procedure, showing the
TV recorder, control monitors and encoding boards for two
observers.

A simplified block diagram of the basic data acquisition program is given in Fig. 7. The program is designed as to allow evaluators to stop and to restart the recorder at their own discretion. So, they can insert rest pauses at any time during the encoding procedure, which is of high workload for the operators.

A control output of the data acquisition program provides the following informations:
- mean fixation time for each target area,
- total fixation frequency for each target area,
- total number of EPR transitions between any two target areas, in both directions.

The reliability of the evaluation procedure was investigated by repeating the encoding of three EPR records after several days. Very good reliability was indicated by the correlation coefficients which were computed for two identical analyses of the same test run regarding mean fixation times and total fixation frequencies:

EPR record of test run No.	Correlation coefficients for mean fixation time/total fixation frequencies	
1	.996	.980
2	.976	.880
3	.976	.960
Average correlation	.987	.952

The ideal case of total congruence of two sets of data from an identical record would have resulted in a correlation coefficient of 1.00, which is only theoretically achievable.

The data generated by the basic data acquisition program may be further subjected to a large variety of statistical procedures. The following features were thought to be essential for the description of pilot scanning behaviour in the context of the study mentioned here as an example.

Per cent dwell time (gives the proportion of time which is spent for fixating the target areas, in relation to the total time of the test runs);

Look frequency per time unit (sets the total number of fixations for any target area in relation to a specified time interval, thus giving for instance the look frequency per minute or per hour;

Average dwell time (computed from the total dwell time and the total look frequency of any target area; gives the average duration of single looks on an item);

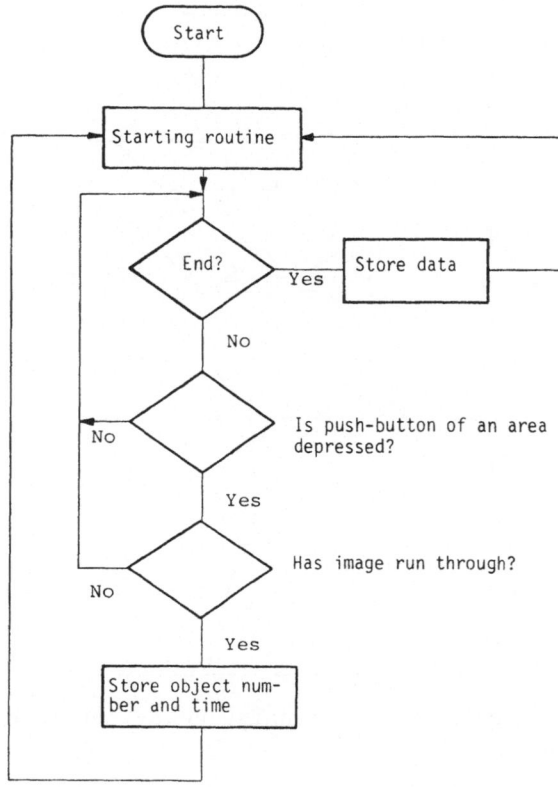

Fig. 7. Simplified block diagram of the data acquisition program.

Per cent transition frequency (gives the proportion of the transition of looks between any two specified target areas, in relation to the total number of transitions occuring in the test runs).

As an example for a comprehensive graphical display of EPR characteristics, Fig. 8 shows the dwell times (circles) and main transitions of looks (arrows) observed from pilots performing a reference flight task during the simulator test series mentioned above.

Apart from the computation of statistical indices, the data generated during the encoding procedures can also be processed in totally different ways. For instance, in a study on pilot performance, when a recently developed airborne link terminal for the communication between pilots and ground control is utilized, time sequences of the pilots' visual activities were the matter of primary interest.

According to· this, the EPR dwell times on specific elements of the new equipment were computed with exact reference to the time elapsed during the test trials. (Again, the test runs, where the eye mark TV records were taken, had been performed in a flight simulator.

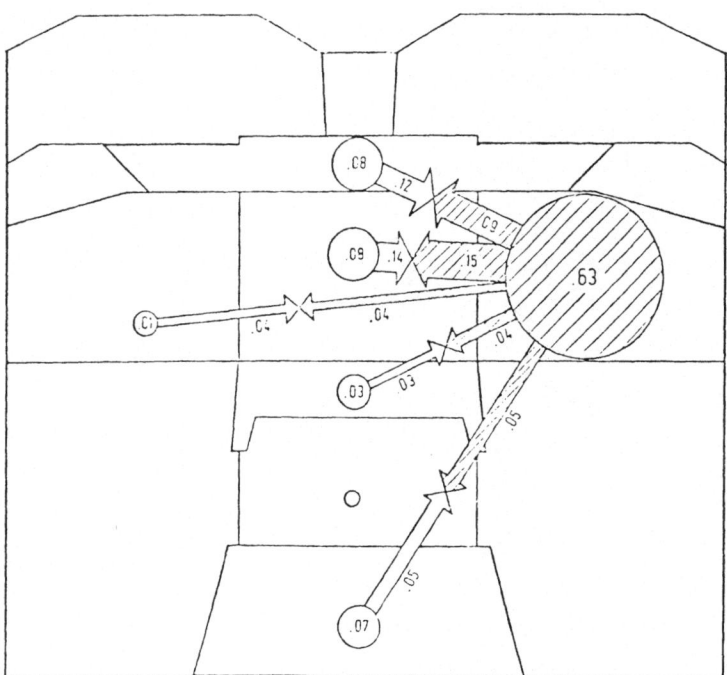

Fig. 8. Graphical display of pilot scanning behaviour (see text for explanation).

The same encoding procedure and the same software for the basic data collection as described above had been used for the evaluation of the records). The dwell times spent on the three components of the airborne link terminal (display, printer and keyboard) were arranged along the time axis and thus put in relation to the information flow, as measured by the number of information changes on the display screen (see Fig. 9).

CONCLUSIONS

The semi-automatic procedure for the evaluation of eye mark records, yet applied for the analysis of the visual behaviour records of two flight simulator studies, has proved to be a practicable method, even applicable on large quantities of data, i.e. sets of records with considerable overall length.

It represents a large step forward, as compared to manual paper and pencil methods. Particularly, the time needed for evaluation is cut down considerably to a justifiable amount.

Some principly existing shortcomings of evaluation-by-tracking procedures are also overcome, since the underlying idea is the direct

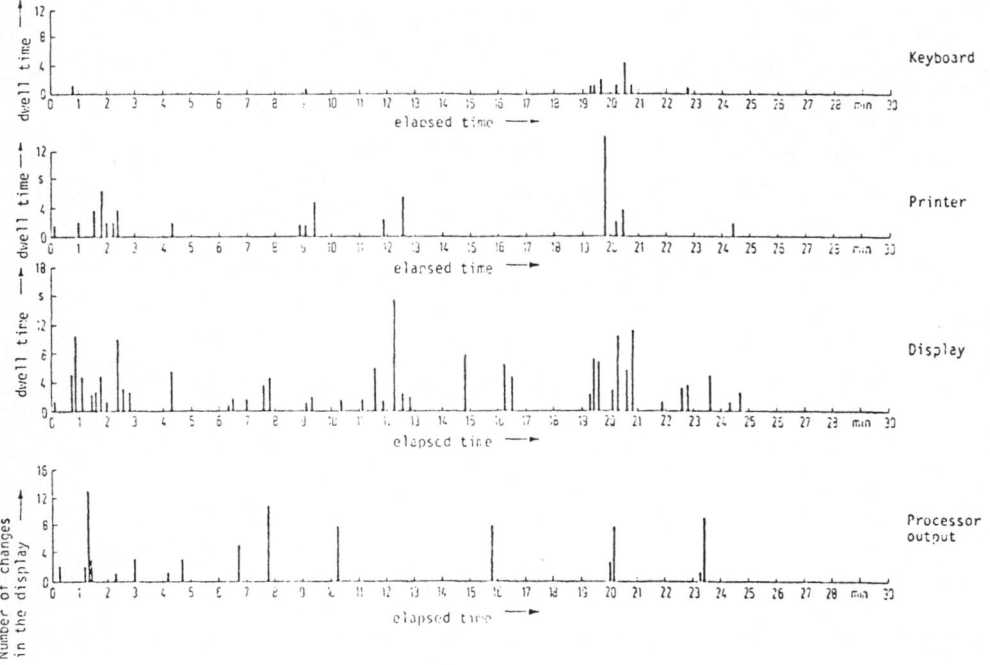

Fig. 9. Eye-point of regard dwell times on the three components of
an airborne link terminal (top) in relation to the information changes of the terminal display (bottom).

classification of targets instead of the identification of targets
by means of computation of visual angle coordinates. This abandons
the necessity to program and use computer stored maps of visual target coordinates, which would anyway suffer from inaccuracies (paral-
lax errors during head movements). Moreover, the semi-automatic classification method is applicable to targets with changing positions
in the visual field, although no use was made of this feature in the
above-mentioned flight simulator studies.

        Finally, the initial transformation of the EPR records into di-
gital data (which can be analysed with a high flexibility of methods
thereafter) came out to be a highly reliable procedure.

REFERENCES

Heinze, W., 1977, Simulatoruntersuchung zur Bestimmung der Piloten-
        beanspruchung bei unkonventionellen, räumlich gekrümmten An-
        flugprofilen, Dissertation, Technische Universität, Braunschweig.
Mackworth, N.H., 1976, Ways of Recording Line of Sight, in: "Eye Move-
        ments and Psychological Processes", R.A. Monty and J.W. Senders
        (eds.), Lawrence Erlbaum Ass., Hillsdale, N.J.

Radke, H., 1979, Ein Verfahren zur teilautomatisierten Auswertung von
    Blickbewegingsmessungen, DFVLR-Mitt. 79-02.
Schick, F.V., Brunner, D., Neumann, E.-H. and Schenk, H.-D., 1979,
    Evaluation of a Central Data Entry System (CDES) for Transport
    Aircraft, DFVLR-FB 79-23.
Young, L.R. and Scheena, 1975, Survey of Eye movement Recording Meth-
    ods, Behav. Res. Meth. and Instr., 7 (5), 397-429.

NARROWING OF THE VISUAL FIELD AS AN INDICATOR OF MENTAL WORKLOAD?

Manfred Voss

INTRODUCTION

The interest in mental workload measuring is growing (e.g. Moray, 1979), because it is an essential element in the adaptation of complex systems to man's capabilities. Several methods have been suggested for the estimation of mental workload: measuring of performance, the secondary task method, measuring of physiological indicators, and the method of subjective judgement.

For the important problem area, the estimation of mental workload related with visual information processing, the special characteristics and tasks of man's visual system should be considered. The cooperation of foveal and peripheral vision plays an important part in the total performance of human vision in man-machine systems. Therefore, based on experimental results on peripheral vision from literature, a measuring method has been developed which uses peripheral visual performance as an indicator of mental workload caused by processing of central visual information (Voss and Bouis, 1979). In the following, this method is described. Furthermore, it is asked whether the decrease of peripheral visual performance is due to the so called tunnel vision effect. Finally, experimental results are discussed and compared with those of other workload measuring methods

Measuring the Mental Workload

Fig. 1 shows the general concept of workload measuring (Moray, 1979; Voss and Bouis, 1979; Jahns, 1973 and Sheridan, 1979): The task demands are the input variables in a given task situation under specific environmental conditions (e.g. a driving task on a definite road under definite weather conditions); partly, the input is

defined as stress. To accomplish these demands, the human operator
has to select and to perceive the information needed, which is often
visual. Furthermore, he has to process the information and to transform it into actions suitable for his task. Some internal variables
play an important part in this context, e.g., the level of training,
circadian periodicity, motivation, actual alertness and attention,
and others, globally described as "internal status". The resulting
output can be divided into two groups of dependent variables:

- Actions of the human operator (e.g. control movements, searching
  for further information), which give a feedback to the timing of
  the input demands;

- Indicators of human operator's workload, measured by different
  methods.

Human operator's workload can be divided as follows (see also Sheridan, 1979):

- Workload caused by operating in a given situation (metabolite effects, physiological workload);

- Workload caused by information processing (perception, decision-
  making, mental transformation into action);

- Workload caused by after-effects of psychohistory (emotional workload); the internal status is the actual state of psychohistory.

   In the following the only aspect considered is mental workload
caused by information processing. - There is no general method of
measuring the mental workload, because the term "workload" is a very
general concept. Rather there exists a plurality of methods (with
specific benefits and disadvantages) of measuring different indicators which are regarded as important aspects of mental workload.
Usual workload measuring methods are (Moray, 1979; Küting, 1976):

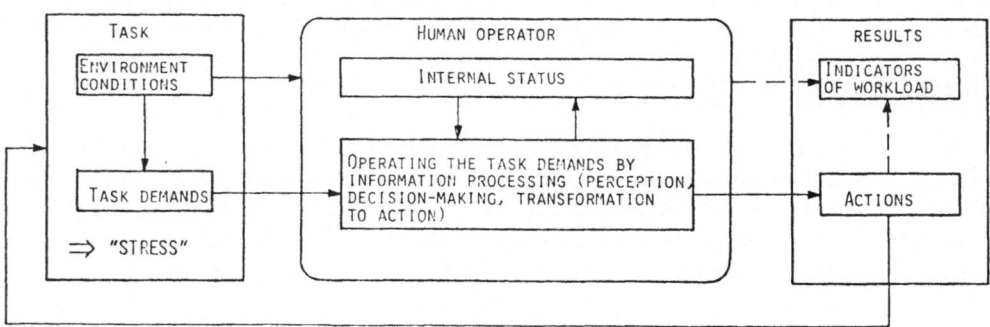

Fig. 1. The general concept of mental workload.

- The measuring of performance criteria of the primary task, e.g. detection or control accuracy;

- Method of secondary task: The measuring of performance criteria of an added (subsidiary or loading) task, e.g. regularity of tapping, solving arithmetic problems;

- The measuring of physiological indicators, e.g. pulse frequency, sinus arrythmia, skin resistance;

- Subjective judgement by a suitable rating scale.

A basic problem of all mental workload measuring methods is the resulting vicious circle: The input variable "stress" is mostly not quantifiable (excepting simple cases). Therefore, the influence of stress on a human operator in a man-machine system has to be judged by the resulting "workload". For this reason, workload must be definable by an established measuring method. On the other hand, it it necessary to know the input "stress" in order to define methods of workload measuring. To solve this problem it seems to be reasonable to look at stress and workload as multidimensional variables, which should be investigated, step by step, by different methods simultaneously (Moray, 1979). This process is not yet completed; in the meantime, the term "workload" and workload measuring should be handled with care.

PERIPHERAL VISUAL PERFORMANCE AS INDICATOR OF MENTAL WORKLOAD

Measurement Method

A method of workload measuring should have the following properties: Little impairment of the subject during the measuring process, time stability, good sensitivity for different workload levels, high temporal resolution of actual measurements taken. For the important problem area of estimating mental workload caused by visual information processing in real situations, the special characteristics and tasks of man's visual system should be considered: The human operator has often to observe continuously and accurately several displays or procedures centered in his visual field. Additionally unpredictable alarm signals or unexpected events have to be detected, interpreted and responded to (e.g. supervisory tasks in a control room, aircraft control or car driving), which are presented more or less in the periphery of the visual field. Peripheral vision comprises most of the visual field; therefore, it plays an important part in the total performance of human vision in man-machine systems.

It is often reported that the performance of peripheral vision is degraded by different influences (e.g. by task-irrelevant information (Mackworth, 1965), by alcohol (Moskowitz, 1974), by danger

(Weltman and Egstrom, 1966), by increased difficulty of or by increased concentration on a visual task presented in the central visual field (Gasson and Peters, 1965)). This effect is often called
"tunnel vision" or "narrowing of the functional visual field", without exactly defining these terms. A detailed discussion of this
point is given in the next section.

Apart from the exact theoretical foundation the detectability
of peripheral lights can be used as indicator of mental workload
(Voss and Bouis, 1979): In addition to the main task, whose influence
on mental workload has to be investigated, subjects have to detect
standard light stimuli presented at constant angular positions in
the peripheral visual field by a special spectacle frame (method of
subsidiary task). Fig. 2 shows the structure of the measurement procedure, and Fig. 3 the lightweighting spectacle frame. The standard
light stimuli were defined as follows: Binary light stimuli (e.g.
realized by LED's) with a duration of 300 ms, the time intervals between two light stimuli statistically distributed with a modified
Gaussian distribution (mean value 5,5 s, standard deviation 3,5 s,
time intervals shorter than 2,7 s were excluded). There was only one
light stimulus presented at once with a presentation angle of 45°,
uniformly distributed on two symmetric positions for the left and
the right eye (e.g. the two temporal positions on the horizontal meridian). Therefore, the influence of horizontal eye-movements on the
detection rate averaged over a measurement period is reduced, and no
special detection strategy is developed by the subjects. Furthermore,
by the use of a spectacle frame, head movements have no influence.
The spectacle frame is individually adjustable; after some training
sessions the subjects reported that they felt no impairment.

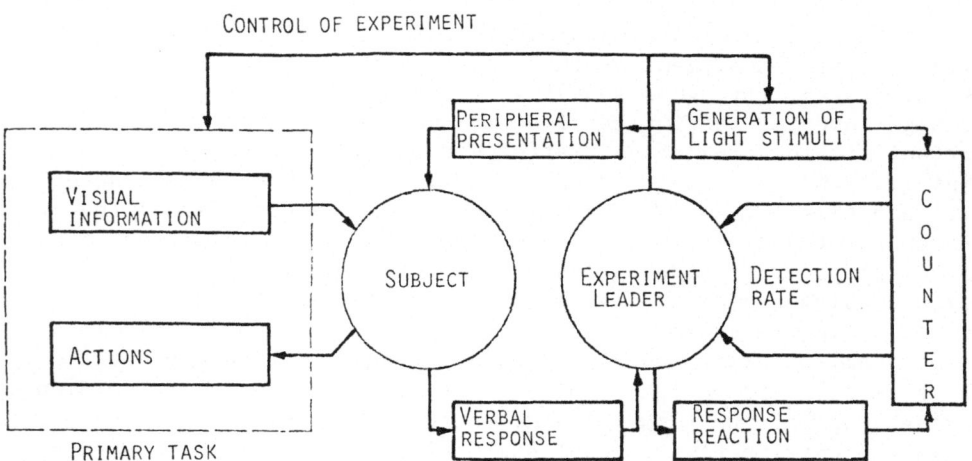

Fig. 2. Schematic diagram of the experimental set-up for measuring
the workload indicator "detactability of peripheral light
stimuli".

Fig. 3. Adjustable, lightweight spectacle frame presenting
peripheral light stimuli.

It is important to assure, that the detection of the light stimuli is a secondary task. Therefore it was added to the main task
step by step in several training sessions. At the beginning of a
measurement procedure, a reference situation with low mental workload was chosen. For this situation the intensity of the light stimuli was adjusted to a detectability of about 80%. Measurements were
carried out in relation to this reference situation. At the end of
a measurement session the reference situation was repeated in order
to estimate the time stability. In most cases the detection rates
measured were stably reproduced; therefore the mean value of both
reference situations was used as a reference value for this session,
and this reference value is suitable for the comparison of different
sessions and subjects.

In order to investigate transient responses of workload in a
main task, it is possible to average the detection rates of the
light stimuli over short time intervals of comparable situations
within one or more measurement sessions. The shorter these time intervals, the more often they have to be repeated for sufficient statistical significance of the mean detection rate. On the other hand,
the consequence of measuring several times repeatedly is an averaging of the situation which is investigated. - As an overview over
the properties of this workload indicator, experimental results are
described in section "Detectability of peripheral light stimuli as
workload indicator: Experimental results".

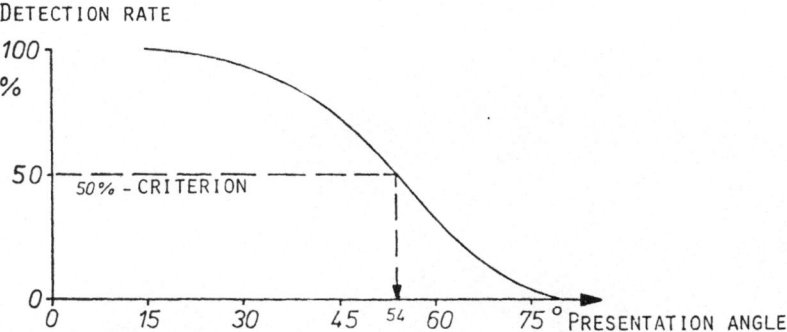

a. Definition of "functional visual field".

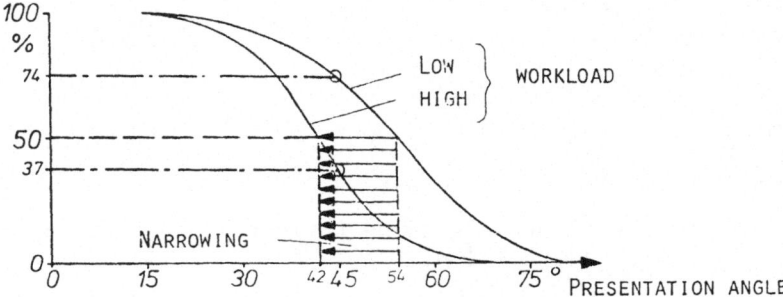

b. Effect of narrowing caused by high workload.

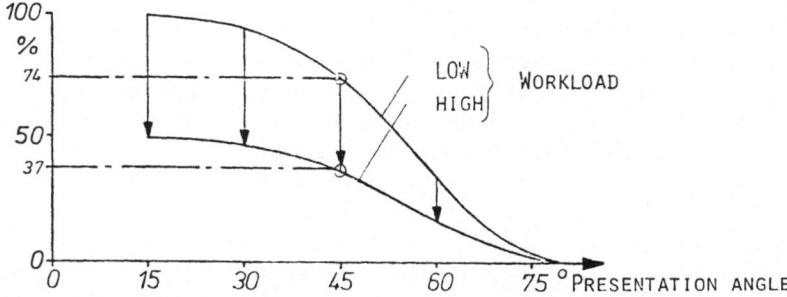

c. Effect of general decrease caused by high workload.

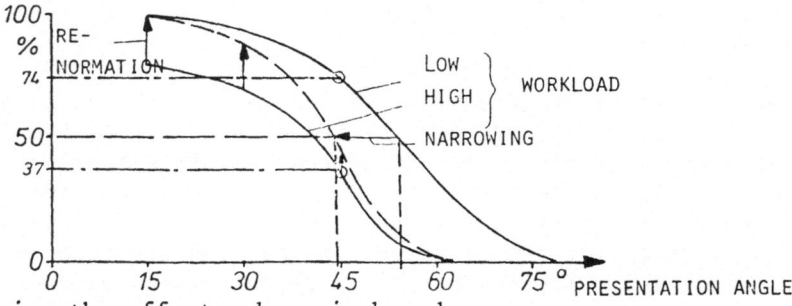

d. Mixing the effects shown in b and c.

Fig. 4. Detection rates vs. presentation angle (hypothetical curves,
        see text).

Tunnel Vision or General Decrease of Peripheral Visual Performance

A decrease of peripheral visual performance, e.g. by one of the influences mentioned above, can be caused differently: Perhaps tunnel vision is the underlying effect often suggested in the literature. Towards an exact definition of the terms "tunnel vision" or "narrowing of the functional visual field", the range of the functional visual field has to be determined by defining a measuring procedure as a first step. The following procedure has been chosen (Voss and Bouis, 1979): Definition of a standard light stimulus (the same as described above), detection rate measurement of this light stimulus as a function of the presentation angle in the horizontal meridian, definition of functional visual field by the presentation angle, where 50% of the light stimuli are detected (Fig. 4a). Thereby the intensity of the light stimuli has to be adjusted to a detectability of just beneath 100% at an angular position located as centrally as possible. In order to give a rough idea, the so determined numerical value can be interpreted as half the aperture angle of a cone, which is forward-opened, and axis of vision centered. This cone represents the functional visual field by definition.

"Tunnel vision" is then characterized as a clear reduction of the aperture angle of this cone, e.g. by increasing workload (Fig. 4b). Another possibility is a general decrease of visual system's performance for the detection task, which is independent of the presentation angle. Consequently it can be measured at any angular position (Fig. 4c(. A third possibility is a combined effect, which can be separated by "renormation" starting at an angular position located as centrally as possible (Fig. 4d).

If detection performance is measured only at one constant presentation angle in the periphery (e.g. at 45° with the spectacle frame), it cannot be decided between these possibilities. Therefore, an experiment was carried out measuring the detection of the standard light stimuli at different angular positions in the horizontal meridian as a function of different workload levels of a tracking task presented in the central visual field (Voss and Bouis, 1979) (see Fig. 5a, b, c). The different workload levels were realized by compensatory control of one, two or four independent control loops. The control errors were simultaneously displayed as vectors on an oscilloscope, the control device was a joystick compatible with the display (Fig. 5c, see also Geiser and Schumacher, 1976). As an example, Fig. 6 shows the results of two different subjects. Further experiments are being undertaken with modifications in presenting the light stimuli; they are not yet finished. As a preliminary result a combined effect is given: Tunnel vision as well as a general decrease of visual system's performance for the detection task; the balance between these two possibilities differs individually without concentration on one of these.

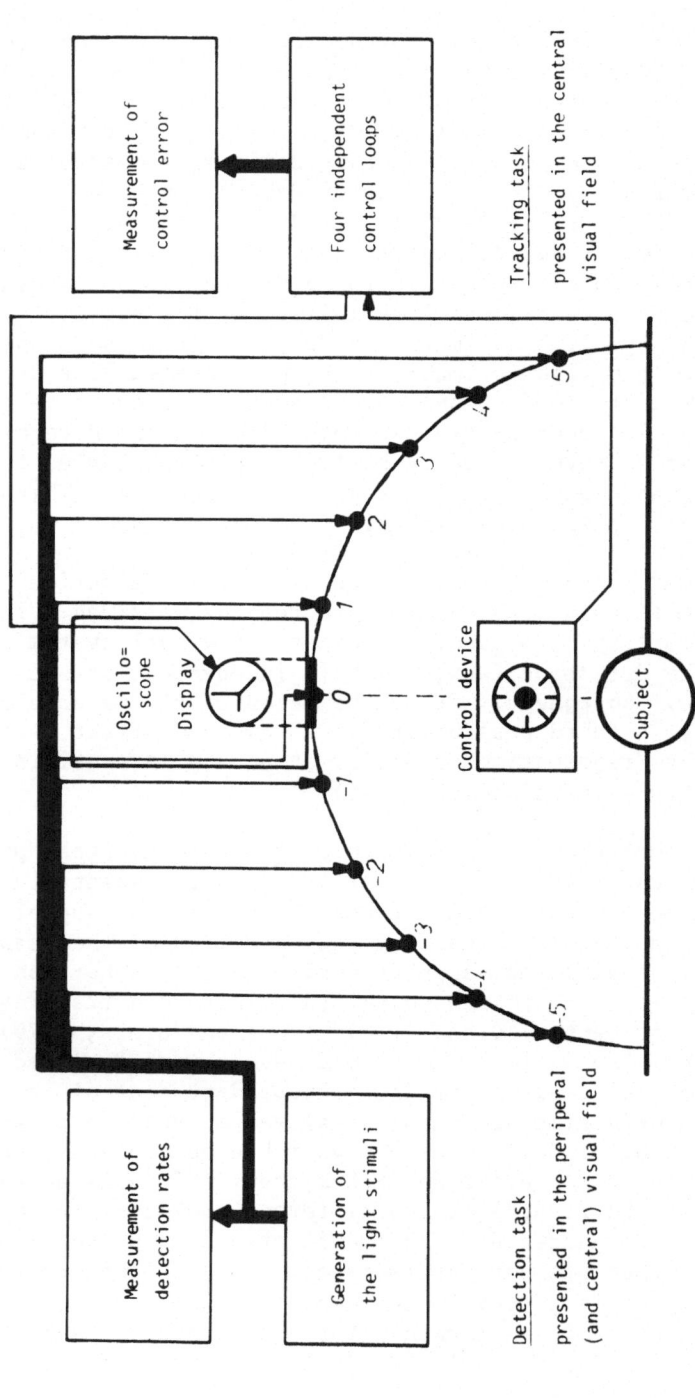

Fig. 5a. Schematic diagram of the experimental set-up for measuring the detection of light stimuli as a function of different workload levels of a tracking task.

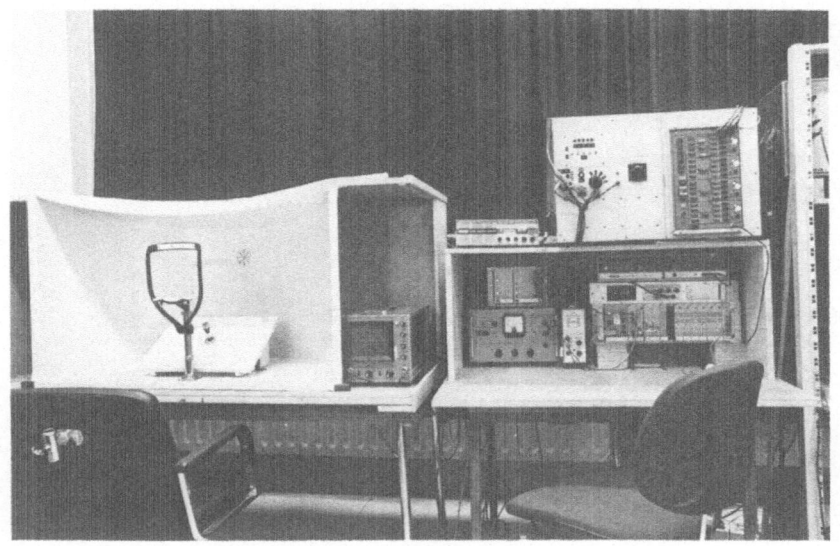

Fig. 5b. Overview over the experimental set-up described in Fig. 5a.

Fig. 5c. Detail of the experimental set-up: Display of the control errors as vectors in the central visual field (with central LED enlightened) and the control device as a joystick compatible with the display.

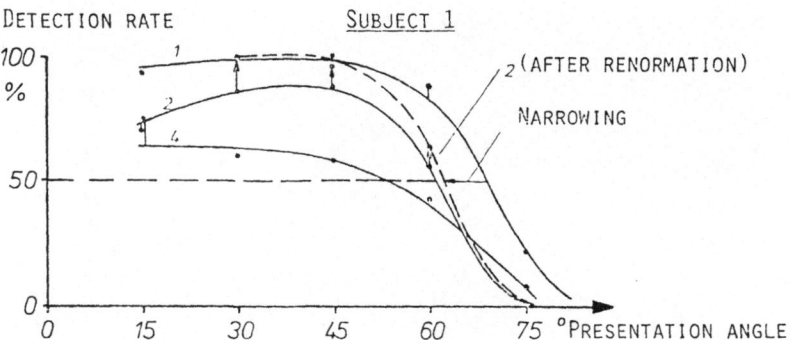

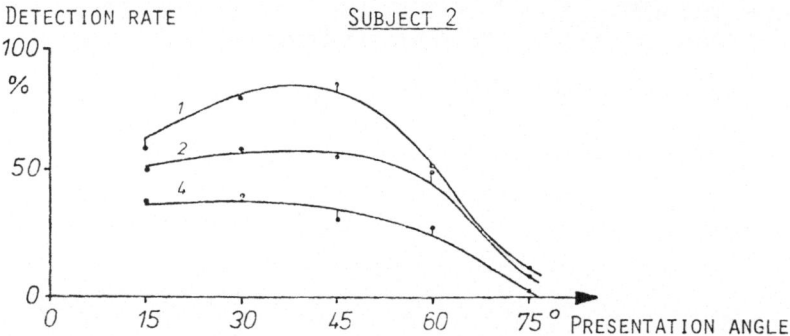

Fig. 6. Detection rates vs. presentation angle; results from 2 sub-
        jects in an experiment with different demand levels defined
        as the number of control loops to be operated (curve para-
        meter).

Detectability of Peripheral Light Stimuli as Workload Indicator: Experimental Results

Several field and laboratory experiments were carried out by
the measuring method described earlier (detection of peripheral light
stimuli, presented by a special spectacle frame, as a subsidiary task
method). As a first example, the influence of workload caused by concurrent discrete demands on the detectability was investigated. The
main task for the subjects was to respond to 8 streams of binary demands, presented by the numbers 1 to 8 on a common numeric display
(Schumacher, 1978; and Schumacher and Geiser, 1978). The arrival pattern of each stream was given by a Poisson distribution. The service
of each demand had to be done in the order of arrival and consisted

in pressing a corresponding push-button during a fixed lapse of time which was indicated by a service time lamp. The demand level of this task is defined by the ratio of the service time and the mean inter-arrival time of the demands, which varied in the range 0.8 to 1.6. Fig. 7 shows the result of an experiment carried out with 6 subjects, 4 experimental sessions each: The detection rate indicates the work-load differences caused by the different demands of the main task, significantly at the 1%-level.

In another experiment different workload indicators were mea-sured simultaneously in a driving simulator (Voss and Bouis, 1979). Due to the use of the cabin of an automobile the simulator employed offers to the subjects a realistic environment. The essential ele-ments of a driver's task course control, speed control, detection of unexpected events and perception of visual and auditory informa-tion are simulated (see also Bouis et al., 1979). In this experiment as the main task the subjects had to carry out the simulated course control task, which is realized as compensatory tracking task of a low-pass filtered white noise. Different demand levels of this task were realized by different cut-off frequencies of the low-pass, ranging from 0.2 to 0.6 Hz. The subjects had to keep a projected light symbol (vertical stripe) within a given range by the steering movements; the symbol was deflected by the noise signal. The experi-ment was carried out with 9 subjects (7 male, 2 female, mixed age and mixed driving experience), several training sessions and 5 ex-perimental sessions each. The following workload indicators were measured:

- Control error of the course control task as an example of a perfor-mance criterion: Fig. 8a shows the control error in percent, re-lated to the case of no control. There is a good sensitivity of this indicator for the different demand levels of the course con-

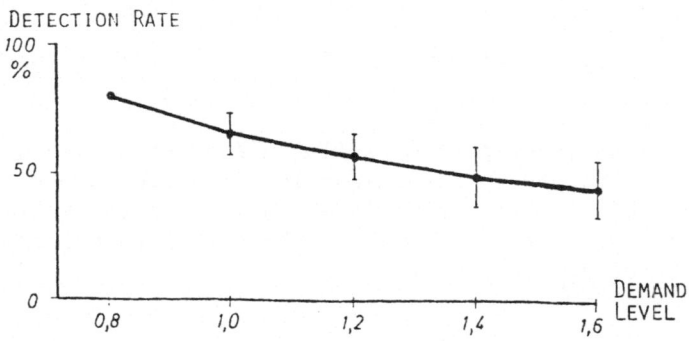

Fig. 7. Detection rate of peripheral presented stimuli vs. demand level defined as ratio of the service time and the mean in-terarrival time of concurrent discrete demands (6 subjects, 4 experimental sessions each).

CONTROL ERROR [%]

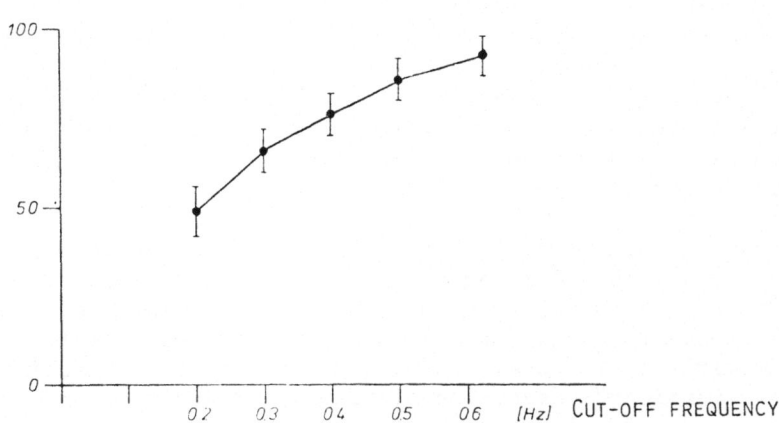

a. Workload indicator: Control error in percent related to the case
   of no control.

DETECTION RATE [%]

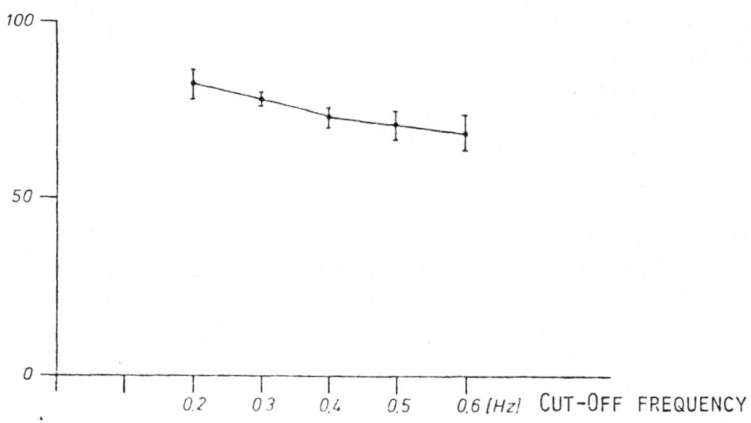

b. Workload indicator: Detection rate of light stimuli peripheral
   presented.

Fig. 8. Results of an experiment measuring different workload indi-
         cators simultaneously vs. demand level defined as cut-off
         frequency of the noise signal, which must be controlled in
         a compensatory tracking task (9 subjects, 5 experimental
         sessions each).

trol task (but in another experiment with the tracking task described earlier the indicator "control error" showed no sensitivity
for the low levels of demand), significant from 0.2 to 0.3 Hz at
the 0.1%-level, from 0.3 to 0.4 Hz and from 0.4 to 0.5 Hz at 1%,
from 0.5 to 0.6 Hz at 5%.

- Detection rate of peripheral light stimuli presented by the spectacle frame developed as an example of a subsidiary task method:
  Fig. 8b shows also a good sensitivity of this indicator, especially for the low levels of demand (significance: from 0.2 to 0.3 Hz
  and from 0.3 to 0.4 Hz at the 0.1%-level, from 0.4 to 0.5 Hz and
  from 0.5 to 0.6 Hz at 6%).

- Some physiological indicators derived from the electrocardiogram
  (pulse frequency and two different measures of arrhythmia (Mulder
  and Mulder-Hajonides van der Meulen, 1973)): There was no systematic variation of these indicators with increasing demand level.
  This may be caused by the lack of emotional workload: In a simulator, no real danger exists.

- Subjective judgements: Different ad-hoc scales, with five categories each, were used to test subjective rating of the demand levels
  given. Some of these were discriminating, others were not. This
  shows that with a validated set of category scales subjective
  judgements are an useful indicator.

Finally, as an example of temporal resolution of the indicator
"detectability of peripheral light stimuli" a field experiment is
described (Voss and Bouis, 1979). The main task was to drive a car
in real traffic along a defined course in Karlsruhe, F.R. of Germany.
A subject performed this 35 times; in addition, he had to detect peripheral light stimuli presented by the spectacle frame. On 3 cross-
ings (without light signals, the subjects having no priority) the
detection rates were measured in 4 successive local areas: approach,
stop, cross and drive away. Often only a few seconds were needed to
stop and cross. Fig. 9 shows the results: In relation to the areas
of approach and drive away the detection rates decreased very strongly in the stop and cross areas. This indicates that driver's mean
ability to detect peripheral events was very poor in this situation.

CONCLUSIONS

The decrease of man's peripheral visual performance caused by
different influences is sometimes reported as tunnel vision. In order to investigate this the range of the functional visual field and
its narrowing have to be defined by a measuring procedure. If it is
done as described above, it must be concluded that there exists a
combined effect: As a preliminary result, tunnel vision as well as
a general decrease of peripheral visual performance occur at the
same time with individual differences, but without a concentration
on one of these. As a practical consequence, the influence of a
general decrease leads in the mean to a lower detectability of e-
vents, which occur in the near periphery of man's central visual
field, as if tunnel vision would be the only influence.

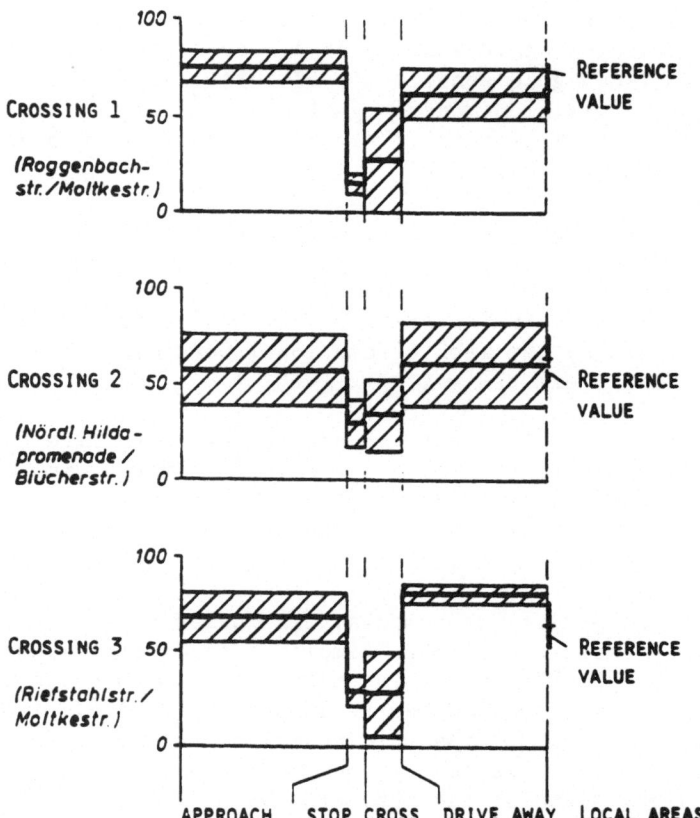

Fig. 9. Detection rates of light stimuli peripheral presented on 3
        crossings in a field experiment (mean value and standard de-
        viation in each local area).

        From this result the narrowing of the functional visual field
(e.g. given as a numerical value of the aperture angle reduction)
cannot be used as derived indicator of mental workload based upon a
model of tunnel vision. On the other hand, the detectability of peripheral light stimuli (as, for example, presented by a spectacle
frame) can be applied to workload measuring in a subsidiary task
method. Numerous field and laboratory experiments were carried out
by this method; some of them are presented in this paper. They all
showed little impairment of the subjects as well as a good discrimination of different demand levels in very different main task situations. The application of this method is limited to situations with
constant background illuminance; in other cases the contrast of the

peripheral lights related to the background had to be adjusted (e.g. automatically). With regard to a temporal resolution this indicator is very useful (by averaging repeated measurement intervals of short duration). The indicator "detectability of peripheral light stimuli" is suitable especially in cases where the co-operation of foveal and peripheral visual performance is investigated. Apart from the problematic term "mental workload" it leads to practical consequences for visual information design in a man-machine system.

This research was supported by the Forschungsvereinigung Automobiltechnik e.V. (FAT) and the Bundesanstalt für Straßenwesen (BASt). The results of this paper are part of a dissertation by the author.

REFERENCES

Bouis, D., Voss, M., Geiser, G. and Haller, R., 1979, Visual vs. Auditory Displays for Different Tasks of a Car Driver, Proceedings of the Human Factors Society, 23rd Annual Meeting, Boston, 35-39.

Gasson, A.P. and Peters, G.S., 1965, The Effect of Concentration upon the Apparent Size of the Visual Field in Binocular Vision, Optician, Part I, 148, 660-665; Part II, 149, 5-12

Geiser, G. and Schumacher, W., 1976, Parallel vs. Serial Instrumentation for Multivariable Manual Control in Control Rooms, in: "Monitoring behavior and supervisory control", T.B. Sheridan and G. Johannsen (eds.), Plenum Publishing Corp., New York.

Jahns, D.W., 1973, A Concept of Operator Workload in Manual Vehicle Operations, Forschungsbericht Nr. 14, Forschungsinstitut für Anthropotechnik, Wachtberg, West-Germany.

Küting, H.J., 1976, Belastung und Beanspruchung des Kraftfahrers, Forschungsbericht der Bundesanstalt für Straßenwesen, Bereich Unfallforschung, Köln.

Mackworth, N.H., 1965, Visual Noise Causes Tunnel Vision, Psychonomic Science, 3, 67-68.

Moray, N. (ed.), 1979, "Mental Workload, Its Theory and Measurement", Plenum Publishing Corp., New York.

Moskowitz, H. and Sharma, S., 1974, Effects of Alcohol on Peripheral Vision as a Function of Attention, Human Factors, 16, 174-180.

Mulder, G. and Mulder-Hajonides van der Meulen, W.R.E.H., 1973, Mental Load and the Measurement of Heart Rate Variability, Ergonomics, 16, 69-83.

Schumacher, W., 1978, Human Operator Strategies in Dispatching Concurrent Demands in Man-Machine Systems, IITB-Mitteilungen, Fraunhofer-Institut für Informations- und Datenverarbeitung, Karlsruhe, West-Germany.

Schumacher, W. and Geiser, G., 1978, Petri Nets as a Modeling Tool for Discrete Concurrent Tasks, Proceedings of the 14th Annual Conference on Manual Control, Los Angeles, 161-175.

Sheridan, Th.B., 1979, Mental Workload in Decision and Control, Proceedings of 18th IEEE Conference on Decision and Control, Fort

Lauderdale, Florida, 977-982.

Voss, M. and Bouis, D., 1979, Der Mensch als Fahrzeugführer: Bewertungskriterien der Informationsbelastung, visuelle und auditive Informationsübertragung im Vergleich, FAT-Schriftenreihe Nr. 12, Frankfurt.

Weltman, G. and Egstrom, G.H., 1966, Perceptual Narrowing in Novice Drivers, Human Factors, 8, 499-506.

EYE MOVEMENT MEASUREMENT IN THE ASSESSMENT AND TRAINING OF VISUAL PERFORMANCE

Heino Widdel and Jürgen Kaster

INTRODUCTION

By analysing visual search behavior knowledge may be guided to improve the performance of search tasks through human engineering design and training. Two ideas have suggested the need for further analysis of visual search behavior. The first is the suggestion of Krendel and Wodinsky (1960) who supposed that visual search strategies could be inferred from the underlying distributions of search times. They postulated that an exponential distribution is indicative of a random strategy where successive fixations are independent of each other and therefore will overlap each other. A linear distribution, on the other hand, indicates a systematic search strategy without overlapping of fixations. This strategy is called an exhaustive search, which may reduce the mean search time in a trial considerably (Howarth and Bloomfield, 1968). The other idea is related to the size of the useful field of view or the visual lobe area. It is defined as the peripheral area around the central fixation point from which specific information can be extracted. The size of the useful visual field depends on the context of a target; e.g. it is reduced when the target is embedded in a complex background or surrounded by irregularly positioned nontarget items (Brown and Monk, 1975).

The attraction of these considerations is that subjects working in the field of visual inspection perhaps can be trained in an exhaustive visual search strategy to reduce visual search time. Also a search task might be modified in such a way that a moving aid scan for exhaustive search is integrated into the search field. Similar problems have been investigated by Lovie and Lovie (1968) who subdivided the display into smaller areas and Townsend and Fry (1960) who used an automatic scanning device. A systematic strategy in this sense could only be successful if the useful field of view for a target is known before subdividing a visual display. But this is mostly not the case in practical situations. Moreover, the exact characteristics of a target including its shape, size, colour, texture, luminance, and the background characteristics embedding the target are not known before the detection of the target. One also has to take into account that a search strategy is not reliable.

PARAMETERS OF EYE MOVEMENTS

From the point of visual search the most relevant eye movements are the saccadic movements and the most relevant characteristics in general for information processing are the fixations. A value of three fixations per second seems to be general for visual search with a minimum near 200 ms for a fixation. Long fixations of more than 500 ms are often found when subjects are confirming the presence of a target (Bloomfield and Modrick, 1976) or may be sometimes artefacts based on the inability to distinguish small saccades of less than 1 deg. The accuracy with which one can assess a fixation and moreover the number of fixations depends on the method used to record the eye movements and on the operational definition of a fixation. The distances between the fixations may reflect the size of the useful field of view (Snyder and Taylor, 1976; Megaw and Richardson, 1979b).

The most relevant eye movements from the point of applied inspection and search tasks of longer search times are the saccadic movements. These are the movements between fixation sequencies to bring different parts of stimulus material on the fovea for detailed analysis. They are executed rapidly and mostly range in amplitude between 5 deg and 40 deg. During a saccade the image becomes very blurred and the amount of detail that can be resolved is greatly reduced. The two stage programming procedure (Hou, 1978) of a saccade needs 200 ms, which represents the average time lapsing between the end of one saccade and the beginning of the following one, usually referred to as the minimum fixation time. Nevertheless saccades exist with smaller amplitudes correcting an over- or undershoot of the primary saccade. Microsaccades last around 20 ms with an amplitude lower than 1 deg and their detection in inspecting tasks is as difficult as the detection of drift-movements and tremor-movements with high frequencies. Eye movements of another type are pursuit movements which have the function of stabilizing the retinal image by matching the velocities of the eye and a moving stimulus. In the scope of search problems their importance depends on moving stimulus material.

STUDIES OF EYE MOVEMENTS

There are only a relatively few eye movement studies related to

practical tasks and jobs in industry, the military or other areas.
One can expect an increase in the rate of publication of practical
studies with recent technical developments. One of the earliest studies of Gerathewohl (1952) dealt with radar operators. He found that
eye movements followed the sweepline of a radar display with a series of saccadic movements and with a general absence of constant
velocity pursuit movements. Similar data, that the eyes generally
tend to follow a rotating sweepline by means of a series of closely
spaced fixations are reported by Wallis and Samuel (1961) and by
White and Ford (1960). Thackray and Touchstone (1980) analysed eye
movements of simulated radar observation with and without a sweepline in an air traffic control situation. They failed to point out
an influence of the sweepline on search performance and a relation
between mean fixation duration and mean detection latency as reported
in other studies. Enoch (1959) found that subjects looking for specific details in aerial maps had most fixations on the center of the
display. Moreover with increasing display size the interfixation distances increased and the fixation times decreased. Stager and Angus
(1978) studied the air-to-ground visual search situation with detection of crash sites. Experienced subjects showed superior performance
but this could not be concluded from eye movement data as, e.g. the
fixation times, coverage or scan-path. Other scopes with practical
importance of eye movements are related to vehicle driving, pilot activities and boat operating (e.g. Miller, 1973; Cohen and Studach,
1977; Gatchell, 1977; Miller et al., 1977).

Eye movement problems are of high importance for examining x-
rays, where Tuddenham (1962) found that up to 30% of certain abnor-
malities were missed by experienced radiologists. A study of Llewellyn-
Thomas and Lansdown (1963) showed that large areas of the x-rays are
unexplored while fixations on edges or borders are favoured. Kundel
and Wright (1969) demonstrated that search strategies are affected
by instructions or feedforward information available to the examiners.
The results indicate that as the tasks become more general or the
stimulus material more ambiguous the complexity of the search pattern increases.

Another research field of eye movements is industrial inspection. Schoonard et al. (1973) analysed the eye movements of experienced inspectors examining slides of integrated circuit chips. The
edges of the chips were neglected and fixations were most frequent
in those areas where stimulus features are most complicated. The distributions of the fixation times were positively skewed and had a
mean value of 275 ms. The number of fixations determined the differences in inspection times between inspectors rather than differences
in fixation times. Moraal (1975) recorded the eye movements of experienced inspectors examining steel sheets moving from left to
right. He found two different search strategies and in agreement with
Ohtani (1969) that mean fixation time was 300 ms and relatively independent from the speed of the examined object. A reason may be the
homogeneous surface of steel sheet. Other studies reported the inspec-

tion of empty bottles (Saito, 1972), of tin plates, woven fabric and electrical edge connectors (Megaw and Richardson, 1979a).

PROBLEM

The purpose of the following investigation is the analysis of eye movements during a series of search tasks in a laboratory situation. The stimulus material presented was meaningless and structure-free in order to eliminate influences of significant parts of a picture or an object as used in real world tasks and the influence of previous experiences with certain types of search material. The targets presented for detection in the search trials are unknown and changed in each trial to realise a similarity with such types of practical search tasks, where faults of unknown pattern appear.

The first question is, if a search training with a forced exhaustive eye movement strategy leads to lower search times than a training without a forced search pattern. The second and third questions are, if fixation duration and fixation frequency are influenced by the mode of training and if they discriminate between subjects of high and low search performance. The fourth question is, if the visual lobe area has a relevant contribution to determine search time. The last question is, if a systematic exhaustive eye movement pattern can be learned and produced after training phases without a visual aid. In all cases the influence of the difficulty of the search task is examined. Additionally methods of measuring eye movements and of estimating the visual lobe area other than that usually used in the reported studies were evaluated.

APPARATUS

The used eye movement recording system of the Honeywell oculometer was developed by Merchant (1969) and its function based on the principles of the pupil-cornea reflection method, also called the point of regard measurement as described in detail in Young and Sheena (1975). The basic operating principle of the oculometer required illuminating the subject's eye with infrared radiation from a single light source. The radiation reflected from the eye is monitored with an infrared sensitive television camera. The optics of the oculometer include necessary lens, mirror and beamsplitters to transmit the infrared radiation from the source to the subject's eye and to return the optical signal to the electro-optical tracker for conversion to an electrical signal. This electric signal generated in the circuitry of the electro-optical tracker is processed by a mini-computer and converted to the desired output format as shown in Fig. 1. The stimulus material was generated by a DEC PDP 11 computer and presented by a graphical display system controlled by a microprocessor (Haeusing et al., 1978). For all task programming, manage-

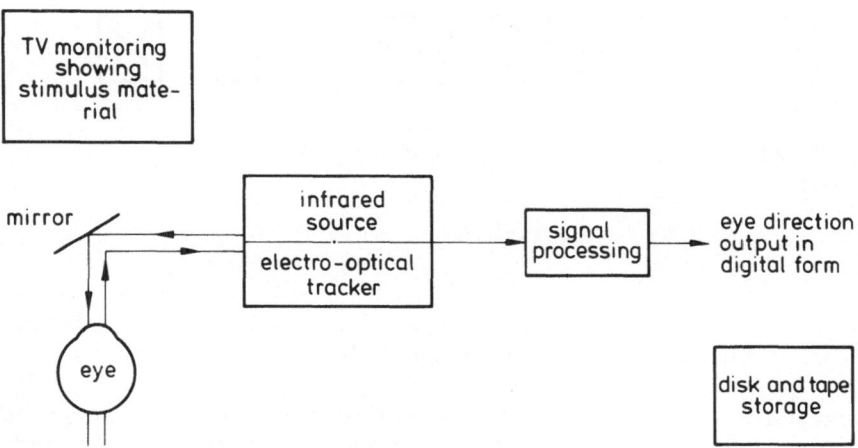

Fig. 1. Schematic illustration of experimental arrangement.

ment of the experimental sequences, and response recording a PDP 11 was used.

EXPERIMENT

Subjects

Six male and two female subjects participated voluntarily in the experimental investigation. All subjects were coworkers of the research institute ranging in age from 25-37 years. They had normal vision acuity without glasses. Subjects who need eye correction had to be excluded because of technical requirements of the oculometer system. Each subject completed all experimental conditions for his group.

Stimulus Material

The visual search material consisted of a composition of a lot of neutral symbols which were spread randomly over the screen representing a stimulus field with an extent of 27 x 22 cm or 30 deg in horizontal and 25 deg in vertical dimension. The symbols were of green colour being on a homogeneous green background and had a size of around 0.5 deg. The small contrast of the luminance of the brighter symbols and the darker background realised the desired high difficulty of the conditions of visual perception. The stimulus material consisted of a field of identical symbols with the exception of one symbol, the target, which had to be detected by the subjects. It differs from the norm-symbols in such a way, that the discrimination is made rather difficult because of its high structural similarity with the nontarget symbols. It has the same size, colour and luminance as all other symbols but differs only in the internal graphical

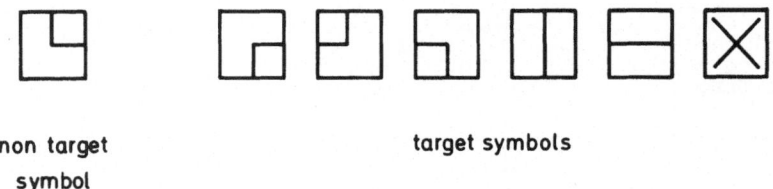

non target   target symbols
symbol

Fig. 2. Stimulus material.

design (Fig. 2). The target was positioned randomly in each presentation of the stimulus field as explained below.

A systematic variation of the difficulty of visual perception
was obtained with three different conditions of density of the symbols. The highest density had the stimulus field with 576 symbols,
the medium density with 461 symbols and the lowest density with 352
symbols. Under one experimental condition the stimulus material presented a marker or a scan which was moving in a systematic exhaustive
pathway over the stimulus field (Fig. 3).

Experimental Procedure

The eight subjects were randomly assigned to two experimental
groups, so each group consisted of four subjects. They had to take
part in nine sessions with fifteen presentations of a stimulus field
each. The time distance between the eight sessions was around one
day and one session lasted around half an hour. The fifteen presentations of one session were comprised of five stimulus fields with

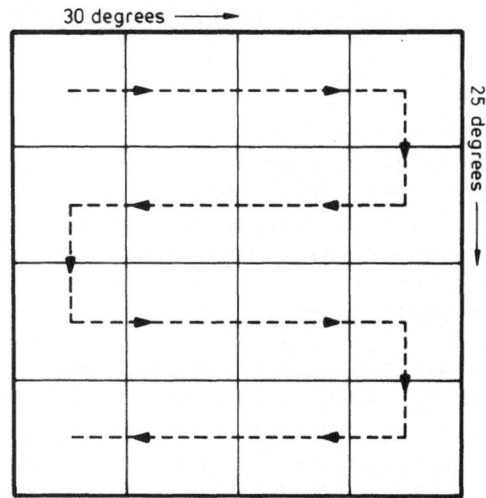

Fig. 3. Schematic pattern of scan moving over display.

three densities each in a random sequence, but this sequence was in-
variant in the comparable experimental parts of the two groups. The
same strategy was used in positioning the targets in a stimulus
field. The field is assumed to be subdivided with an implicit grid
of sixteen squares (Fig. 3), in which the targets were randomly generated, and the random sequence order of the squares selected for
target positioning was fixed in comparable parts of the two groups.
After each presentation the subjects had to fixate successively three
points in the scope of the target with distances of 5 deg., in order
to get the fixation coordinates of the oculometer data for the following linearization procedure.

In the first session of the investigation subjects were instructed what to do and had an opportunity to become familiar and
gain practical experience with the experimental procedure. The second session was a pretraining experimental session, called pretest-
session, to get data of search performances and eye movements before
training. In the next five sessions both groups received training.
For one group the eye movement task was to follow a scan which was
systematically moving over the stimulus field as shown in Fig. 3,
while the other group had a free individual search condition. The
eighth session was a post-training experimental session, called
posttest-session, to get after-training data. In the ninth session,
called transfertest-session, new stimuli were presented to determine
whether learning effects could be transferred to new stimuli situations (Fig. 4). The collected data were search time and eye movement
parameters. The search time was fixed by the subjects pressing a
button which stopped the time and the procedure of collecting eye
movement data.

RESULTS

Search Time

The time of presenting the stimulus material was limited corresponding with the density level. The limitation of time for presenting the stimuli was determined by the speed of the scan marker
moving over the picture during the training sessions of the experi-

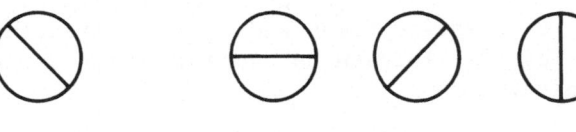

non target   target symbols
symbol

Fig. 4. Stimulus material of transfer task.

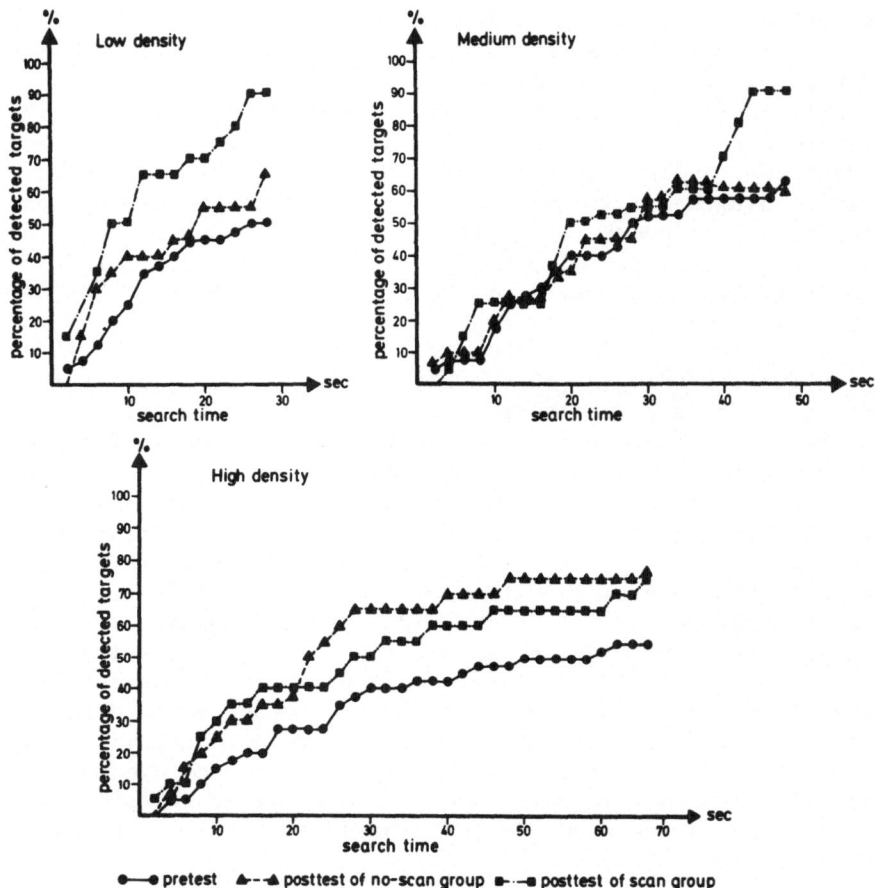

Fig. 5. Cumulated percentage of detected targets with regard to
        search time at three levels of density.

mental group. The stimuli have been shown for 30, 50 and 70 sec for
the low, medium and high densities in both groups. Especially at the
pretraining session the targets often had not been detected in the
given time limit. Therefore a direct comparison between the densities
had to be restricted up to 30 sec and 60 sec, respectively. An evaluation of the search time in this scope has to be realised in estimating the differences of the polygons illustrated in Fig. 5. The subsequent investigations of fixation times and fixation frequencies are
analyzed with the U-Test for independent data and with the Wilcoxon-
Test for dependent data.

    No difference in search ability exists between the scan and no-
scan groups at the pretest phase. This result indicates that the forming of random groups was successful and the training procedure began
with approximately the same level of search performance. Comparison

of pretest and posttest results would show any learning effect. The
scan group shows a clear learning effect for the stimulus material
with low density and a moderate effect for medium density. In the
high density situation there exists a moderate training effect for
both learning groups. These results are valid in a similar way for
the transfertest sessions. An expected relationship between search
time and level of difficulty of the perceptual conditions could be
verified in the posttest session, where the higher densities correspond with longer search times. In the other sessions a difference
of search time exists only between low density and medium and high
densities but not between the latter two.

Fixation Duration

There are no significant differences between the fixation times
in the pretest- and posttest-situations for the scan-learners regardless of the density of the stimulus pictures. That means in general
there is no systematic influence of the learning procedure on fixation duration.

The analysis of search performance showed that a group of three
subjects with a high performance level could be clearly separated
from a group of five subjects with a lower performance level. A comparison of fixation durations for these two subject groups and for
the three stimuli densities are shown in Fig. 6. The illustration represents the results of the posttest-situation. In all cases as well
as in the transfertest-situation not illustrated the subjects with
high search performance have a shorter fixation duration than the
subjects with a lower search performance. These differences are, with
the exception of the posttest-situation combined with low density,
statistically significant (p<.01).

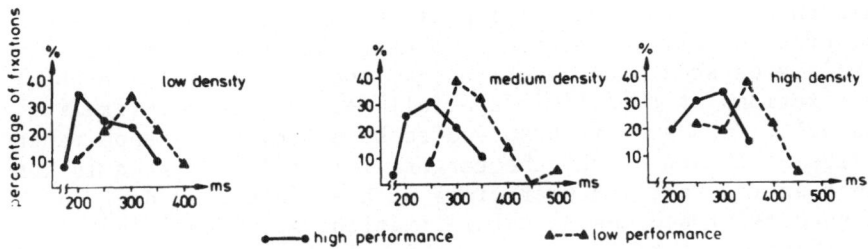

Fig. 6. Distributions of fixation duration in posttest-situation.

Number of fixations

There is a clear influence of the learning activities on the
fixation frequencies in that the number of fixations increases when
the training sessions have been ended (p<.01). These consequences of

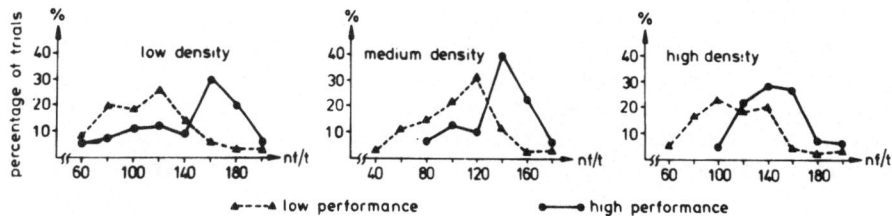

Fig. 7. Distributions of number of fixations per unit time (n/t) in posttest-situation.

the learning procedure are independent of the mode of training because the subjects with and without scan learning have an increasing fixation frequency; no difference was found in the posttest- and transfertest-situations between the two learning groups. The subject groups with high and low search performances, as described above, were compared for fixation frequency in the posttest- and the transfertest-situations separated for the three stimuli density levels. Fig. 7 shows generally higher fixation frequencies for superior searchers than for lower performance searchers (p<.01) and this result is identical with the transfertest-situation not illustrated.

Visual Lobe Area

Usually the visual lobe area or the useful field of view are defined by the value of the interfixational distance. Use of this value in the present study indicated that no differences occurred between subjects of low and high search performance, between different stages of the training procedure, between modes of training, or between different levels of stimuli density with regard to the size of interfixational distances. Another definition of the visual lobe area takes into consideration only the search area near the target. The outer limit of this area should be determined by the fact that the probability that fixations at this limit are followed by a target-fixation approaches null. This area is supposed to have a circular appearance with a radius of 10 deg around the target. This critical area is subdivided into equidistant circular rings with a width of 2 deg. In this area successful and unsuccessful fixations are identified. A successful fixation is that fixation, which is followed immediately by a fixation of the target. When no target fixation followed it is considered to be an unsuccessful fixation. The philosophy of this definition is that during a successful fixation the target is detected in the peripheral area, determining a parameter of the visual lobe area, so that the following saccade reaches the target for fixating.

The distributions of frequencies of unsuccessful fixations in the different distances of 1 deg to 8 deg from the target are shown in Fig. 8. They are separated for different densities of the stimulus material and point out the differences between the two subject

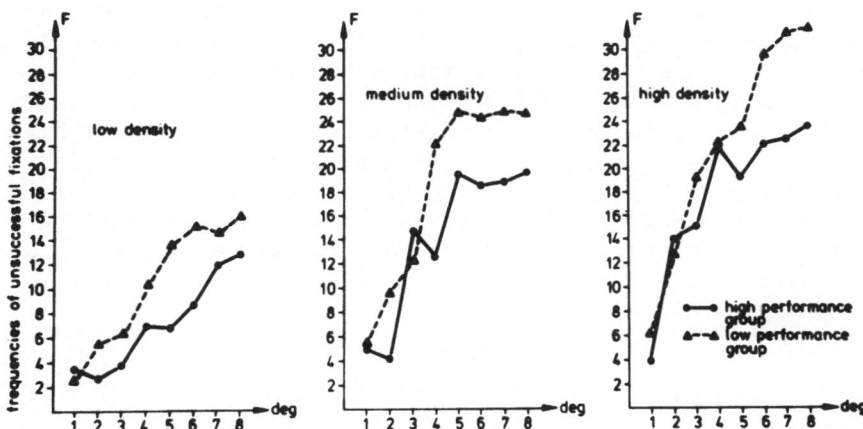

Fig. 8. Distributions of unsuccessful fixations (F) of subjects with
        high and low search performance with regard to distance (deg)
        to target.

groups of high and low search performance. In the case of low density the high performance group has a lower rate of unsuccessful fixations over the whole distance than the lower performance group
(p<.01). In the cases of medium and high density this effect appears
in a larger distance of around 4 to 5 deg from the target (p<.05).

Analysing the visual lobe area in a further step the number of
successful fixations in the different areas were counted and the frequency distributions were illustrated in Fig. 9. The total amount of
successful fixations is relatively small, the maximum number of successful fixations in a trial is only one, and statistical analysis
failed to point out a significant difference between the two subject
groups. Nevertheless, Fig. 9 shows in two conditions of density a
tendency that subjects with lower search times have a higher rate of

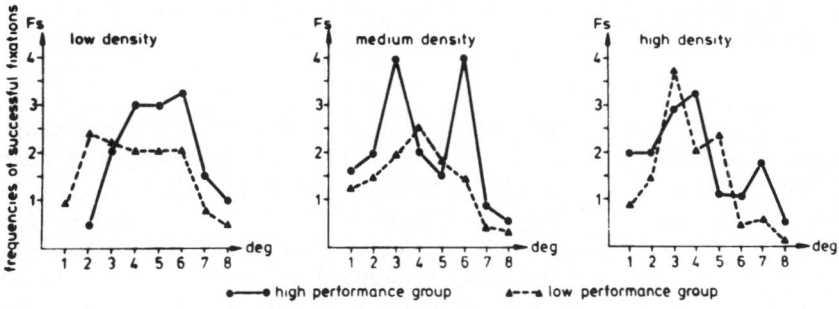

Fig. 9. Distributions of successful fixations (Fs) of subjects with
        high and low search performance with regard to distance (deg)
        to target.

successful fixations in larger distances from the target than subjects with longer search times. To illustrate the analytic procedure
more clearly a two-dimensional distribution of the positions of the
successful fixations of all subjects in the display plane is presented in Fig. 10 corresponding to the distribution in Fig. 9 for high
density. One can see an accumulation of fixations in the circular
ring of 2-5 deg around the target. The general accumulation of fixations in the region above the target is a consequence of the eye
movement strategy so that information about the shape of the visual
lobe area could not be inferred from this illustration.

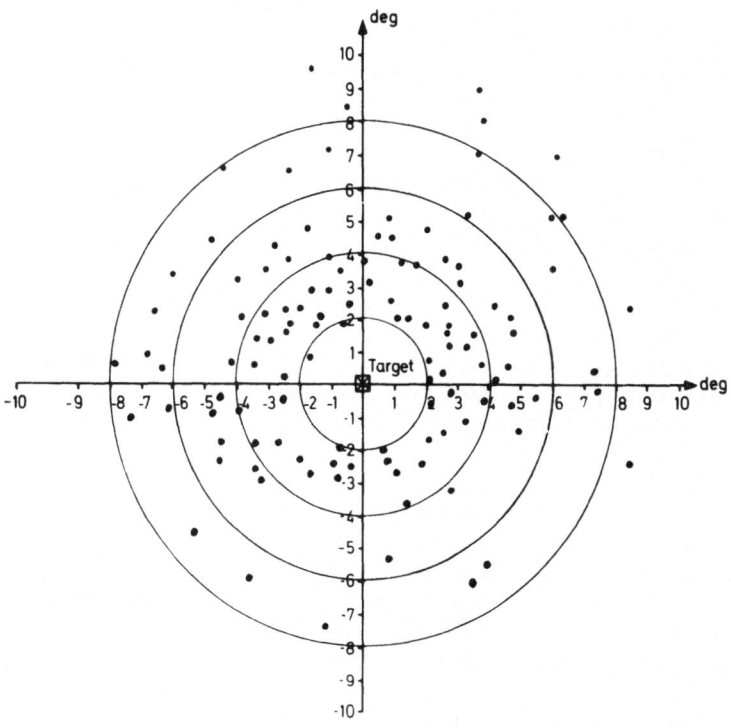

Fig. 10. 2-dimensional distribution of successful fixations at high
         density condition.

Finally, an index of the probability of success in detecting a
target during a fixation in the critical area is defined as a quotient of the number of successful fixations divided through the total number of fixations. The distributions of these indices (Fig. 11)
show a clear difference for the low and medium density conditions between the high and low performance groups. The subjects with lower
search times have a higher probability of detecting the target especially at larger distances from the target than subjects with longer search times.

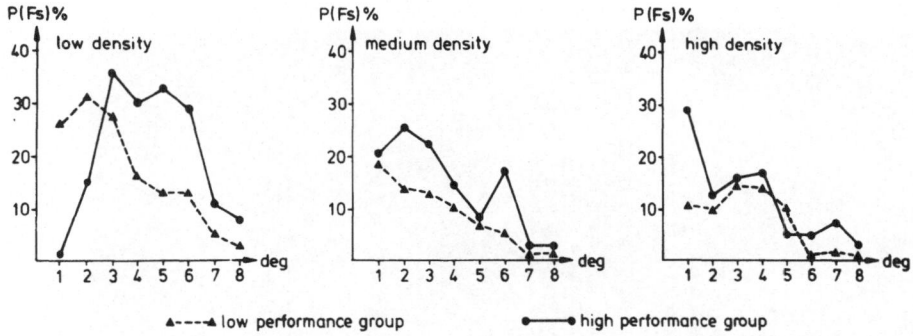

Fig. 11. Probability of successful fixations (P(Fs)) of subjects with
         high and low performance with regard to distance (deg) to
         target.

        The three modes of estimating parameters of the visual lobe area,
the frequencies of unsuccessful fixations, the frequencies of successful fixations and the probability of success, are not independent from
each other and certianly include some redundant information. They
point out a more efficient visual lobe area for subjects with higher
search performance than for subjects with lower search performance,
but no difference of its size.

        The dependence of the size and efficiency of visual lobe area
from density of the stimulus material is shown in Fig. 12. The cumu-

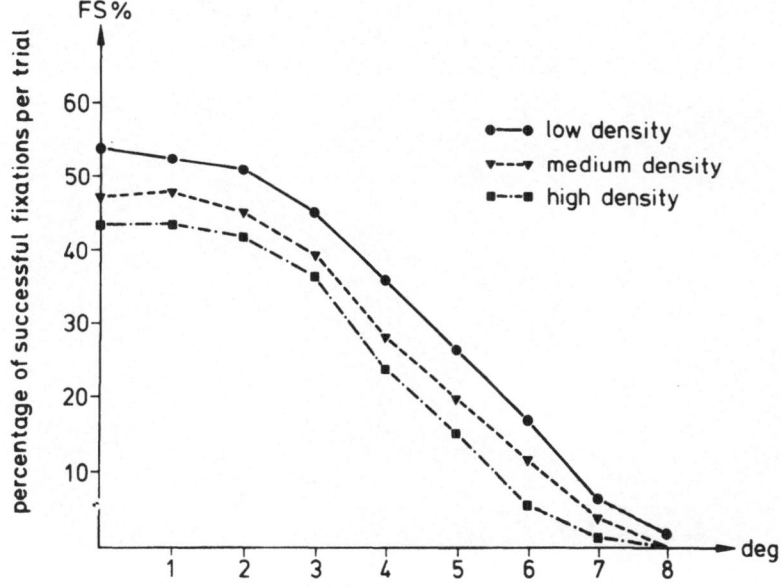

Fig. 12. Probability of successful fixations with regard to density.

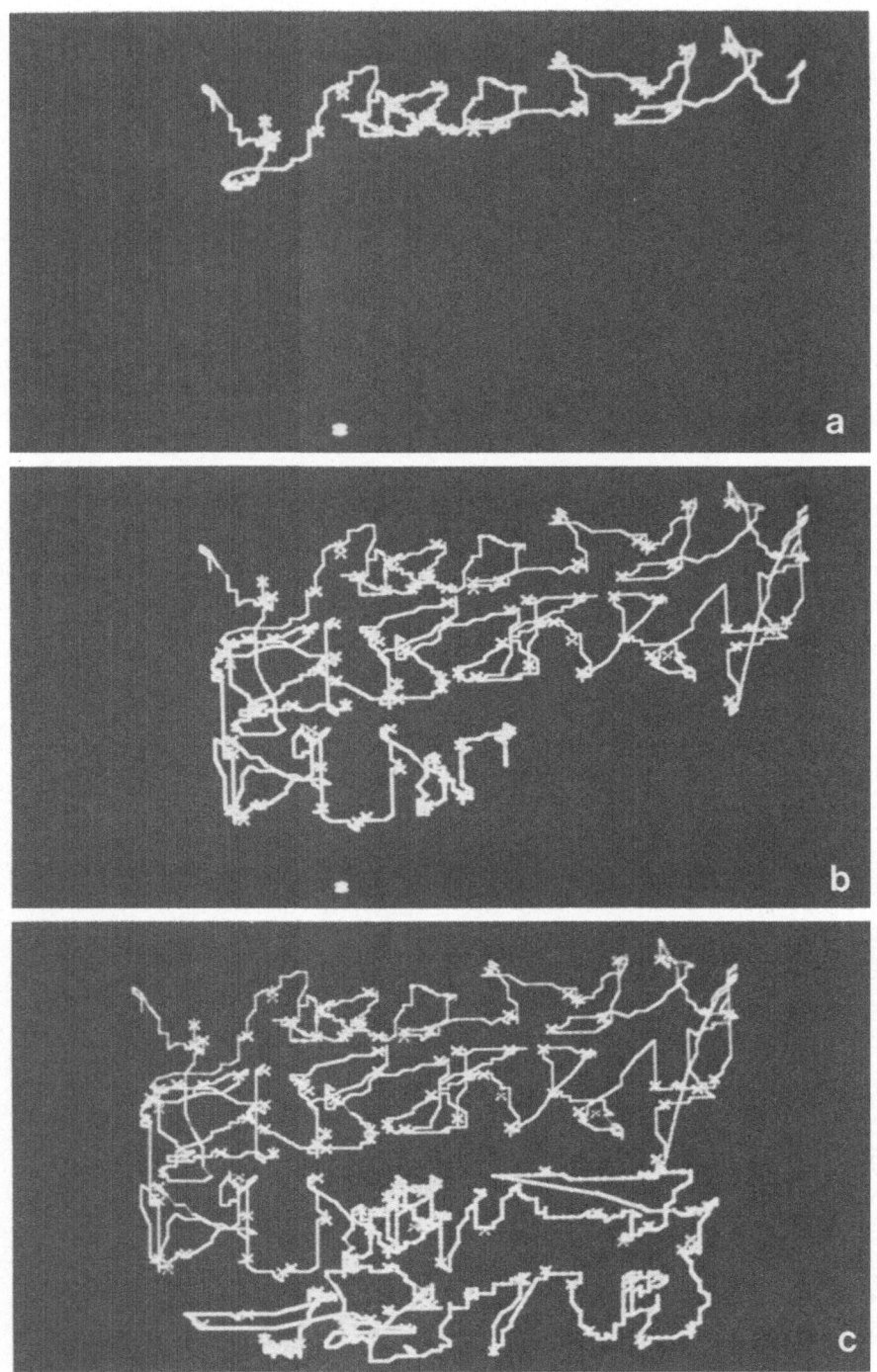

Figs. 13a-13c. Replayed pattern of eye movement learned with scan
marker in three stages of development.

lative polygons are based on the data of all sessions and represent
the distributions of the probability of successful fixations. At low
stimulus density fixations in the visual lobe area have a higher
probability to be followed by the target fixation than at medium or
high density of the stimulus material (p<.01).

Eye Movement Strategy

An important question of the present investigation concerns the
problem of learning an exhaustive search strategy and, whether appropriate training leads to superior search performance in comparison with the training effect of an individual, self-determined and
free scanning pattern. The results of search times show a superiori-
ty of subjects in that group which learned systematically with material of low and medium density and furthermore that this learning
effect is transferable to a new search task.

The analysis of relevant differences of eye movement pattern in
posttest and transfertest-situations of the two groups with different modes of learning was performed by qualitative methods. The eye
movements of each subject and trial could be replayed and a time history could be formed and shown on the videograph display which also
was used to present the stimulus material during the experimental
procedure. The eye movements then can be evaluated subjectively with
regard to an exhaustive scan pattern. Such a scan pattern resembles
the one illustrated in Figs. 13a-13c which show successive stages of
the exhaustive eye movement pattern typically of the one learned by
the scan group. In comparison to this example Fig. 14 shows an eye
movement pattern of the no-scan group which was selected as the one

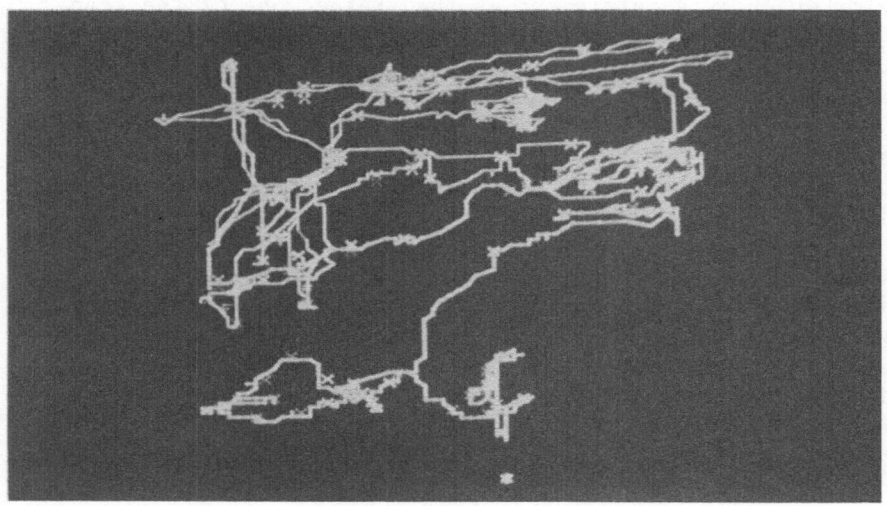

Fig. 14. Replayed pattern of eye movement learned without scan marker.

with the highest similarity to a scan group pattern. Nevertheless, it is evident that the search pattern used was also somewhat systematic, but it can be seen that the eye movement pattern missed greater parts of the stimulus picture. During further progress of the search procedure the subject has to remember the missed areas which later on must be scanned. This process of remembering and of relatively unsystematic scanning will be accompanied by errors and redundant eye movements which all lengthen the search time for detection of a target.

DISCUSSION

The results of the present investigation are for the most part easily interpreted and coincide with the results of other studies but not entirely. The density of the stimulus pictures is of determinative relevance for the search time and a linear relationship as Snyder and Taylor (1976) reported exists only in the pretraining session. After the learning sessions the difference between search time for medium and high density disappeared and one may suppose that a visual search training minimizes the influence of task difficulty especially at extreme difficulties.

Another result is that search time decreases after training for the scan group at low and medium density in relation to the no-scan group but not at material of high density. The individual no-scan training procedure did not show an improvement on search performance for the two lower densities which may lead to the supposition that individuals without a systematic training further on are searching randomly. On the other hand subjects of the no-scan group found an individual systematic search strategy for search tasks of very high difficulty which is as efficient as the trained one of the scan group. It can be argued that the training with a forced exhaustive eye movement strategy is effective for search tasks of high difficulty and this effect can be transfered on other and more difficult tasks. Of course a great number of different eye movement patterns are possible to implement which have exhaustive characteristics and it seems to be a desirable question to compare the training effects of different exhaustive patterns on search performance.

At the group level, the training sessions did not affect the mean fixation duration. This result seems to be contrary to other findings of Ford et al. (1959), White and Ford (1960) or Gould and Schoonard (1969). The consensus is that experienced inspectors produced lower fixation durations than unexperienced and naive observers, but a fair comparison is problematical because Gould and Dill (1969) pointed out a dependency of fixation duration on the task. This conclusion is valid only for search material of generally low density as Moffitt (1980) pointed out. He analysed 11 studies and found that fixation duration is influenced by experimental manipu-

lation only in the case of low display density but not in the case
of high display density.

In agreement with the present results Snyder and Taylor (1976)
did not find differences in fixation duration before and after practice with search tasks. This implies that fixation duration is not
a variable which could be easily influenced by practice. In the
three studies quoted before, a corresponding relation was found between high search performance and low fixation duration; superior
searchers were identical with experts and low searchers with un-
trained subjects. In the present investigation this finding occurred in different task situations, posttest- and transfertest-situations, where high and low performance is correlated with short and
long fixation times, respectively.

A number of previously quoted authors and others, e.g. Neisser
et al. (1963) or Weber et al. (1970) found that experienced and
trained inspectors and searchers which showed normally high search
performance require fewer fixations to detect a target than subjects
with lower search performance. The present investigation presents
results which are distinctly contrary to those studies with the exception of Krebs (1975) who also reported higher fixation rates for
experienced FLIR operators. In the present study, regardless of
whether subjects were trained with or without forced eye movement
scan markers, their fixation frequency increased. Furthermore, superior subjects have a higher fixation rate than subjects of low performance.

The lower fixation rates of experienced searchers, as reported
usually in literature, correspond with larger interfixational distances from which it was concluded that the visual lobe area was
larger for experienced than for unexperienced searchers. The present
investigation shows that interfixational distances, which have values around 6-13 deg and seem to be similar in value to those already
reported by Ford et al. (1959), do not differ in a systematic way.
The novel method of defining the visual lobe area described earlier
restricts the area to the target region and leads to the result that
subjects of high search performance have not larger, but more efficient visual lobe areas than subjects of low performance. Additionally the dependency of visual lobe area on nontarget density was
found too. The experiments do not permit conclusions about the shape
of the visual lobe, which seems to be wider than it is high as Enoch
(1959) already pointed out because the distribution of fixations in
the target area is intensively influenced by the eye movement strategy.

The new method of estimating the visual lobe area gives facilities to collect data not only with regard to its size but also to
the density of probability for successful fixations in this area.
The present results lead to the conclusion that the latter parameter

is of higher relevance to target detection than the size, which seems
to be equal for subjects with low and high search performance in the
present study. In terms of the new definition of visual lobe area it
represents characteristics of attentional factors. Further investigations of search behaviour should take these considerations into account. It is projected to evaluate this method as measurement corresponding to mental workload and to use it for analysis of the con-
spicuousness of colors in order to develop strategies for color coding procedures.

REFERENCES

Bloomfield, J.R. and Moderick, J.A., 1976, Cognitive processes in vi-
    sual search, Proceedings of the 6th International Ergonomics As-
    sociation Conference, Maryland, 204-209.
Brown, B. and Monk, Th., 1975, The effect of local target surround
    and whole background constraint on visual search times, Human
    Factors, 27, 81-88.
Cohen, A.S. and Studach, H., 1977, Eye movements while driving cars
    around curves, Perceptual and Motor Skills, 44, 683-689.
Enoch, J.M., 1959, Effect of the size of a complex display upon vi-
    sual search, Journal of Optical Society of America, 49, 280-286.
Ford, A., White, C.T. and Lichtenstein, M., 1959, Analysis of eye
    movement during free search, Journal of Optical Society of
    America, 49, 287-292.
Furst, C.J., 1971, Automatizing of visual attention, Perception and
    Psychophysics, 10, 65-70.
Gatchell, S.M., 1977, Power boat operators' visual behavior patterns,
    Proceedings of the 21st Annual Meeting of the Human Factors So-
    ciety, Los Angeles, 179-183.
Gerathewohl, S.J., 1952, Eye movement during radar operations, Avia-
    tion Medicine, 23, 597-607.
Gould, J.D. and Dill, A.B., 1969, Eye movement parameters and pattern
    discrimination, Perception and Psychophysics, 6, 311-320.
Gould, J.D. and Schoonard, J.W., 1969, Eye movement during visual in-
    spection of integrated circuit chips, IBM Report.
Haeusing, M., Holzhausen, K.P. and Wolf, P., 1978, Einsatz eines Mi-
    kroprozessors bei der Übersichtsdarstellung und Manipulation
    großer Datenmengen auf Farbfernsehmonitoren, Wachtberg, FAT
    Forschungsbericht No. 39.
Howarth, C.I. and Bloomfield, J.R., 1968, Towards a theory of visual
    search, AGARD Conference Proceedings, No. 41, London.
Hou, L.R., 1978, The programming strategy of the saccadic eye move-
    ment control system, doctoral thesis, California Institute of
    Technology, Pasadena.
Krebs, M.J., 1975, Scanning patterns in real-time FLIR displays, Pro-
    ceedings of the Human Factors Society 19th Annual Meeting, Dal-
    las, Tex., 418-422.
Krendel, E.W. and Wodinsky, J., 1960, Visual search in unstructured

fields, in: "Visual search techniques", A. Morris and E.P. Horne
    (eds.), National Research Council, Washington D.C., 151-169.
Kundel, H.L. and LaFollette, P.S., 1972, Visual search patterns and
    experience with radiological images, Radiology, 103, 523-528.
Kundel, H.L. and Wright, J.D., 1969, The influence of prior knowledge
    on visual search strategies during the viewing of chest radio-
    graphs, Radiology, 93, 315-320.
Llewellyn-Thomas, E. and Lansdown, E.L., 1963, Visual search patterns
    of radiologists in training, Radiology, 31, 288-292.
Lovie, A.D. and Lovie, P., 1968, The effect of a horizontally struc-
    tured field and target brightness on visual search and detection
    time, Ergonomics, 11, 359-367.
Merchant, J., 1969, Laboratory oculometer, NASA CR-1422.
Megaw, E.D. and Richardson, J., 1979a, Eye movements and industrial
    inspection, Applied Ergonomics, 10, 145-154.
Megaw, E.D. and Richardson, J., 1979b, Target uncertainty and visual
    scanning strategies, Human Factors, 21, 302-315.
Miller, J.M., 1973, Visual behavior changes for student pilots fly-
    ing instrument approaches, Proceedings of the 17th Annual Meet-
    ing of the Human Factors Society, Santa Monica, 208-214.
Miller, J.M., Gatchell, S.M. and Dykstra, D.R., 1977, The visual be-
    havior of recreational boat operators and its relationship to
    boat collisions, University of Michigan, Dept. of Industrial and
    Operations Engineering, CO-31-77.
Moffitt, K., 1980, Evaluation of the fixation duration in visual
    search, Perception and Psychophysics, 27, 370-372.
Moraal, J., 1975, The analysis of an inspection task in the steel in-
    dustry, in: "Human reliability in quality control", C.G. Drury
    and J.G. Fox (eds.), Taylor and Francis, London, 217-230.
Morris, A. and Horne, E.D. (eds.), 1960, "Visual search techniques",
    National Academy of Sciences - National Research Council,
    Washington D.C.
Neisser, V., Novick, R. and Lazar, R., 1963, Searching for ten tar-
    gets simultaneously, Perceptual and Motor Skills, 17, 955-961.
Ohtani, A., 1969, Eye movements during visual inspection task, Pro-
    ceedings of the 16th International Congress on Occupational
    Health, Tokyo, 88-90.
Saito, M., 1972, A study on bottle inspection, Journal of the Science
    of Labour, 48, 395-400.
Schoonard, J.W., Gould, J.D. and Miller, L.A., 1973, Studies of vi-
    sual inspection, Ergonomics, 16, 365-379.
Snyder, H.L. and Taylor, D.F., 1976, Computerised analysis of eye
    movements during static display visual search, Aerospace Medi-
    cal Research Laboratory, Report AMRL-TR-75-91, Ohio.
Stager, P. and Angus, R., 1978, Locating crash sites in simulated
    air-to-ground visual search, Human Factors, 20, 453-466.
Stern, J.A. and Bynum, J.A., 1970, Analysis of visual search activi-
    ty in skilled and novice helicopter pilots, Aerospace Medicine,
    41, 330-305.
Thackray, R.I. and Touchstone, R.M., 1980, Visual search performance

during simulated radar observation with and without a sweepline, Aviation, Space and Environmental Medicine, 51, 361-366.

Townsend, C.A. and Fry, G.A., 1960, Automatic scanning of aerial
photographs, in: "Visual search techniques", A. Morris and E.P.
Horne (eds.), National Research Council, Washington D.C., 194-
210.

Tuddenham, W.J., 1962, Visual search, image organisation and reader
error in Roentgen diagnosis, Radiology, 78, 694-704.

Wallis, D. and Samuel, J.A., 1961, Some experimental studies of radar operating, Ergonomics, 4, 155-168.

Weber, R.J., Patrick, C. and Perry, W., 1970, Visual search through
random walk number fields, Psychonomic Science, 18, 207-209.

White, C.T. and Ford, D., 1960, Eye movements during simulated radar
search, Journal of Optical Society of America, 50, 909-913.

Young, L.R. and Sheena, D., 1975, Survey of eye movement recording
methods, Behavior Research Methods and Instrumentation, 7, 397-
429.

PHYSIOLOGICAL MONITORING AND THE CONCEPT OF ADAPTIVE SYSTEMS[1]

Frank E. Gomer

INTRODUCTION

My intent is to briefly review current trends in the development
of manned systems that are produced by the aerospace industry. I will
describe these trends within the context of a generic tactical aircraft. Moreover, I will attempt to delineate the benefits which may
be realized if we are successful in applying new methods for monitoring human performance in near real-time. I will begin by presenting
the challenge to which my colleagues and I (see Gomer, 1980; Gomer
et al., 1979; Gomer and Youngling, 1978) have been responding[2].

We continue to see greater automation built into the design of
large, computer-based systems. In the case of tactical aircraft, direct pilot participation in those tasks which comprise so-called
housekeeping functions (i.e., navigation, systems monitoring, communications, and flight control) has been markedly reduced. This is due,

---

[1] The work described herein was supported, in part, by the Cybernetics Technology Office of the Defense Advanced Research Projects
Agency (DARPA) through contract MDA-903-78-C-0181.

[2] I have drawn heavily from the pioneering research conducted at the
University of Illinois by Professor Emanual Donchin and his associates. To large degree, their studies have made it possible for
me to focus on a rather novel means of enhancing manned system effectiveness.

in part, to the assumption that overall system performance can be dramatically improved by lessening dependency on the most unreliable component - the operator. Although one might argue that there is merit in this position, the importance of the human operator must not be underestimated if new systems are to retain the capacity for anticipating and responding to unpredictable events. Consequently, I believe that the human operator will remain an essential ingredient of evolving computer-based systems that are being incorporated onboard tactical aircraft to accomplish difficult mission functions (i.e., **detection**, location, identification, decision, execution and assessment).

It is important to note that the technological advances which have resulted in increased automation also allow us to consider the possibility of building adaptive properties into computer-based systems. That is, because of the tremendous flexibility afforded by the computer, we can envisage that the distribution of responsibilities between the operator and the hardware or software elements of a particular system might be modified as circumstances change. As I will describe in greater detail a bit later, one such redistribution might take the form of a series of computer-initiated actions to unburden or assist the pilot during periods of peak demand.

Of course, we must first define the requirements for implementing adaptive procedures. The computer must be made aware of 1) the current status (or health) of the hardware and software components of the system and 2) the current status of the operator. In order to act upon these inputs, the computer also must store information which will allow it to determine whether all components of the system (including the operator) are performing acceptably. Digital avionics programs have provided the means for real-time transmission of data to the computer about the operating characteristics of the hardware and software elements of each system. Furthermore, the computer can store detailed profiles which define normal performance limits for these elements. For many system functions then, the pilot need intervene only if alerted of a particular hardware or software problem. While the communication link from system to computer is impressive indeed, we must ask whether comparable means are available for the pilot to convey status information to the computer. As I will attempt to show in reviewing configurational changes in crew station displays and controls, "pilot status" at this time refers principally to momentary fluctuations in attentiveness or in the ability to process information and make appropriate decisions. I think it is quite apparent that an operator usually communicates with a computer via manual responses or perhaps a small vocabulary of verbal commands. In the sections that follow, I also hope to demonstrate that these forms of input are not sufficiently sensitive to transient changes in mental function and if we are to attain our goal of implementing adaptive procedures, a new communication channel must be added from the operator to the computer.

MULTIFUNCTION DISPLAYS AND CONTROLS AND THE ASSESSMENT OF MENTAL WORKLOAD

The evolution of computer-based systems brings with it an attendant concern for improvements in display/control technology. Not surprisingly, growing alarm over the increasing task demands imposed on the pilot and severe space limitations within the crew stations of advanced tactical aircraft have forced engineers to discard a design philosophy which prescribes the use of dedicated, single purpose instruments and controls (see Fig. 1). An accelerated obsolescence of this conventional approach to crew station design and a new emphasis on multipurpose displays and keyboards have been brought about by developments in two areas: digital avionics (as mentioned

Fig. 1. Dedicated instruments and controls within the cockpit of a modern fighter/bomber aircraft.

above) and programmable electronic display devices. Large scale integration (LSI) has produced the necessary microcircuit/microprocessor electronics to create an extremely flexible display generation capability. Raster-scan graphics can be used to generate realistic and varied synthetic visual scenes, and they afford an ease of mixing display symbology with the outputs of imaging sensors such as radar, low-light-level TV, and forward-looking infrared. Display presentations can be limited (in theory) to those sources of information which are most relevant at a specific time period in the mission. Moreover, the sequencing of presentations is controlled by the particular computer programs in operation and by the crew member's keyboard inputs.

Fig. 2 illustrates this transition from a preponderance of single-purpose indicators, switches, and warning devices to the incorporation of multifunction displays and keyboards. The virtues of a digital design concept are that it offers flexibility, reliability, ease of maintenance, and low-life-cycle cost. Moreover, as mentioned earlier, by automating many of the usual manual control functions, a management or decision-making role is created for the pilot, since he now is assigned greater responsibility for monitoring displayed events. Some design engineers believe that an intelligent integration of information when coupled with a logical routing of sequential switching operations will dramatically reduce pilot workload (McIver and Hatfield, 1978). Although new generation aircraft will certainly challenge the pilot's cognitive abilities to a much greater extent than his motor skills, there is a concern that mental workload problems will indeed remain significant. In many development programs, there has been a tendency to misuse the flexibility afforded by programmable display generation techniques and to continue a policy of saturating the pilot with much more information than he can possibly assimilate (Lovesey, 1977).

Having introduced the term "mental workload", I would like to offer a definition which is consistent with our goal of implementing adaptive procedures. I stated earlier that we needed to develop a communication channel which would be responsive to momentary fluctuations in attentiveness or in the ability to process information and make appropriate decisions. Although many definitions of immediate workload refer either to the demands imposed on the pilot (cf. Gerathewohl, 1976) or to the expenditure of mental effort required to meet these demands (cf. Jenny et al., 1972; Roscoe, 1978a), I favor a more integrated approach which recognizes both input variables and pilot response capabilities. Man has a finite capacity to process information and reach timely decisions for action. When confronted with multiple task demands he must allocate limited resources and function in a time-sharing mode.

The purpose for which workload is measured largely determines the method of assessment that is chosen. Again, our stated goal is

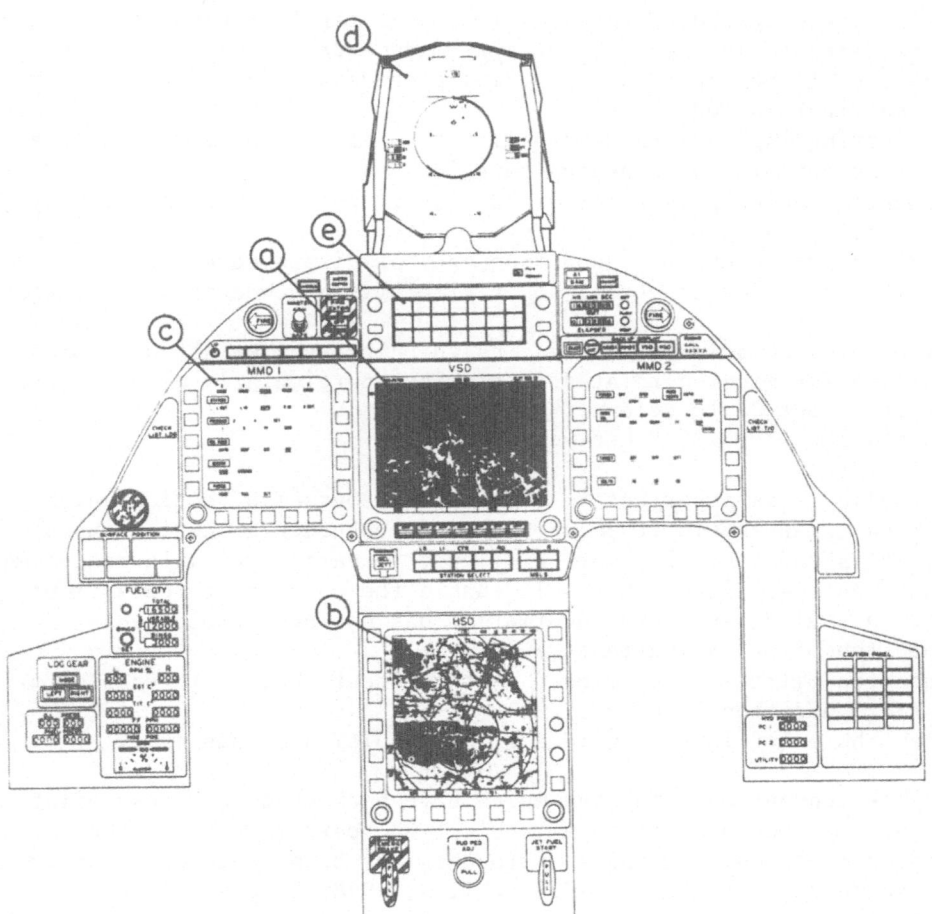

Fig. 2. Integrated crew station concept for a post-1985 tactical aircraft. Essential features are labeled and include: a) a vertical situation display which presents attitude and position
information in the vertical plane, as well as predictive symbology for all flight phases, b) a horizontal situation display which provides geographic data in the form of terrain
images or computer generated maps, as well as tactics; c)
multipurpose displays which present system malfunctions, sensor imagery, as well as pilot-selected engine parameters,
check-lists, weapon status, etc.; d) a head-up display which
presents (as a virtual image) weapon delivery and flight information, and e) a multifunction control unit which allows
the pilot to sequence events via keyboard inputs.

to more fully utilize the evolving adaptive properties of computer-based systems. As mission requirements change from moment-to-moment, so do the task demands imposed on the pilot. If the controlling computer were provided with reliable estimates of time-varying workload levels (which according to my definition refer to the mental resources required to meet particular system demands), then adaptive procedures could be implemented. During peak demand periods, these procedures might include:
- redistributing task responsibilities by effecting greater automation of certain housekeeping functions,
- reducing the complexity of or "decluttering" information displays, especially the HUD,
- modulating the dynamics and frequency of displayed events,
- cueing the pilot to attend to critical flight, weapons, and target data,
- displaying adaptive decision aids which present weighted recommendations for mission-related strategies, particularly with respect to fire control functions,
- furnishing remedial "checklists".

While it is important to outline the courses of action which might be taken should it be necessary to unburden or assist the pilot, it is substantially more difficult to define reliable real-time measurement techniques that will enable the computer to determine:
- when visual or auditory information has not been processed,
- when the pilot is inattentive,
- when the pilot is task-loaded to the extent that he is unable to perform additional duties,
- when the pilot lacks confidence in a decision he has made.

The conventional methods which have been used to assess pilot workload may be categorized as subjective, performance-related, or physiological. Each of these approaches has been reviewed recently in some detail (cf. Ellis, 1978; Chiles, 1978; Roscoe, 1978b). I will only attempt to describe the limitations of current measurement techniques within the context of adaptive systems.

Subjective reports, because they are introspective, suffer from uncontrollable changes in response criterion. Moreover, those pilot rating procedures which have received support in system development programs are directed primarily at the expenditure of physical effort or at the aircraft's handling qualities. Although Sheridan (1980) would argue that mental workload is best defined in terms of subjective experience, the multidimensional scaling required to measure such impressions with precision does not qualify as a real-time analysis technique.

Similarily, as the pilot's role changes from one of an active controller to that of a monitor and interpreter of displayed information, behavioral indices of workload necessarily become less sensitive to momentary fluctuations (cf. O'Donnell and Hartman, 1979). The cognitive events of interest usually take place when the operator is not engaged in discernible overt activity. Thus, if we are attempting to uncover transient mental processes which may ultimately threaten the successful completion of complex tasks, a strict dependency upon antecedent behavioral data may not allow us to make accurate predictions of impending performance failures.

Physiological measures of pilot workload would appear to be the most appropriate for application to adaptive systems. However, Roscoe (1978b) has stated that "there is little evidence available to indicate the value of physiological measures in assessing pilot workload"(p. 39). I attribute this to the fact that almost without exception in these studies, mental workload has been equated with the state of arousal (Wierwille, 1979). Therefore, most applied investigations have examined peripheral physiological responses (e.g., electrodermal, cardiovascular, electromyographic, and respiratory) mediated by the autonomic nervous system - responses which are primarily affected by anxiety, tension, or stress. I have stated elsewhere (Gomer et al., 1979) that recordings of brain electrical activity are more appropriate for monitoring specific mental functions. With respect to our need for a real-time communication link from the operator to the computer, Donchin (1979) has stated that:

"A channel carrying electrophysiological data (unobtrusively) acquired from the operator can supply the adaptive controller with at least part of the necessary information. We thus assume that mental activities manifest themselves in electrophysiological signals. We further assume that it is possible to make strong inferences about mental activity from such signals" (p. 3).

MOMENTARY CHANGES IN MENTAL FUNCTION AND THE ANALYSIS OF BRAIN ELECTRICAL ACTIVITY AND EYE MOVEMENT BEHAVIOR

To evaluate brain function while a crew member performs a demanding task, as suggested above, we must rely upon electrophysiological techniques to observe the underlying interactions within and between populations of cortical cells. Donchin (1979), John (1977), and Thatcher and John (1977) have suggested that cognitive operations can be conceptualized in terms of coherent patterns of activity within distributed cell groups. And importantly, orderly neural behavior gives rise to rhythmic voltage fluctuations in the two types of scalp-recorded electrical activity which must be analyzed - electroencephalographic activity and event-related potentials.

Electroencephalographic (EEG) activity consists of spontaneous or on-going voltage fluctuations (see Fig. 3). Event-related potentials (ERPs), on the other hand, are transient voltage fluctuations

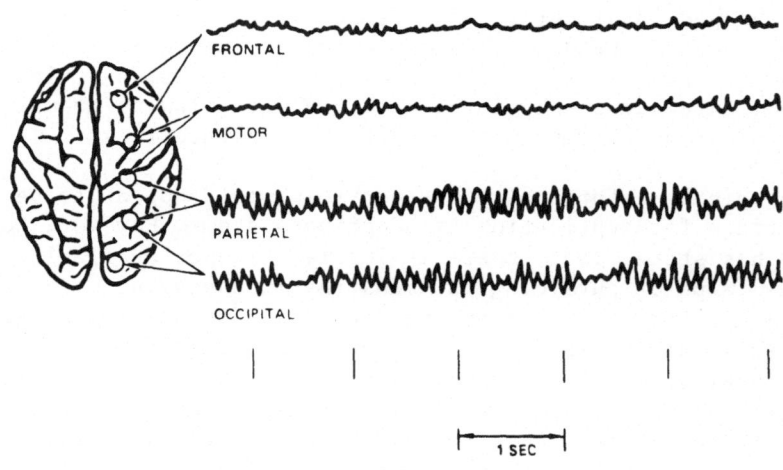

Fig. 3. Electroencephalogram of a normal human adult.

which are associated with a critical inducing event (i.e., a sensory stimulus or a cognitive operation) and which are imbedded in the
EEG activity. Since the EEG is generally more pronounced in amplitude, it usually obscures the waveform of the ERP. However, there are
several strategies for extracting and measuring ERPs (or selected
"components") in real-time (John et al., 1978).

If a discrete visual stimulus is presented to an observer who
must classify it and then signify the decision he has reached with
a behavioral response, the resultant ERP is composed of successive
positive and negative deflections continuing for up to 750 msec post
stimulus. Fig. 4 depicts such a waveform in which the major deflections or components have been labelled by a character-number desig-
nation. The character refers to the polarity of the component (P =
positive, N = negative), while the number indicates the temporal delay or latency between the eliciting event and the peak voltage of
the component.

The morphology of the components occuring within 200 msec after
stimulus onset is influenced markedly by the physical attributes
(e.g., wavelength, intensity, or contrast) of the evoking stimulus
(cf. Regan, 1972). Consequently, these so-called early components
are referred to as exogenous. If the input must be processed and a
decision reached, the prominence of late or endogenous component activity, particularly P_{300}, is affected (Gomer et al., 1976). The
term "endogenous" signifies that these components are not affected
by the sensory qualities of external events. It is interesting to

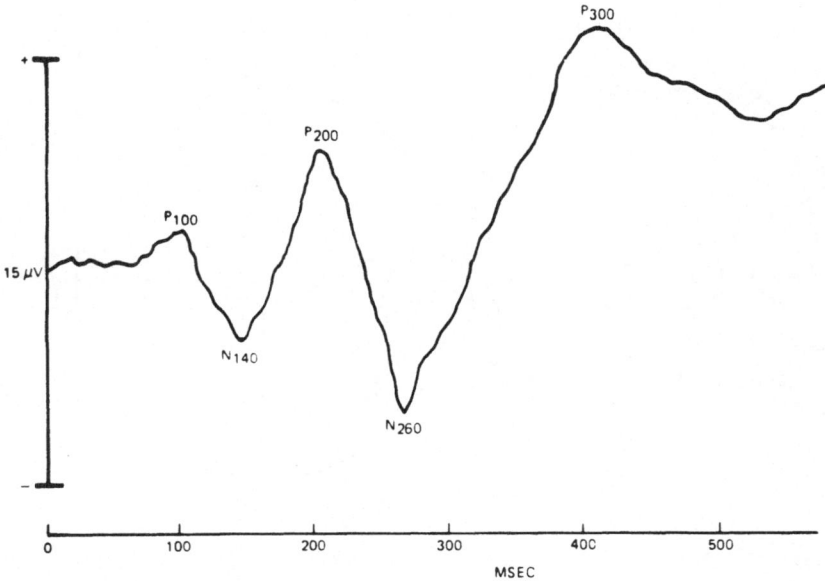

Fig. 4. Example of vertix event-related potential elicited during a
display recognition task.

note that Roscoe (1978b) cites the work of Spyker et al. (1971) when
concluding that ERPs are unsuitable for measuring pilot workload.
Unfortunately, Spyker et al. analyzed only the early components of
ERPs that were evoked by a flashing light source to which the subjects paid little attention.

The location of recording electrodes must be given careful consideration when denoting the amplitude and latency of ERP components.
This follows from the previously stated position that neural representations of information processing and decision-making involve the
coordinated behavior of disparate cell populations. Thus, the spatiotemporal distribution of late component activity, when referenced to
the International Electrode Placement System (see Fig. 5), is a most
important indicator of cognitive function (cf. Adam and Collins,
1978; Courchesne et al., 1975; Goff et al., 1978; Thatcher, 1976).

One may question whether it will be possible to evaluate ERP
components during tactical missions, since discrete visual events
are rarely presented to the pilot (other than in the form of checklist items). Rather, alphanumerics, symbolic characters, and sensor
imagery are displayed continuously, and they change in content or

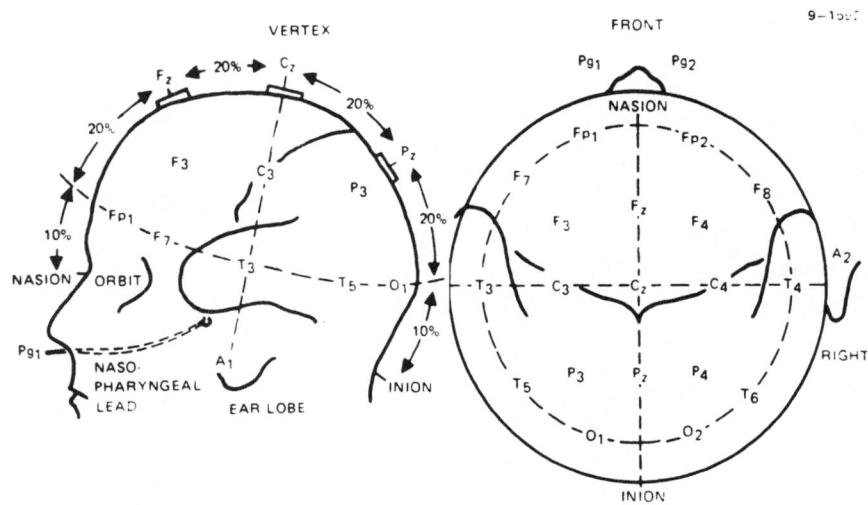

Fig. 5. Lateral and Superior views of International Electrode Placement System.

value dynamically. Donchin (1980) has shown that discrete probe stimuli (in this case auditory) can be introduced artificially into
those situations in which dynamic visual events predominate. By analyzing "background" responses (principally P_{300}) to these probes we
can infer how well the operator is performing "foreground" tasks.

For example, Isreal et al. (1980) recently simulated an air
traffic control task of monitoring symbolic representations of aircraft which moved in straight-line paths across a display. The observers were required to respond to sudden changes in the displayed
trajectories of certain aircraft. Mental workload was manipulated by
varying the number of aircraft to be monitored simultaneously. Auditory probes were presented in the background as a Bernoulli series
of low- and high-pitched tones. The high tones occurred quite infre-
quently, and the observers were instructed to maintain a running
count of these events and to report the count at the end of the experiment. It was found that the P_{300} component of the high-tone ERP
exhibited a monotonic reduction in amplitude as the number of displayed aircraft was increased from zero to eight. Thus, these investigators have demonstrated an instantaneous measure of residual processing capacity during visual loading.

The probe technique used to elicit P_{300} changes can also be used
to generate another endogenous waveform referred to as the contingent
negative variation or CNV. The CNV is a slow, negative shift in EEG

baseline which develops in the interval between successive presentations of discrete visual events (see Fig. 6). Donchin (1979) and
others have shown that the time course, amplitude, and scalp distribution of this waveform provide a general index of attentiveness,
with latency increasing and amplitude decreasing during periods of
waning attention.

    "Foreground" activities can be assessed directly, however. That
is, we can evaluate the EEG changes with accompany information processing and decision-making operations that are integral to the tasks
performed by the pilot. The EEG is categorized with respect to two
basic dimensions, frequency and amplitude. Usual frequency bands are:
- delta (.5 - 4 Hz),
- theta (5 - 7 Hz),
- alpha (8 - 12 Hz),
- beta (19 - 30 Hz).

Energy distributions can be measured within these and more restricted
frequency bands at each recording site. In fact, energy asymmetries
in homologous leads (left vs. right hemisphere) may be quite sensitive to subtle differences in cognitive function (Rebert, 1980).

    I must remind the reader of the conditions which must be satis-
fied before electrophysiological signals will be valuable in the context of adaptive systems. With respect to monitoring pilot status,
clearly delineated and unambiguous patterns of brain electrical activity must be associated with information processing, attentiveness,
and decision-making. Further, optimal patterns of activity (profiles)
must be defined for these cognitive functions as the pilot successfully performs housekeeping- and mission-related tasks. The profiles
must be continuously updated and stored in computer memory onboard
the aircraft. Current electrophysiological data which are recorded
during the various stages of the mission must then be evaluated in
comparison with the profiles for deviations from acceptable levels.
Progress has been made in defining patterns of brain electrical ac-

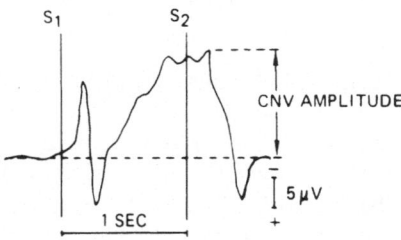

Fig. 6. Example of contingent negative variation (CNV) recorded from
        vertex electrode placement. S_1 and S_2 denote successive sti-
        mulus events.

tivity that are related to the efficacy of various cognitive functions (Donchin, 1979; Thatcher and John, 1977; John, 1977). Further,
very significant improvements have been reported regarding the techniques used to extract (from noise) and classify these "messages" in
real-time (John et al., 1978). These advances notwithstanding, a
great deal more must be accomplished (in computer technology, software development, and the design of physiological monitoring equipment) before it is both feasible and practical to implement adaptive
procedures in dynamic, operational environments. Nonetheless, I remain hopeful that the necessary breakthroughs will continue to occur.

We also have not determined conclusively whether a single physiological measure can serve as a reliable indicator of operator
status, or, rather, if several physiological response systems must
be monitored to reveal transient changes in cognitive processes which
may threaten system performance. I suggest that ocular measures be
taken to supplement electrophysiological measures of mental function.
It is generally accepted that eye movements reflect the observer's
distribution of attention within visual space. However, the dynamics
of these movements (i.e., timing, velocity, and pattern) appear to
be quite sensitive to shifts in information extraction and processing strategies.

A comprehensive discussion of the types of eye movements and
their relation to perception and cognition is presented by Cumming
(1978). We are most concerned with the dynamics of large eye movements that occur as the pilot scans displays and the visual scene
outside the aircraft. Two-stage models of visual perception hold that
events of interest are located initially via peripheral vision; this
leads to fixation and, therefore, more detailed analysis. As mentioned above, many investigators believe that fixation implies attention.
However, the relationships between the velocity and the amplitude
(angular distance) of saccadic eye movements may provide additional
insights concerning fluctuations in attentiveness or in decision-
making (Stern, 1978). Equally as important in this regard are temporal parameters, such as fixation duration (dwell time), since unusu-
ally long or short fixation pauses may reflect periods in which visual information is not being processes effectively.

PHYSIOLOGICAL MONITORING DURING FLIGHT SIMULATION

Several research programs have begun to evaluate aspects of eye
movement behavior during simulated (Krebs et al., 1977) and actual
(Gerathewohl et al., 1978) flight. However, virtually all of the investigations of brain electrical activity that I cited earlier took
place in university laboratories, and the task demands imposed upon
the subjects were somewhat constrained from an operational point of
view. It is apparent that the progression of a program which considers electrophysiological signals as input to adaptive systems will

require that these investigations be extended to appropriate part-task and full-mission simulations. In this regard, I must take exception with those who claim that brain electrical activity is a better in-flight measure of crew workload than are the measures currently being used by the manufacturers of commercial aircraft (Air Line Pilot, Vol. 47 (10), p. 56, 1978). Unhappily, real-time monitoring of brain electrical activity has not reached the point in its development where unambiguous interpretations of in-flight recordings are possible.

Therefore, my colleagues and I are conducting preliminary flight simulations in which we have challenged the pilot's ability to process information from several sources and to make multiple decisions within short periods of time. Our intent in these investigations is to further define the features of brain electrical activity and eye behavior which, when analyzed on-line during the simulation, may forewarn imminent deteriorations in system performance. I believe that there is a fundamental difference between our work and that of the Honeywell group (North et al., 1979; Wolf, 1978). The Honeywell group also has recorded physiological data within the context of flight simulation. Their purpose has been to show that changes in certain physiological variables correspond to changes in conventional flight performance during approach and landing. However, since they have chosen to study muscle tension, respiratory activity, and cardiovascular patterns, one might argue that their physiological measures of pilot workload actually reflect the expenditure of physical effort which is required to achieve control of the aircraft.

In contrast, we are simulating certain segments of an Air-to-Ground Strike mission, consisting of low altitude ingress, a pop-up maneuver for target acquisition and weapon release, and then egress. The pilots in our studies are responsible for maintaining a commanded flight profile, for systems monitoring, for threat detection/selection of countermeasures, and for target acquisition/weapon release. The difficulty of each of these tasks is manipulated independently. With respect to eye behavior, we are using electro-oculographic and oculometer methods to measure instrument dwell times, fixation duration, saccade amplitude, saccade velocity, blink rate, and pupil diameter. The aspects of electroencephalographic activity with which we are concerned include frequency distributions of energy, energy ratios in each scalp derivation, and energy asymmetries in homologous leads. Further, we are using Donchin's probe technique for generating auditory event-related potentials, and we are analyzing the spatiotemporal distribution of P_{300} elicited by infrequent tonal events during the mission.

In conclustion, while we have a great deal more to accomplish before brain electrical activity is recorded routinely during flight simulation (see Fig. 7), nonetheless, I think that we are moving steadfastly toward the goal of implementing adaptive procedures (Reising, 1979).

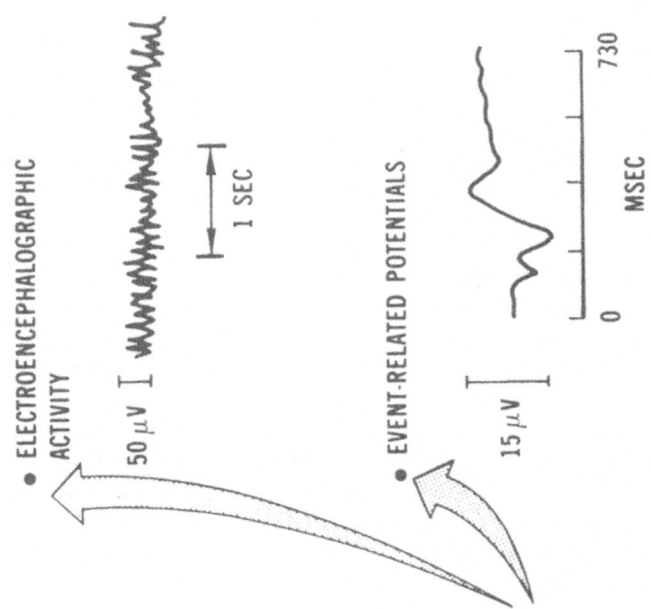

- ELECTROENCEPHALOGRAPHIC
  ACTIVITY

  50 μV | 1 SEC

- EVENT-RELATED POTENTIALS

  15 μV   0   MSEC   730

Fig. 7. Toward the concept of adaptive systems.

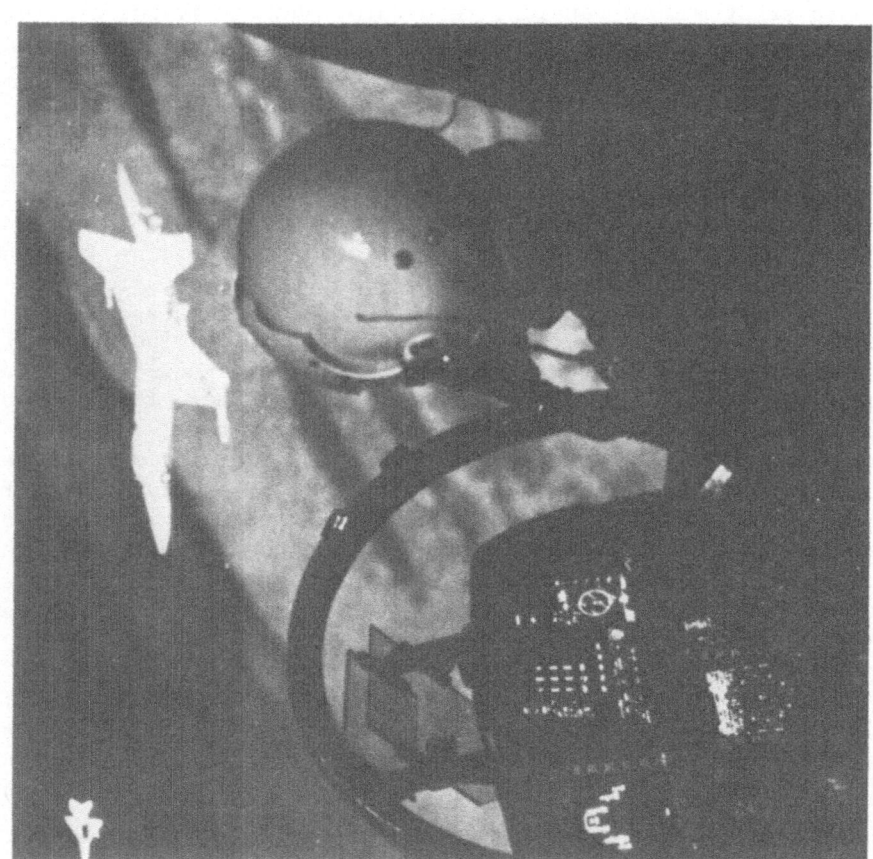

REFERENCES

Adam, N. and Collins, G.I., 1978, Late components of the visual e-
    voked potential to search in short-term memory, Electroencepha-
    lography and Clinical Neurophysiology, 44, 147-156.
Chiles, W.D., 1978, Objective methods, in: Assessing Pilot Workload,
    A.H. Roscoe (ed.), AGARD Conference Proceedings 233.
Courchesne, E., Hillyard, S.A. and Galambos, R., 1975, Stimulus novel-
    ty, task relevance and the visual evoked potential in man, Elec-
    troencephalography and Clinical Neurophysiology, 39, 131-143.
Cumming, G.D., 1978, Eye movements and visual perception, in: "Hand-
    book of Perception", E.C. Carterette and M.P. Friedman (eds.),
    Vol. IX, Academic Press, New York.
Donchin, E., 1979, Event-related potentials: A tool in the study of
    human information processing, in: "Evoked Brain Potentials and
    Behavior", H. Begleiter (ed.), Plenum Press, New York.
Donchin, E., 1980, Event-related potentials: Inferring cognitive ac-
    tivity in operational settings, in: Biocybernetic Applications
    for Military Systems, F.E. Gomer (ed.), Proceedings of the DARPA
    Conference, Chicago, 1978, McDonnell Douglas Astronautics Com-
    pany, St. Louis Division, Technical Report MDC E2191.
Ellis, G.A., 1978, Subjective assessment: Pilot opinion measures of
    workload, in: Assessing Pilot Workload, A.H. Roscoe (ed.), AGARD
    Conference Proceedings 233.
Gerathewohl, S.J., 1976, Definition and measurement of perceptual and
    mental workload in aircrews and operators of Air Force weapon
    systems, in: Higher Mental Functioning in Operational Environ-
    ments, AGARD Conference Proceedings 181.
Gerathewohl, S.J., Brown, E.L., Burke, J.E., Kimball, K.A., Lowe,
    W.F. and Stackhouse, S.P., 1978, In-flight measurement of pilot
    workload: A panel discussion, Aviation, Space and Environmental
    Medicine, 49, 810-822.
Goff, W.R., Allison, T. and Vaughan, H.C., 1978, The functional neu-
    roanatomy of event-related potentials, in: "Brain Event-Related
    Potentials in Man", E. Callaway, P. Tueting and S. Kaslow (eds.),
    Academic Press, New York.
Gomer, F.E. (ed.), 1980, Biocybernetic Applications for Military
    Systems, Proceedings of the DARPA Conference, Chicago, 1978,
    McDonnell Douglas Astronautics Company, St. Louis Division,
    Technical Report MDC E2191.
Gomer, F.E., Beideman, L.R. and Levine, S.H., 1979, The application
    of biocybernetic techniques to enhance pilot performance during
    tactical missions, McDonnell Douglas Astronautics Company, St.
    Louis Division, Technical Report MDC E2046.
Gomer, F.E., Spicuzza, R.J. and O'Donnell, R.D., 1976, Evoked poten-
    tial correlates of visual item recognition during memory-scann-
    ing tasks, Physiological Psychology, 4, 61-65.
Gomer, F.E. and Youngling, E.W., 1978, Electrophysiological applica-
    tions to human factors problems in military settings, Human Fac-
    tors, 21 (8), 1-3.

Isreal, J.B., Wickens, C.D., Chesney, G.L. and Donchin, E., 1980, The event-related brain potential as an index of display-monitoring workload, Human Factors, 22, 211-224.

Jenny, L.L., Older, H.J. and Cameron, B.J., 1972, Measurement of operator workload in an information processing task, NASA Contractor Report CR 2150.

John, E.R., 1977, "Functional Neuroscience, Volume 2: Neurometrics", John Wiley & Sons, New York.

John, E.R., Ruchkin, D.S. and Vidal, J.J., 1978, Measurement of event-related potentials, in "Brain Event-Related Potentials in Man", E. Callaway, P. Tueting and S. Kaslow (eds.), Academic Press, New York.

Krebs, M.J., Wingert, J.W. and Cunningham, T., 1977, Exploration of an oculometer-based model of pilot workload, NASA Contractor Report CR 145153.

Lovesey, E.J., 1977, The instrument explosion - A study of aircraft cockpit instrument, Applied Ergonomics, 8, 23-30.

McIver, D. and Hatfield, J.J., 1978, Coming cockpit avionics, Astronautics and Aeronautics, 16, 54-63.

North, R.A., Stackhouse, S.P. and Graffunder, K., 1979, Performance, physiological and oculometer evaluation of VTOL Landing Displays, NASA Contractor Report CR 15081.

O'Donnell, R.D. and Hartman, B.O., 1979, Contributions of psychophysiological techniques to aircraft design and other operational problems, AGARD Technical Report AG-244.

Rebert, C.S., 1980, Electrocortical correlates of functional cerebral asymmetry: Relevance to performance in operational environments and to personnel selection, in: Biocybernetic Applications for Military Systems, F.E. Gomer (ed.), Proceedings of the DARPA Conference, Chicago, 1978, McDonnell Douglas Astronautics Company, St. Louis Division, Technical Report MDC E2191.

Regan, D., 1972, "Evoked Potentials in Psychology, Sensory Physiology and Clinical Medicine", Chapman and Hall, Ltd., London.

Reising, J., 1979, The crew adaptive cockpit: Firefox, here we come, Proceedings of the 3rd Annual Conference on Digital Avionics Systems, Dallas, Texas.

Roscoe, A.H., 1978a, An introduction to the topic of pilot workload, in: Assessing Pilot Workload, A.H. Roscoe (ed.), AGARD Conference Proceedings 233.

Roscoe, A.H., 1978b, Physiological methods of assessing pilot workload, in: Assessing Pilot Workload, A.H. Roscoe (ed.), AGARD Conference Proceedings 233.

Sheridan, T.B., 1980, Mental workload - What is it? Why bother with it?, Human Factors, 23 (2), 1-2.

Spyker, D.A., Stackhouse, S.P., Khalafalla, A.S. and McLane, R.C., 1971, Development of techniques for measuring pilot workload, NASA Contractor Report CR 1888.

Stern, J.A., 1978, Eye movements, reading and cognition, in: "Eye Movements and Higher Psychological Functions", J.W. Senders, D.F. Fisher and R.A. Monty (eds.), Lawrence Erlbaum Associates,

Hillsdale, N.J.
Thatcher, R.W., 1976, Electrophysiological correlates of animal and
    human memory, in: "Neurobiology of Aging", R.D. Terry and S.
    Gershon (eds.), Raven Press, New York.
Thatcher, R.W. and John, E.R., 1977, "Functional Neuroscience, Volume
    I: Foundations of Cognitive Processes", John Wiley & Sons, New
    York.
Wierwille, W.W., 1979, Physiological measures of aircrew mental work-
    load, Human Factors, 21, 575-593.
Wolf, J.D., 1978, Crew workload assessment: Development of a measure
    of operator workload, Air Force Flight Dynamics Laboratory,
    Technical Report AFFDL-TR-78-165.

ELECTROPHYSIOLOGICAL MEASUREMENT TECHNIQUES

Günter Rau

INTRODUCTION

In ergonomics, the use of physiological parameters is especially attractive since, as such, they are objective in contrast to the psychological response parameters. Physiological parameters are utilized in the field of ergonomics to obtain indicators for functions of the living system and especially for strain of single organs, organ systems or the total organism. A special interest is directed to the observation of physiological signals in order to detect underlying psychological states and to obtain a correlation between the stress put on a subject (e.g. human operator in a man-machine system), and the subject's strain as a consequence.

Before considering the different physiological signals which are of interest in this context, a serious restriction in the application to ergonomics has to be taken into account: In nearly all ergonomic applications only non-invasive techniques can be used on a broader scale for many medical and psychological reasons. E.g., a pilot will not agree to have a certain number of EMG wire or needle electrodes inserted into his arm muscles or to have an intra-arterial catheder for bloodparameter detection implanted. As a consequence, especially for electrophysiological signals, their detection by means of surface electrodes is the mostly desired practicable solution.

When utilizing surface electrode methods two aspects have to be taken into consideration: (1) The electrophysiological sources within the body as a volume conductor produce a resulting electrical field at the skin surface which forms the boundary of the volume. It is shown by Helmholtz already that the relation between the source configuration within a conducting volume and the resulting electrical

field at any closed boundary plane is not unequivocal. This is e.g.
the reason for many difficulties in ECG diagnosis because changes in
the ECG leads even at the thorax can be related to changes in the
heart function only based on indirect methods of correlation. (2) The
detection of electrophysiological signals at the skin surface has to
be designed according to the knowledge of the electrical properties
of the skin, the electrodes, the amplifier specifications etc. as
well as the signal characteristics. The distortion and the disturbances can be reduced by a proper design.

The implications and the possible improvements are in principle
similar in all kinds of electrophysiological measurements. Therefore,
the single steps in the design procedure of a complete measuring system will be shown and the improvements will be demonstrated by adequate examples. A comprehensive review of all commonly used psychophysiological techniques and discussion of the most relevant physiological parameters with respect to ergonomics is to be found in the
recently published AGARDOgraph by O'Donnell (1979). Therefore, this
paper is focussed on the technical aspects of these measurements.
Considering the different stages of the measurement channel we cut it
down into sections as indicated in Fig. 1.

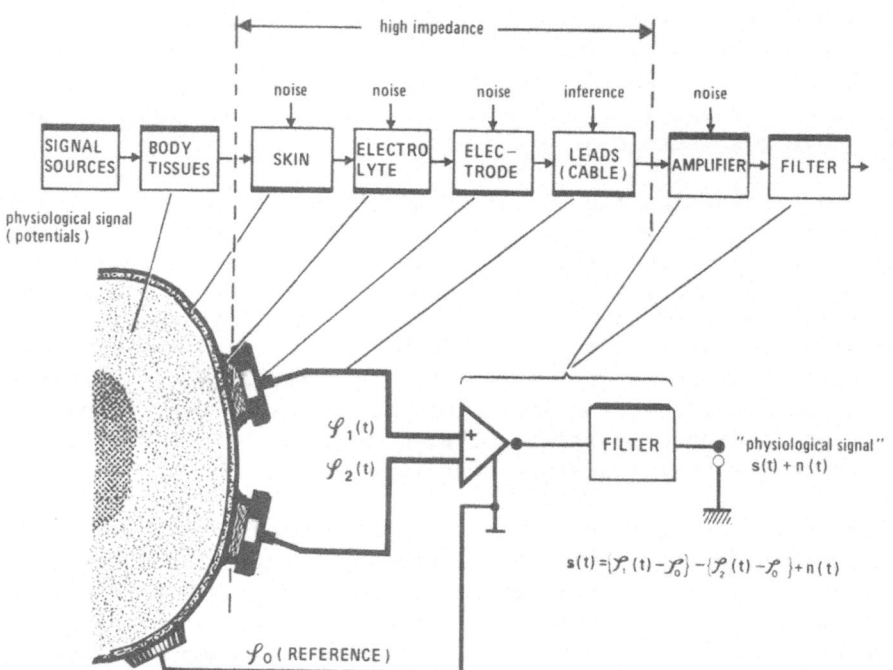

Fig. 1. Different stages of a set up for measurement of electro-phy-
siological signals, schematically.

SIGNAL SOURCE AND SIGNAL CHARACTERISTICS

First of all, the signal in question has to be specified by the most relevant parameters, which are frequency range and amplitude distribution. Some of the signals often used are summarized in Table I. Most of the values are valid for surface electrode applications. Considering the small values of amplitude it is obvious that low noise figures are indispensible in order to obtain a reasonable signal/noise ratio. In addition, the active area of the electrodes influences the frequency range: Very small electrodes shift the signal frequency band towards higher frequency components.

Table I. Characteristics of electrophysiological signals

Parameter and measuring technique	Principle measurement range (max. amplitude)	Signal frequency range (Hz)	Electrode type
Electrocardiography (ECG)	5 mV	0.1-250	skin
Electroencephalography (EEG)	500 µV	0.1-40	(skin) scalp
Electromyography (EMG)	10 mV	1-2000	skin needle wire
Nerve potentials (ENG)	5 mV	1-2000	skin
	5 mV	1-10.000	needle wire
Eye potentials			
EOG	5 mV	dc-100	skin
ERG	1 mV	dc-100	"contact"
Galvanic skin response (GSR)	500 kΩ	0.01-10	skin
Visual evoked potential (VEP)	100 µV	0.1-40	(skin) scalp

COMPLEX SKIN RESISTANCE

The resistance of the unprepared skin is complex and varies within a wide range (Rau, 1973) depending on (a) skin site, (b) subject, (c) time, and (d) skin treatment. As worst case we have to cope with values up to several M Ohms at 0.5 Hz and about 1 M Ohm at 10 Hz. This is not only valid for dorsal skin sites at the extremities but also on the human thorax as shown in Fig. 2.

In short, the electrical properties of the skin according to Fig. 2 shows the characteristics of a complex electrical filter which

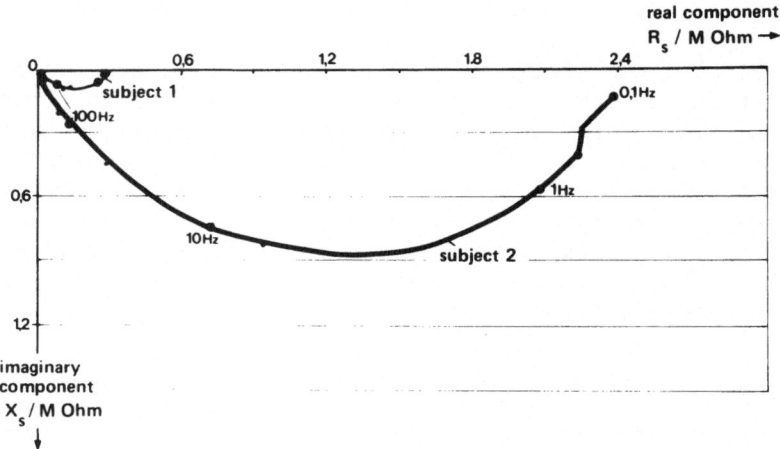

Fig. 2. Skin impedance loci of two subjects. Parameter: measuring
     frequency (from Silny and Rau, 1977).

has to be passed by the signal. Therefore, the filter can introduce
undesired and unknown changes to the signal. It is possible to suppress the influence of this filter as will be shown later on.

ELECTRODE AND ELECTROLYTE

     The complex electrical resistance of the electrode and of the
electrode/electrolyte transition depends on the electrode material
and the electrolyte (sweat, also electrode paste, jelly or cream) in
use. The measurement of active electrophysiological signals (ECG,
EEG, EOG, EMG, etc.) permits the use of electrolyte in contrast to
passive electrophysiological parameters (skin resistance = SR, skin
resistance reaction = SRR, skin conductance = SC, etc.), where so
called "dry electrodes" are to be preferred (Faber and Schubert,
1979).

     When detecting a signal from the body surface the active area
of the electrodes has to be selected accordingly. EMG signals picked
up locally from small muscle tissues demand small sized electrodes
(finger muscles: $\emptyset \approx 2...3$ mm; biceps muscle: $\emptyset \approx 8$ mm). For oculogra-
phic signals the active areas have to be small compared to the com-
monly used ECG electrodes. The size and the localisation (with respect to the source as well as the interdistances) have to be adapted
to the specific measuring problem; this aspects will not be discussed
here. However, the smaller the electrode area the higher the electrode resistance. In addition, small electrodes contact only small
areas of skin, therefore the resulting skin resistance is also high,
since the skin resistance increases with the decrease in size of the
skin area.

The complex resistance of the electrode/electrolyte transition depends on the electrode material and the kind of electrolyte used. However, the values are about two orders of magnitude smaller compared to the skin impedance values; roughly speaking, it can be neglected with respect to the skin resistance.

The electrolyte improves the contact between electrode and skin. Simultaneously, it decreases the skin resistance by intruding into the skin tissues. Many of special electrolytes as jelly, cream or paste are commercially available. The reduction in skin resistance is very pronounced when "aggressive" electrolytes are used while other types are of less effect. But the "aggressive" type can not be used in long term application because skin irritation is one consequence (Zipp et al., 1977). All contact electrolytes, as far as known, cause a gradually decrease in skin resistance which is still not stabilized even after 24 hours.

Closely related to the effect of decrease in skin resistance, there is a reduction in movement artefact susceptibility to be found. Another important practical specification is the rate of water loss of the paste or jelly, especially in long term application (e.g. 24 h). Good experiences are made with electrolyte paste (Beckmann), neptic electrolyte jelly (Sander Ltd), electrolyte cream (Hellige), as indicated by Zipp et al. (1977). As electrode surface Ag/AgCl is recommended, but for frequencies above 10 Hz stainless steel can be also used. Low polarization voltage is observed at Ag/AgCl electrodes. Unfortunately the Ag-content varies dependent on the manufacturer, resulting in different polarization; this parameter should be tested before selecting a definite electrode type.

"Dry" electrodes are used without a contact electrolyte but the normal skin is wet with sweat which act as natural electrolyte forming a galvanic connection. In skin conductance and skin reaction measurements the skin properties must not be changed by adding electrolytes because then the effects under investigation decrease in amplitude. For this type of measurements black platinum surface is recommended by Faber and Schubert (1978), since it showed the best electrical properties which were constant within the frequency range between 5 Hz and 15 kHz. With respect to long term stability black platinum is superior to other materials, but its boundary layer produces a slightly higher noise than gold surface.

Especially measurements with "dry" electrodes are sensitive to movement artifacts. They can be reduced markedly by pressing the electrode to the skin surface (Zipp and Schad, 1979). The artifact amplitudes are decreased to $1/10$, when increasing the pressure from 40 p/cm^2 to 350 p/cm^2 according to Zipp et al. (1978). - It is clear that a compromise has to be accepted taking into account all the parameters mentioned.

CABLE, CONNECTING ELECTRODES TO AMPLIFIER

In a recently performed study on cable properties (Silny et al., 1979) it has been shown that some cable produced movement artifacts of several millivolts in voltage, when being moved as indicated in Fig. 3.

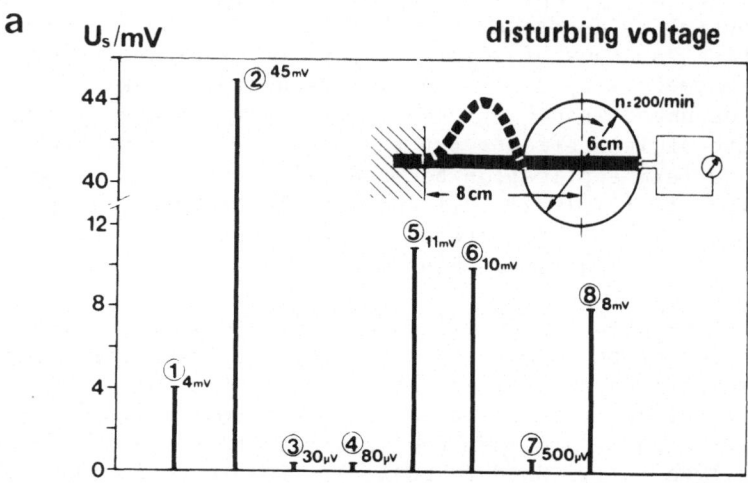

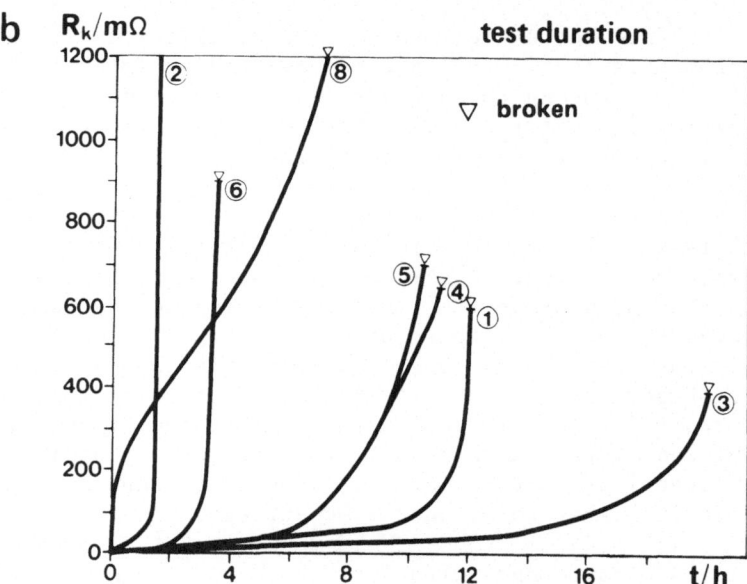

Fig. 3. Electrical and mechanical properties of cable, tested during mechanical movement. The lose end is moved in a circle (∅ = 6 cm, 200 revolutions/minute).
   a. 8 different cable types; no. 3 and no. 4 showed the smallest voltages.
   b. In addition, no. 3 showed the longest mechanical test life time (18 hours) (From Silny et al., 1979).

Even a "special ECG cable" as offered on the market showed very bad results; in contrast to that we found a cable which produced 10^2 times less voltage artifact compared to the cable mentioned above. It is remarkable that the movement can generate up to 45 mV amplitude as shown by Fig. 3 (cable A4). Before making use of a cable it should have been tested.

AMPLIFIER STAGE

The specifications of the differential amplifier can be adapted to the measuring problem very easily when utilizing modern semiconductor technology. Old instrumentation gives rise to some problems since not only the input impedance but also the input offset current and the noise are given.

The desired and recommended specifications for newly designed differential amplifiers which somebody is going to buy now are summarized in Table II.

Table II. Amplifier specifications.

Input impedance	> 10^{10}	Ohm
Input offset current	< 50	pA
Common mode rejection ratio CMRR	> 100	dB
Noise level	< 5	μV_{rms}
(f = 0,1...1000 Hz, source resistance 100 k Ohm)		

The noise level as given in the manufacturer's specifications list, is mostly measured while the amplifier's input is short circuit; that is a non-relevant figure in this context. This noise level should be specified with a source resistance (e.g. 100 k Ohm) since here the "current noise" plays an important role. A further improvement in S/N ratio is achieved by adapting the filter (upper and lower limit) to the signal frequency band (Table I).

The influence of the complex skin resistance can be suppressed effectively by a high input impedance (Rau, 1974) as indicated by the attenuation ratio shown in Fig. 4, schematically.

By these means, also variations in skin impedance dependent on time, skin site, subject, etc., can be neglected. A preparation of the skin as formerly applied, which has to decrease the skin resis-

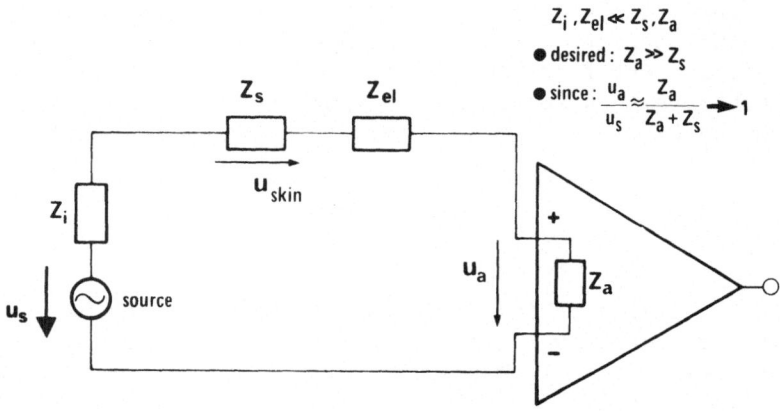

Fig. 4. Influence of the amplifier input impedance, simplified.

tance to several k Ohms, is not necessary any more. The low input
offset current is necessary to reduce a type of artifact which may
be caused by modulation of the skin resistance. It is known that the
value Z_S (Fig. 4) varies depending on mechanical pressure (Zipp and
Schad, 1979). Time varying pressure occurs when movements are induced to the electrode relative to the skin surface, e.g. via cable
motion. With a change ΔZ_S = 50 k Ohm and an input current of 50 nA
the generated voltage amplitude is Δu = 2,5 mV. By restricting the
input offset current to 50 pA, the amplitude is reduced to 2,5 µV.

In addition, pressure applied to the skin produces a skin voltage which can, by no means, be separated from the desired signal
voltage. This voltage may reach values of up to 5 mV. It can be reduced only by such a pick up design where both electrodes, connected
to the differential amplifier, are pressed to the skin equally and
simultaneously.

Electromagnetic disturbances (50 Hz noise etc.) have been a superior problem in the past. Microminiaturization provide small sized
complete differential amplifiers fulfilling the desired specifications. A small amplifier (V = 1000, filter fu = 1 Hz, F_o = 1000 Hz)
used for an EMG pick up (developed in our group) has the dimensions:
20 x 10 x 5 mm^3. By means of this technique it is possible to incorporate electrodes and amplifiers in a complete pick up arrangement
or, at least, to reduce the cable length between amplifier and electrodes to several centimeters. A further advantage of this design is
a low cable capacity.

The amplifier delivers a signal of several volts which can be
connected to a main unit by a long cable since the output resistance
is very low. Three leads of the cable are used to supply voltages
and reference voltage (simultaneously shielding of the cable).

FURTHER DESIGN RECOMMENDATIONS FOR IMPROVEMENT OF SIGNAL/NOISE RATIO

Optocoupling

Most recently the progress in optoelectronic devices opened the possibility to reliable and inexpensive optocoupling of analog signals. The principle is shown in Fig. 5.

First, the subject's security is improved which might be critical whenever the signal is fed to a computer. Second, the electromagnetic inference is reduced. The galvanic decoupling of the amplifier is possible when the supply voltages are taken from a battery set or a special DC-DC-converter. The galvanic decoupling should be strongly recommended as standard.

Filter Techniques

Digital filter techniques are of rapidly increasing importance since digital processing becomes less expensive and easily available.

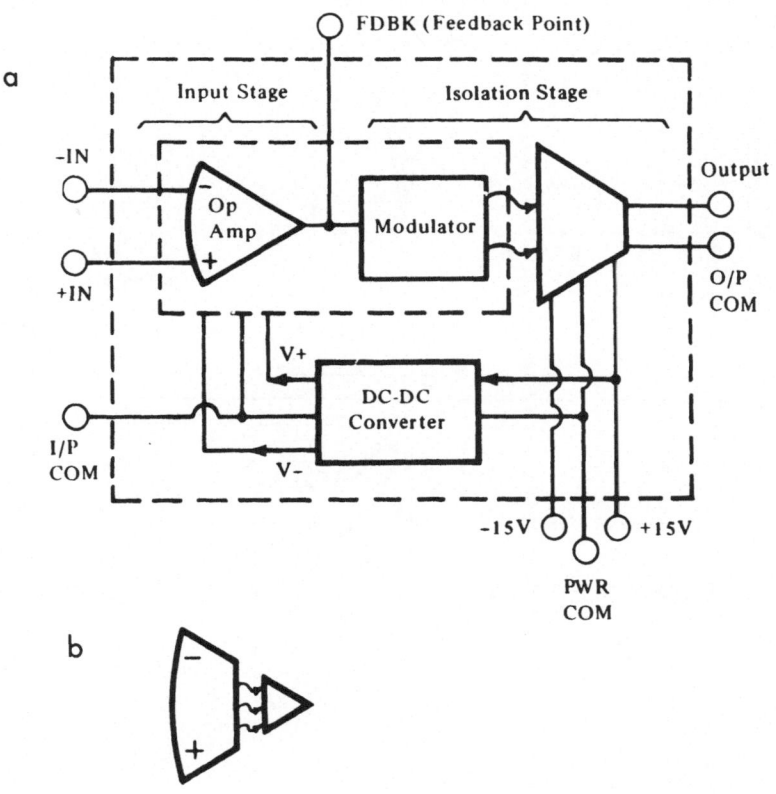

Fig. 5. Isolation amplifier arrangement for galvanic decoupling;
        a. block diagram; b. symbol.

(As an introduction e.g. see: Rabiner and Gold, 1975.) Here, only a
few remarks on filter procedures still in use seem to be necessary.
50 Hz disturbance is sometimes "eliminated" by notch filter application. This filter type has the disadvantage of phase distortions introduced to the signal. The amplitude and phase characteristics are
shown in Fig. 6.

   A different kind of filter for improving the cut off characteristics of a filter is achieved by digital or hybrid technology which
is superior to analog techniques (Murr et al., 1980). As a very im-

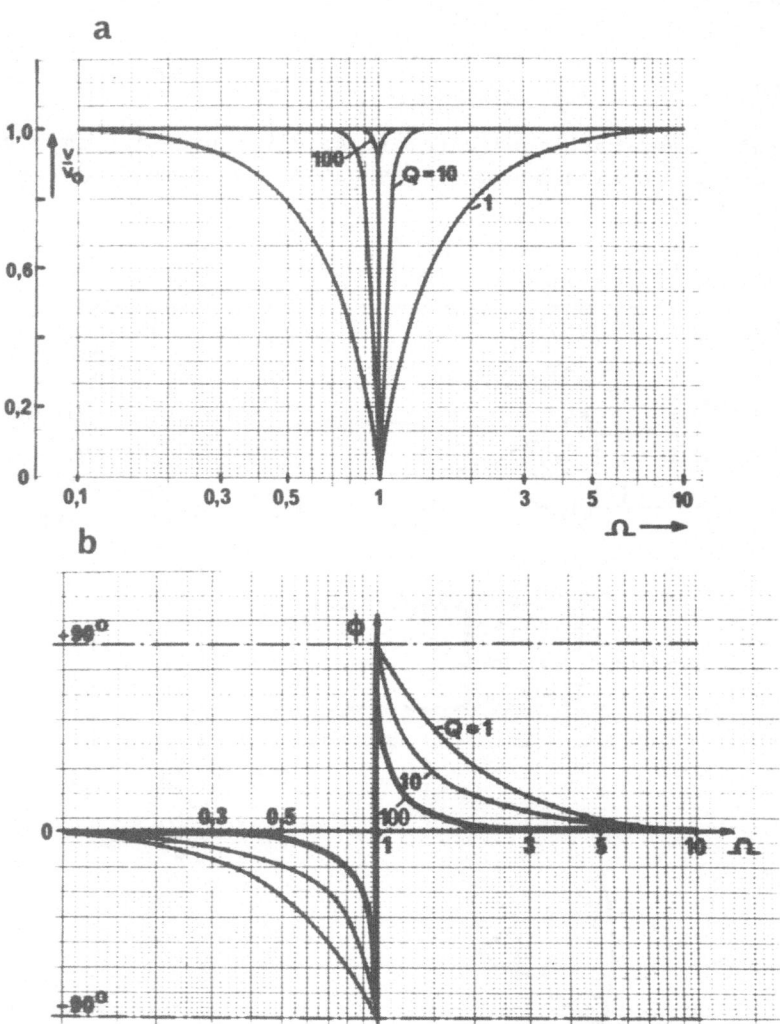

Fig. 6. a. Amplitude and b. Phase ($\emptyset$) characteristics of a notch fil-
     ter; Ω = normalized frequency.

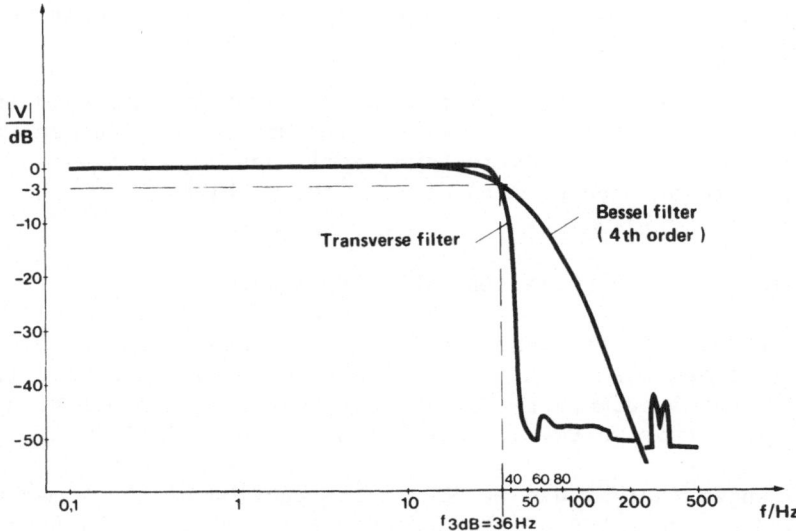

Fig. 7. Amplitude characteristics of a transverse filter and a Bessel filter.

portant example, a low pass filter of the 4th order Bessel type is compared to a new transverse filter type in Fig. 7.

The cut off frequency selected is $f_{3\,dB}$ = 36 Hz which is appropriate e.g. for EEG and VEP signals. The advantage of the transverse filter is the sharp cut off (150 dB/oct); in addition, the influence to the phase of the signal can be neglected. This new type of filter will become very important in the future.

Artifact Detection and Suppression

Strictly speaking, this matter is a big problem of pattern recognition which we will not go into detail in this context. Only one item should be picked out. Very small signals demand special methods for improvement of signal/noise ratio. In evoked signals which are synchronous to a stimulus or to an other defined event, a very common method is to calculate an average from repeated measurements. A typical example is the Visual Evoked Potential (VEP) which will be illustrated later on. By this procedure, all signal components correlated with the reference event are enhanced relative to all uncorrelated components as e.g. undeterministic noise. Because of the low amplitude level, the signal is highly susceptible to disturbances as movement artifacts or muscle potentials. If one disturbed sample is taken to the set of samples contributing to the average, it takes an additional number of samples to suppress the influence of this disturbance. E.g. in VEP, one disturbed sample makes additional 20-50 samples necessary; sometimes the measurement has to be started again from begin. This means a prolongation of the measurement duration which sometimes is not tolerable. Other specific consequences are

not mentioned here, but a considerable loss of information is one
other effect.

For these reasons, the rejection of a disturbed sample is appropriate since a correction is too difficult. Marked disturbances can
be detected easily as e.g. over-range amplitudes. Other types of disturbance have to be treated according to the problem.

IMPROVED VISUAL EVOKED POTENTIAL (VEP) PROCEDURE

The suggested improvements in design of a measuring set up lead
to marked improvements in quality of the electrophysiological signals.
Two examples can be demonstrated: EMG as shown by the paper presented
by Gärtner (see this volume), and VEP (Murr et al., 1980).

It is known that by use of VEP the refraction of the human visual apparatus can be detected. The principle is described by Regan
(1979) as well as in the review by O'Donnell (1979). A changing
checker-board is displayed to the subject, and the potentials are detected from the skull surface by electrodes as indicated by Fig. 8.

The improvements are 1. the electrode with small pins are fas-
tened by an elastic ribbon to the skull while the pins are protrud-
ing the hair layer making good skin contact; 2. the complete amplifier is placed above the electrodes ("active electrodes") achieving
very small electromagnetic disturbances; 3. amplifier specifications
according to Chapter "AMPLIFIER STAGE"; no skin preparation is neces-

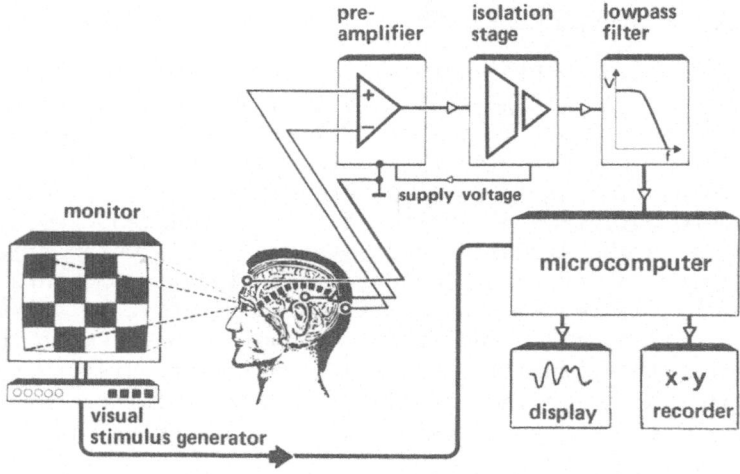

Fig. 8. Experimental set up for detection of visual evoked potential
        (VEP) (from Murr et al., 1980).

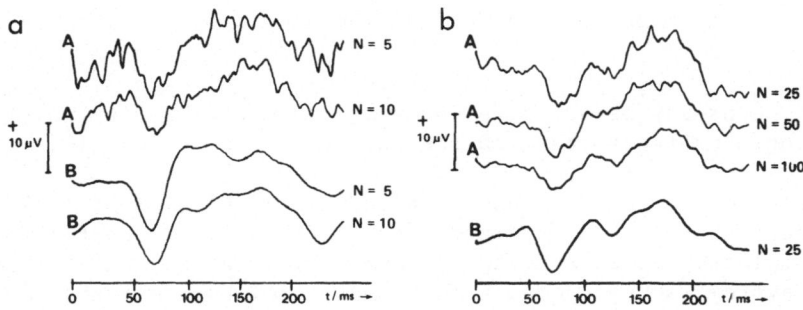

Fig. 9. Visual evoked potentials – comparison between curves, A: conventional processing, and B: processing by means of the suggested new concept. All signals were taken under identical
conditions (from Murr et al., 1980).

sary, the use of electrolyte is optional; movement artifacts are minimized; (4) optocoupling and transverse filter according to Chapter
"FURTHER DESIGN RECOMMENDATIONS .... RATIO"; (5) averaging by a microprocessor.

The result of the measurement performed with conventional procedures and our concept (Murr et al., 1980) is shown together in Fig. 9.

Fig. 9a shows an example of VEP curves where a high quality VEP
signal is obtained if only 5 averages are taken. In Fig. 9b one can
see that the quality in conventional techniques does not improve by
increasing the number N of samples included in the average; in our
set up the quality is already satisfying when N = 25 or less is selected.

By the consequent and proper design applied to the VEP procedure,
we now can measure the evoked potentials even in unshielded rooms,
while the normally used number of averaging steps N = 100 can be reduced sometimes to only N = 5. We hope that we can detect the refraction with an accuracy of ± 0,25 dioptries completely objective and in
a reasonable observation time (Murr et al., 1980) which will be a
marked progress.

SMALL LABORATORY SIGNAL ACQUISITION AND PROCESSING SYSTEM

Acquisition and processing of electrophysiological signals in
a laboratory has to be planned and designed in such a way that different kinds of experiments can be performed without long preparations. In the medical and clinical surrounding a system is necessary
which can be handled by technically innocent staff members. Some of
the desired properties are: transportable, small, safe, reliable,
easy to operate. In addition, the signals have to be checked in a

play back display mode, and the evaluation should be possible in an interactive mode.

These needs can be satisfied by a microcomputer system which is hardware configurated similar to the example shown in Fig. 10.

We selected for our purposes the DEC-LSI 11/03 (11/02; 11/23) running under RT 11. With the interfaces (A/D. D/A. I/O ports) and periphery (plotter, display graphic terminal, printer, dual floppy disk) the system is very compact on a small trolley. Only the signal conditioning (preamplifier, output for stimulus conditioning, etc.) is added as an extension according to the problem.

The hardware as outlined is not very original without a comprising software package. Therefore, a special language has been designed on the basis of software modules which can be called by short single-line commands. Of course, the modules can also be incorporated in assembler and basic programs, or they can be connected to each other in a "batch stream" very easily. A complete experiment can be controlled e.g. by a medical-technical assistant after an introduction of less than one day. As an example, in cardiology such a system is used in an intensive care unit for detection and evaluation of precordial mapping in acute infarction patients (Silny et al., 1979). The measurement as well as the evaluation is performed in an interactive mode as indicated by Fig. 11.

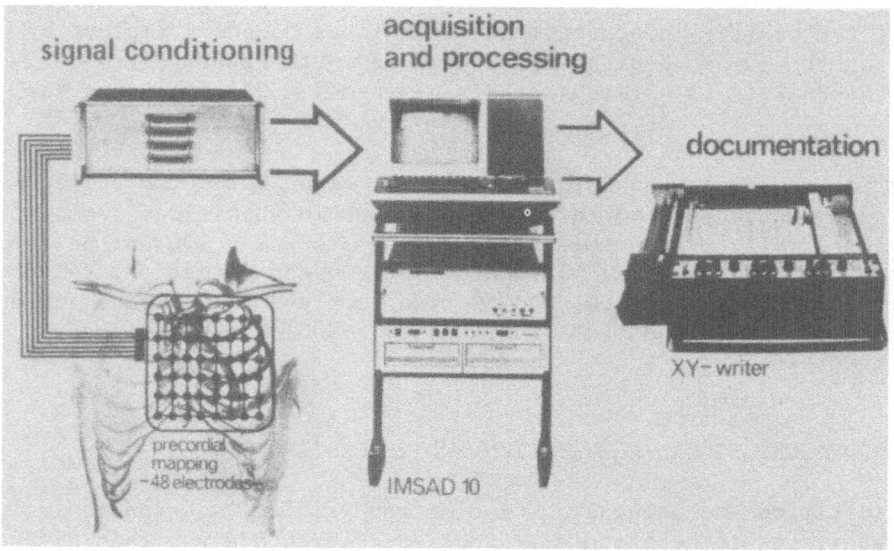

Fig. 10. Small microcomputer system for electrophysiological signal acquisition and processing (example: precordial mapping; 52 channels, ·500 samples/second). Signal conditioning is restricted here to ECG acquisition channels.

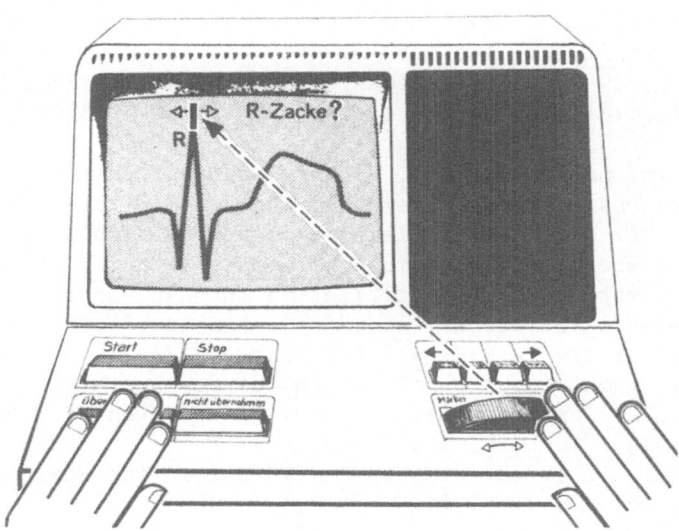

Fig. 11. Interactive computer supported signal processing. Example: evaluation of ECG (from Rau, 1979).

The user (physician) is guided and supported by the program sequence according to his wishes which have to be specified together with the computer specialist.

CONCLUSION

In electrophysiological measurements, the experiences and the improvements of all components of the measuring channel from the electrode/skin interface to the evaluation and display of the signal according to the above suggestions, a marked progress in overall quality can be achieved. However, the selection of certain parameters in hardware (e.g. electrode size and location) or of signal parameters by software (e.g. time durations, amplitudes, correlations etc.) are related to each individual problem and have to be worked out specifically. A hard- and software package has been described which enables its user in a laboratory to solve measurement problems in medicine and ergonomics. Since the small size and the modular concept make the system easy to handle and to service, it can also be utilized in most field studies.

REFERENCES

Faber, S. and Schubert, R., 1978, Neue Elektrode für den Einsatz bei Hautleitfähigkeitsmessungen, Biomed. Technik, 23, 40.

Murr, G., Silny, J. and Rau, G., 1980, Verbesserte objektive Refraktionsbestimmung mittels visuell evozierter Potentiale (VEP), Biomed. Technik, 25, 182-186.

O'Donnell, R.D., 1979, Contributions of Psychophysiological Techniques to Aircraft Design and other Operational Problems, AGARD-AG-244.

Rabiner, L.R. and Gold, B., 1975, "Theory and application of digital signal processing", Prentice-Hall, Englewood Cliffs, USA.

Rau, G., 1973, Der Einfluß der Elektroden- und Hautimpedanz bei Messungen mit Oberflächenelektroden, Biomed. Technik, 18, 23-27.

Rau, G., 1974, Improved EMG quantification through suppression of skin impedance influences, in: "Biomechanics IV", R.C. Nelson (ed.), University Park Press, Baltimore, 322-327.

Rau, G., 1979, Ergonomische Überlegungen bei der Gestaltung komplexer medizinischer Instrumentierung unter Einsatz von Mikro-prozessoren, Biomed. Technik, 24, 10-15.

Regan, D., 1979, Electrical Responses Evoked from the Human Brain, Sc. American, 12, 108-117.

Silny, J. and Rau, G., 1977, Messung des komplexen Hautwiderstandes, Biomed. Technik, 22, 409-410.

Silny, J., Hinsen, R., Rau, G., v. Essen, R., Merx, W. and Effert, S., 1979, Multielektroden-Meßeinrichtung zur elektrokardiografischen Verlaufskontrolle des akuten Myokardinfarktes, Biomed. Technik, 24, 106-112.

Zipp, P., Gessner, H. and Faber, S., 1977, Elektrische und Biologische Eigenschaften von Elektroden, Biomed. Technik, 22, 395-396.

Zipp, P. and Schad, E., 1979, Quantifizierung der Bewegungsempfindlichkeit von Oberflächenelektroden, Biomed. Technik, 24, 76-81.

ELECTROMYOGRAPHY AND APPLICATIONS

Klaus-Peter Gärtner

INTRODUCTION

There are a number of problems which require information about the effort humans have to exert in various tasks and work situations and what effect this exertion has on normal and handicapped subjects. Subjective measures can be rather unreliable. This is especially true with muscular activity because of large individual differences and because subjects frequently are unable to judge questions of effort and fatigue. What is needed is a reliable objective method for measuring muscular activity and a way of relating such measures to fatigue, workload etc. One such approach is electromyography (EMG).

The EMG signal provides data on changes in muscle activity and on changes in the interaction between muscles or muscle groups at rest or during movements of the body or parts of the body. Today it is generally accepted that electromyography is useful for the assessment of muscular loading in medicine. It is a valuable tool in medicine for diagnostic purposes and for testing of rehabilitants. It is also thought that electromyography may be usefully applied in the wide field of ergonomics, i.e. to man at work. For example, the introduction of machines has reduced heavy physical work loadings on man, but at the same time has led to repetitive tasks which has increased the duration and number of situations in which stresses are concentrated on small parts of the body. This can cause disorders such as muscle tensions which are measurable with EMG that degrade the performance of man in the system or worse, lead to significant injurys or reduce one's capacity for work.

BASIS OF ELECTROMYOGRAPHY

In normal anatomic description a distinction is made between muscles and tendons. The tendons connect muscles and bones and serve as the actual power transmitting elements. The contracting unit is the muscle cell or muscle fibre, best described as a very fine thread. A muscle fibre has a length of up to 30 mm and a diameter about 40μ to 150μ depending on the muscle function. It can shorten in length by contracting up to 57% of its resting length.

The Motor Unit

Todays thinking about the electrophysiology of the muscle assumes that a muscle fibre never contracts as an individual but rather several fibres contract at the same time. All these fibre groups are stimulated by the terminal branches of one single nerve fibre, called the axon. The axon cell body is located in the anterior horn of the spinal grey matter of the spinal cord. A motor unit can be seen in Fig. 1. It consists of the nerve cell body, the long axon with terminal branches connected to the end plates and the single muscle fibres. The size of the motor unit, i.e. the number of muscle fibres activated by the same axon, can be quite different for different muscles. It seems to be certain, that one motor unit can serve 5 to 2000 muscle fibres (Basmajian, 1967; Strong, 1979). The force produced by a motor unit ranges from 0.1 to 250 grams weight. The muscle fibres connected and activated by one axon do not lie together, as shown in Fig. 1, but rather the muscle fibres of different mo-

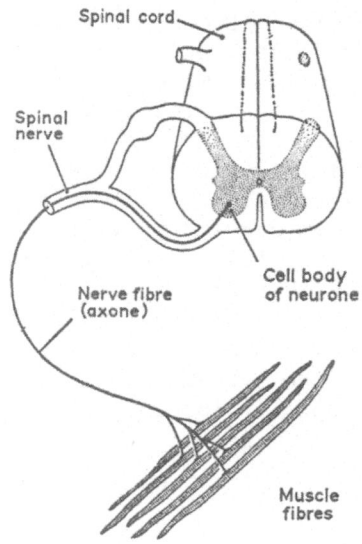

Fig. 1. Scheme of a motor unit (Basmajian, 1967).

tor units are interlaced. The number of motor units varies among the
different muscles of the body, but as a criterion, one could say,
the longer the muscle, the more motor units will form that muscle.

Muscle Action

The muscle cell membrane has a different electrical charge on
both sides. The actual internal cell potential, the so called rest
potential, or normal cell potential is polarized at 90 mV. The fir-
ing signal comes as a traveling wave from the motor neuron along its
axon to cause a depolarisation of the motor end plates. This in turn
depolarizes cells inbedded in the muscle fibres and generates an action potential of about 120 mV causing certain physical-chemical processes. The result is a wave of contractions spreading over the fibres
which produces a brief twitch. The duration of the twitch ranges
from 1 to 4 ms. After that excitation a rapid and complete relaxation follows. This process is shown in Fig. 2. Normal muscular activity is characterized by steadnessm precision, and smoothness of mo-

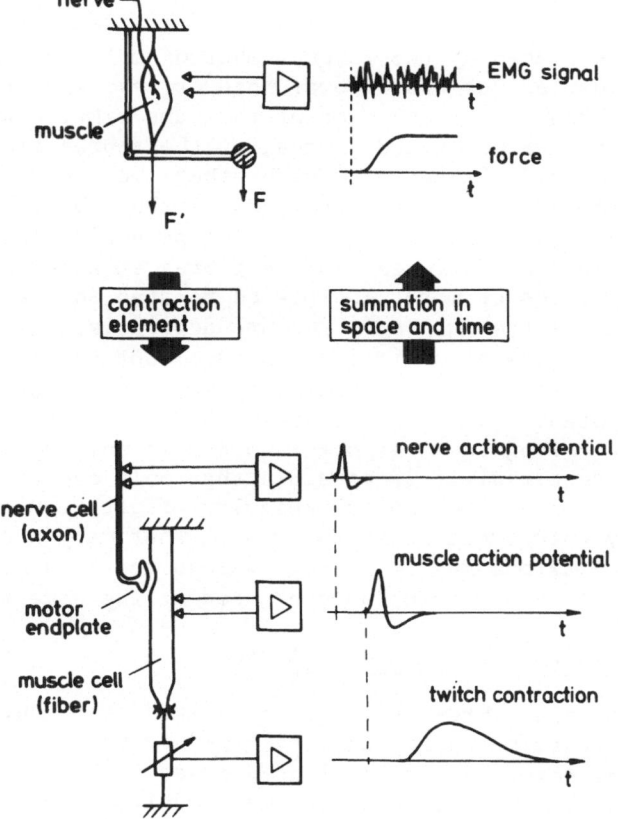

Fig. 2. Electrical and mechanical processes activating a skeletal
muscle (Rau, 1973).

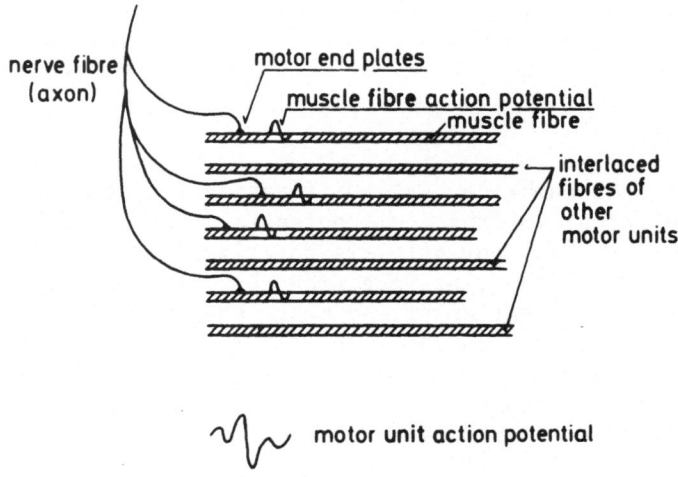

Fig. 3. Constitution of the motor unit potential.

vement. This is explaned by the large number of motor units constituting any one muscle. But also fibres more or less are stimulated
by the terminal branches of one axon at the same time. However, the
stimuli sites i.e. the motor end plates, of the fibres of a single
motor unit are not all aligned with each other i.e. they are not all
located at the same distance from the ends of the fibres (Buchthal,
1959). Therefore a single motor nerve action potential (see Fig. 2)
depolarizes a number of different muscle fibres at a number of different local positions of motor end plates related to the fibres,
shown in Fig. 3. All the muscle fibres are depolarized at the same
time and the spikes travel a similar distance. But they are out of
phase with each other. The sum of these polyphasic spikes constitutes
the motor unit potential. In Fig. 3 those muscle fibres shown without
motor end plates belong to other motor units. If only a small muscular activity is required, it is possible that only one motor unit
will be brought into action. Normal muscular effort requires the activation of many motor units with the result that many muscle fibres
contract. If the maximum muscle effort is required then all motor
units belonging to this muscle would be used at the same time.

The Potential Obtained during Muscle Action

Since all muscle fibres of one motor unit do not contract at the
same time, the electrical potential developed by the single twitch of
all fibres of the motor unit is prolonged by about 2 to 12 ms. Because of the sum of the s pikes of one motor unit action potential
(see Fig. 3) and the many superimposed motor unit potentials only an
interference pattern of the electrical activity of the muscle can
be measured. By inserting microelectrodes, fine needle or wire elec-

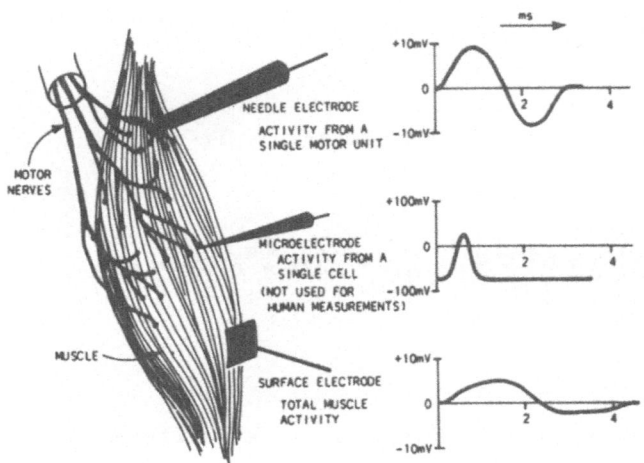

Fig. 4. The muscle potential traces obtained with various electrode
      types is known as the electromyogram or the EMG (Strong,
      1979).

trodes, into the muscle it is possible to record action potentials
generated by a single motor unit. If the electrode is very well
placed it can detect the depolarisation and repolarisation of one
single cell fibre. The duration of the total process will take a-
bout 2 ms (see Fig. 4). If a needle electrode is placed near the
muscle cells, activity of many fibres of the corresponding motor u-
nit can be observed. Because of the small time variation between the
different fibres, as shown in Fig. 3, the excitation wave is slightly ahead in some fibres compared with others and as a result, a superimposed potential will be built-up with a time duration of about
2 to 12 ms.

     The amplitude of the actual electromyogram (EMG) is considerable
and typically shows the total activity of many superimposed motor
units having random relationships to one another. A recording of the
electromyographic (EMG) activity obtained by a surface electrode at
different levels of muscular tension is illustrated in Fig. 5. The
relaxed muscle is at a low or resting electrical level i.e. EMG trace
is nearly zero with small deviations (top of Fig. 5). During low muscular tension (center of Fig. 5) the signal becomes irregular and
with swelling tension the measured electrical activity also increases
(bottom of Fig. 5). When a muscle is to contract, a large number of
motorneurons must fire, contracting many motor units. A fire rate
greater than 10 Hz causes several individual twitches, and an intermediate amount of contraction. Rates of 50 Hz and higher generate a
complete contraction of the muscle. In every stage of excitation the
motorneurons must fire in asynchronous volleys to produce smoothness
of muscle tension over a certain period of time. If the volleys become synchronized, a tremor would be produced by the muscle, as sometimes happens in spinal diseases.

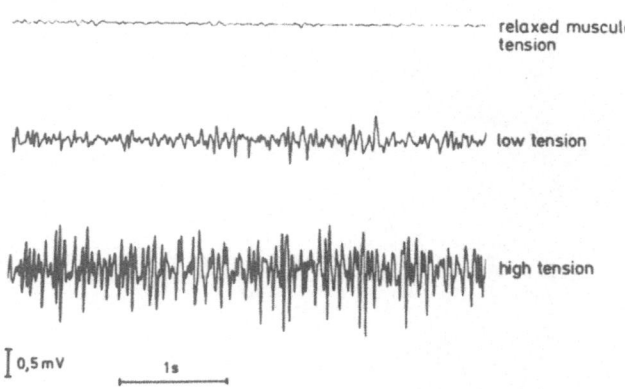

Fig. 5. Electromyographic (EMG) activity at different levels of muscular tension.

SURFACE ELECTROMYOGRAPHY

Microelectrodes, needle or wire electrodes are best avoided in
human factors or human engineering research; they are mainly used
for electrodiagnostic purposes in medicine. Because they are invasive
it requires a lot of experience to place them and they produce pain
unnecessarily and risk infections. For practical applications in human engineering or ergonomics surface electrodes are used. There
will be no risk of infection by preferring that type of electrode in
industrial or research laboratories or in work place situations. The
only principle argument against surface electrodes is that the gathered signal is mostly a representation of the contractile activity
within a relative large volume of the muscle and that sometimes
"cross-talk"-effects from neighbouring muscles are measured. This,
however, it in fact a specific advantage of surface myography when
studying the activity of whole muscle.

To sense the EMG signal present at the skin, surface electrodes
have to be brought in contact with the skin at a place with the
shortest possible distance to the muscle. Most commonly in use are
permanent or disposable electrodes with disks of silver coated with
silver chlorides. More recently, conductive adhesive electrodes have
become commercially available. A disadvantage with all type of electrodes is that electrical interference in certain environments can
occur when electrodes are connected with relatively long cables to
the amplifier input. A modified type of surface electrode is the suction-cup electrode (Moore, 1966; Vredenbregt, 1969) and a successful
further development was described by Rau (1973). A small case supports two electrodes at a fixed distance (Fig. 6). Each small silicon

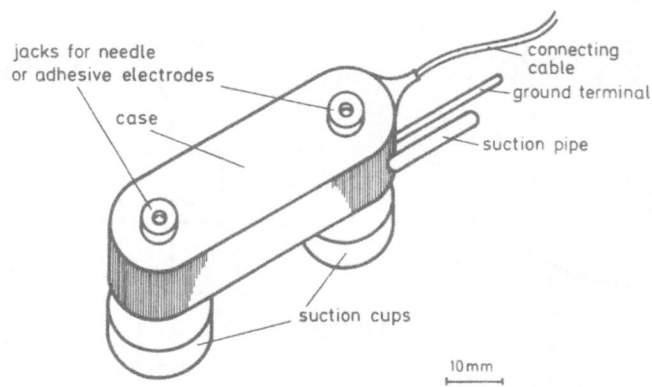

Fig. 6. Suction-cup electrode (Rau, 1973).

suction-cup incorporates a concave rounded steel electrode. The air
is evacuated from the cup after it has been applied with saline electrode paste to the skin. To reduce the influence of jamming signals,
a preamplifier is built into a case which provides mechanical pro-
tection as well as electrical shielding. A very high input impedance
(greater than 200 MOhm) is chosen in order to suppress the complex
influences of skin resistance on the EMG signal and artifact signals
evoked through movements. The assembly of two suction-cup electrodes
with the integrated electronic input stage and case form a small and
handy device. With it and reasonable care in placing it the EMG can
be easily detected and used for direct inspection of the muscles in
action. Every one who has ever worked with EMG electrodes realizes
the enormous advantages in using that special and convenient type of
electrode. Obviously it can also be used in connection with all the
other available electrodes.

ANALYSIS TECHNIQUES

Rectifying, Integrating and Averaging

     Analysis of EMG signals is more difficult than obtaining the
signal. To convert the EMG signal into a measure of muscular force
a number of different procedures are used. These are discussed by
Rau (1973). One important processing procedure is shown in Fig. 7.
The procedure involves double-wave rectifying, integration of the
total power under the curve and averaging of the EMG signal. This
can be done with a digital computer or with analog electronic networks with sufficient accuracy. The analog solution has the advantage that during long running experiments the huge quantity of data
can be pre-processed or reduced at the time and place the experiment
is performed before the data is recorded.

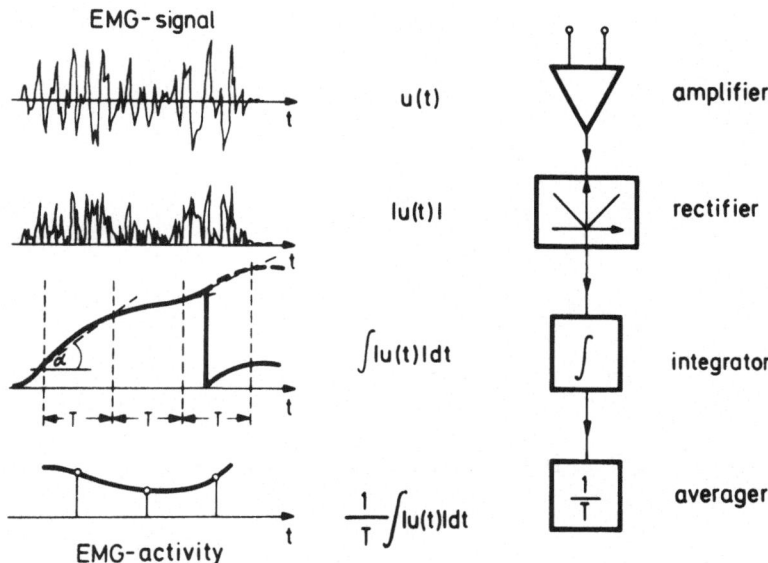

Fig. 7. Processing of the EMG: Rectifying, integrating and averaging (Rau, 1973).

Frequency Analysis

Another important method of processing is frequency analysis of the EMG signals. As demonstrated by Kogi and Hakamada (1962) the power spectrum of the myoelectric signal changes with increased muscle fatigue. According to the literature it is known that under fatigue conditions the high-frequency content of the myoelectric signal decreases and the low-frequency content increases (Kadefors et al., 1968). EMG signals measured during a fatiguing muscle contraction are constantly changing and the related power spectrum must be expressed as a function of both frequency and time. By having those two continuous variables one has the choice of making step increments with either frequency or time in performing a power spectrum analysis. Broman et al. (1973) developed a method for on-line frequency analysis, as shown in Fig. 8. In the left of the figure the intensity level of the myoelectric spectra is separated into four frequency bands and occur as continuous outputs. In the second case illustrated the power spectra frequencies are calculated continuously for a sequence of time intervals. An explanation of the changing EMG signals is that the power spectrum shifts towards lower frequencies (Lindström et al., 1970). The spectral changes of the EMG are associated for the most part with a decrease in conduction velocity of the depolarizing wave in the individual muscle fibre which can be reduced to nearly 50%.

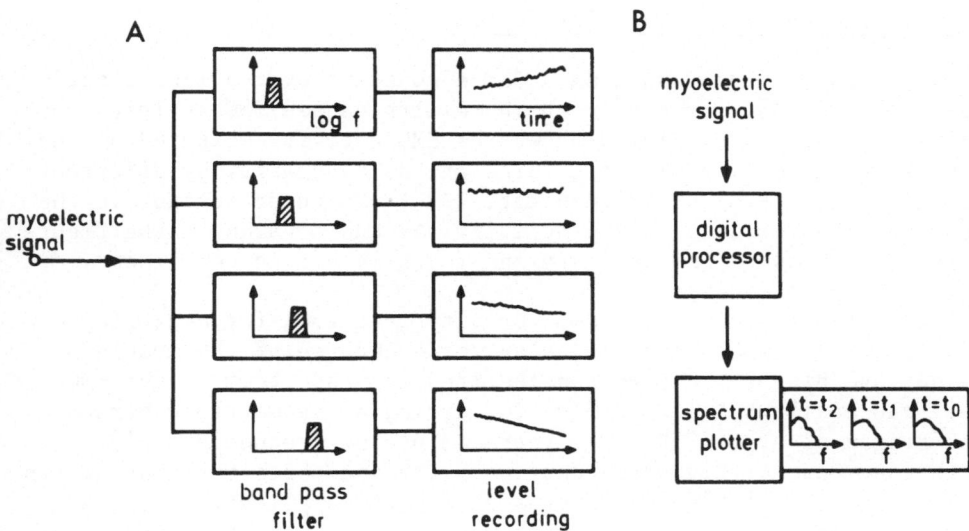

Fig. 8. Power spectrum analysis of myoelectric signal (Broman et al., 1973).

Analysis of Further EMG Parameters

Grieve and Cavamagh (1973) pointed to the advantages of analyzing the number of polarity reverses of the signal and extreme values for calculating the corresponding amplitude and time histograms. That kind of data processing is useful e.g. for investigating fast dynamic muscle contraction with and without limb movement.

APPLICATIONS

In modern man-machine systems high muscle strain is very seldom especially since his role has been transformed more and more towards that of a supervisory controller. The amount of force he needs for manual work decreased and consequently muscular strain of large muscle groups also decreased. However, the number of working situations in which muscular strain occurs in <u>small</u> muscles and muscle groups has increased, such as, e.g. in repetitive working movements. Muscular strain also occurs with static or isometric activation as discussed earlier and rapidly leads to overstrain of small muscle groups (Rohmert, 1967). General body fatigue which sometimes serves as a kind of safety alarm is largely missing because of the small sizes of strained muscles. Industrial groups have tried to use the EMG as an objective way to verify decreases in worker performance because of muscle fatigue and to find ways of improving work methods (Anderson et al., 1973; Radl et al., 1979; Rau and Radl, 1979).

EGM as an Indicator for Muscular Strain

When controling or holding a force constant over a longer period of time the muscular strain which results causes EMG activity to
increase continuously. Fig. 9 shows the EMG activity measured on the
M-biceps while holding a static force at the hand wrist by different
constant forces (Rau and Vredenbregt, 1970). Muscular fatigue is the
reason for the increase in EMG activity. As can be seen in the figure
the relation between EMG activity and force is not longer linear.

In most practical cases EMG activity is a tool for measuring
muscular strain rather than muscular force because of the muscle's
previous use history. Changes in the frequency spectrum of the EMG
signal can easily be seen in Fig. 10, during an isometric contraction (Broman et al., 1973). The increase of low-frequency activity
and the decrease of high-frequency activity indicates muscular strain
which indicates a fatiguing process during the application of static
muscular force. The contraction period in the illustrated case is a-
bout 30 sec. The pairs of brisk contractions provide a means of studying power-spectrum recovery.

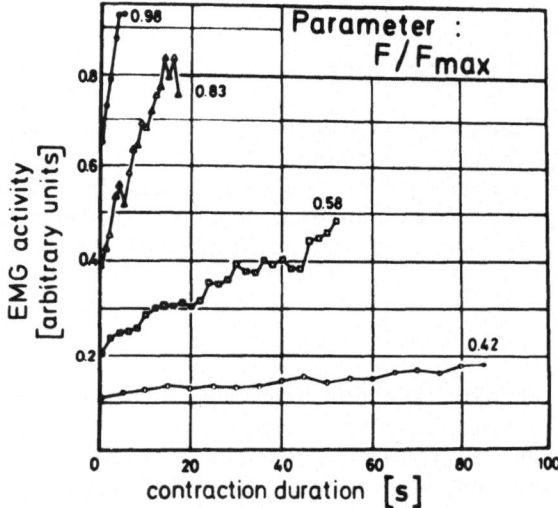

Fig. 9. EMG activity of different static continuous muscular con-
        tractions as a function of contraction duration (Rau and
        Vredenbregt, 1970).

EMG used as Biofeedback

Recent developments in EMG analysis are related to the phenomenon of biofeedback. Budzynski et al. (1970) demonstrated the clinical
utility of EMG feedback with tension headache patients. Much of their

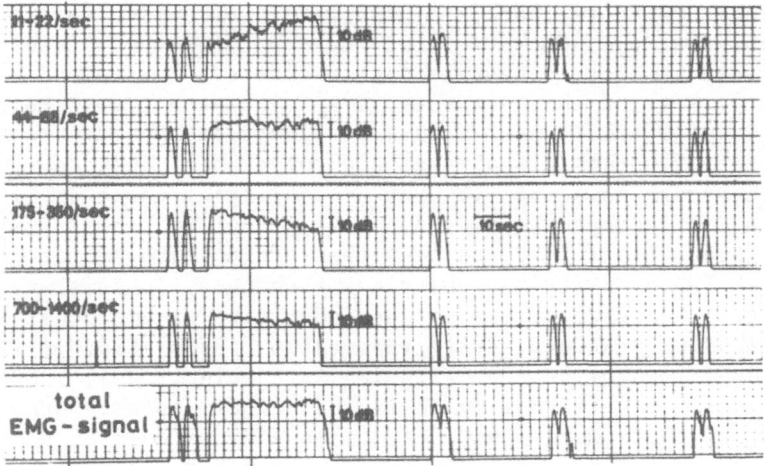

Fig. 10. EMG activity showing the events appearing during a maximal
voluntary isometric contraction analysed with a four-band
filter spectrum analyzer (Broman et al., 1973).

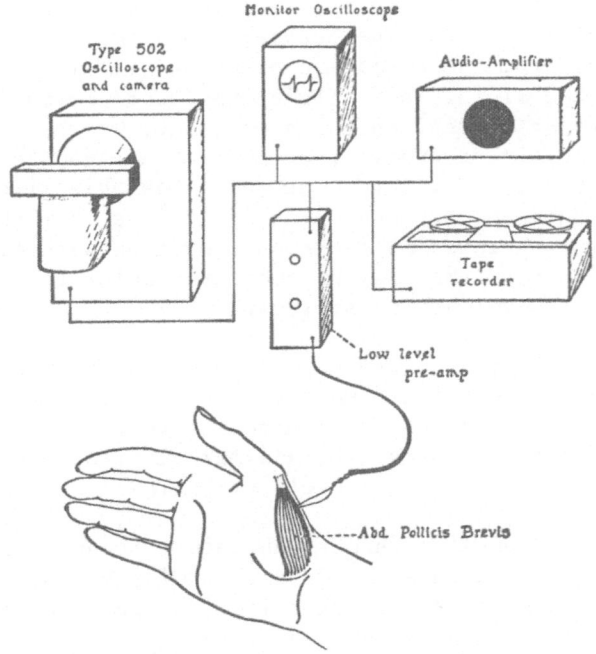

Fig. 11. Technique of recording from abductor pollicis brevis (Basmajian, 1979).

work combined EMG biofeedback training with progressive relaxation, behavior therapy, and home practice. The hypothesis behind this is that for every thought or mental state there is a corresponding muscle activity. Most of this activity is so minute that it remains beneath our level of awareness. With wire electrodes one can sense the change of action potentials of a single motor unit. This EMG technique is therefore so sensitive that by amplifying these small changes the EMG feedback becomes a powerful tool for investigating mind-body interactions. The trainees then are able to see or listen to their own EMG activity and observe any changes. In this way one can learn to make associations between cognitive or emotional states and corresponding muscle activity (Basmajian, 1979). Fig. 11 illustrates an experimental set-up. With implanted electrodes, individual motor units are identifiable by their individual shapes. Here, the subjects heard their motor unit potentials and saw them on monitors. The main muscle tested in all subjects was the right abductor pollicis brevis. In that experiment subjects learned to activate different single motor units by an effort of will. Some of them could control them so sensitively that they produced rhythms of contraction in one unit, e.g. imitating drum rolls.

EMG used as an Ergonomic Tool

A further approach with the EMG technique is the evaluation of controls. Here also one prefers to have objective criteria and to obtain reproducible relationships between EMG signals and the manual control situation. From the ergonomic point of view it can clearly be seen that the EMG signal is a helpful tool for estimating whether a muscle is active or in rest during a movement; whether contraction is continuously, static or dynamic; or whether a low or high force is generated. However, the interpretation of the objectively measured results are uncertain in most measuring situations. For that reason the meanings of the various EMG techniques must be determined through further research so that they may become more useful for solving all kinds of problems in man-machine systems.

In the last 20 years the use of muscle potential measures in aviation research was restricted mostly to a few attempts to measure the effects of flight stresses; determine the relationship between muscular and psychological stresses; and make limited attempts to measure alertness and effort (O'Donnell, 1979).

An attempt has been made in our laboratory to apply EMG measures to ergonomic problems. In this study, a control yoke which requires two-hand operation was tested to determine its operating ranges (Gärtner and Rau, 1978). The intention of this investigation was to find out the optimal form of the control yoke and the maximum permissible operating range in both rotating axes. In these experiments controls had no spring resistance. The control yoke has two rotating axes. Vehicle direction changes to the left or right accomplished by

turning the yoke as with a steering wheel of an automobile, called
here roll motion. Vertical vehicle direction changes are accomplished
by rotating the yoke handles towards or away from the operator which
will be called pitch motion. In the left of Fig. 12 is to be seen the
neutral position and in the middle and right pictures the extreme excursion during the pitch movement with a 45° roll angle position.
These two pictures illustrate the biomechanical position limits of
the hand when rotating the yoke towards and away from the operator.
The pitch motion of the hand towards the operator is accomplished by
radial abduction; pitch motion away from the operator is accomplished
by ulnar abduction.

With 0° pitch angle and roll motion to the right, radial preab-
duction will have occurred in the right hand and some ulnar pre-abduction in the left hand, thereby restricting the available amount
of further abduction for pitch command purposes. It can be shown
that with increases in roll motion to the right pre-abduction will
increase until biochemical limitations make pitch commands impossible
or very difficult. Similar pre-abduction occurs with left roll motions.

EMG activity for the right hand in a number of different roll
angle positions are illustrated in the upper picture of Fig. 13 as
a function of pitch angles for roll angles in the right direction.
The upper EMG value of "1" unit was arbitrarily given to the EMG level obtained when wrist joint pain was experienced after repeatedly
holding an angle position for a few seconds. The maximum value of the
curves (approximately .75 units) is obtained at the maximum pitch
angle which was measured. The maximum pitch angle measured was selected after experimentally determining the maximum pitch angle at which
no wrist pain build-up occurred during fairly long measuring sessions.
For any given pitch angle there is a tendency for pre-abduction to be

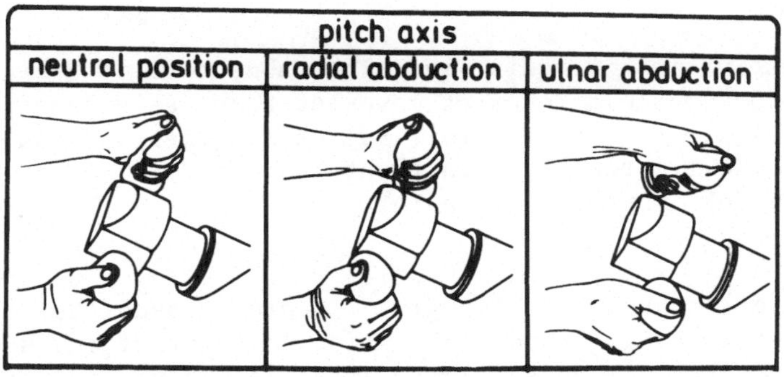

Fig. 12. Influence of roll axis rotation of a twin grip control yoke
on radial and ulnar abduction angles of both hands (Gärtner
and Rau, 1978).

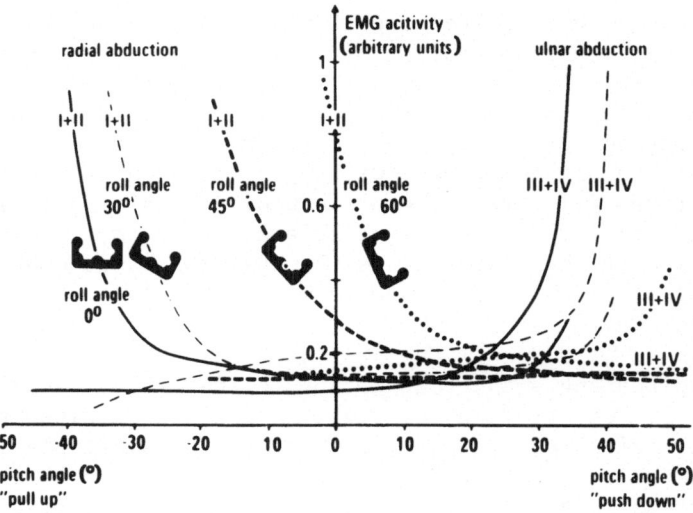

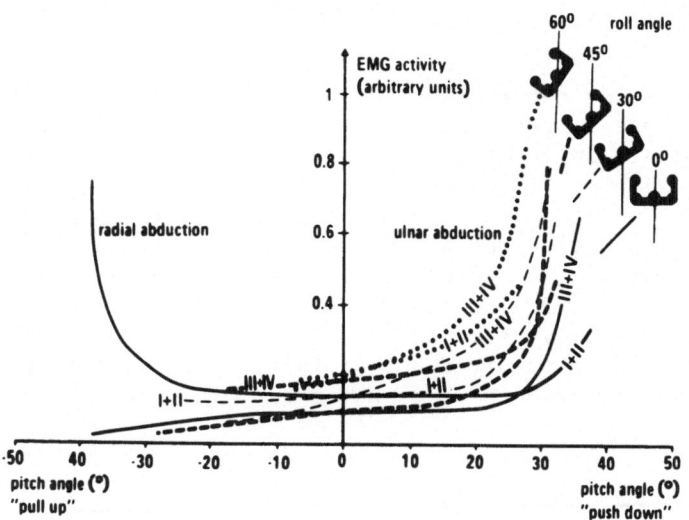

Fig. 13. EMG activities of radial and ulnar abductors of a right hand
with 90th percentile wrist movement range as a function of
pitch angle for various roll angles (Gärtner and Rau, 1978).
upper: Roll movement is in right;
lower: Roll movement is in left.

larger with larger constant roll angles. It can be seen that at 0°
roll angle the full range of possible wrist movement of the subject
can be used for pitch commands in both directions because there is no
pre-abduction. At 60° roll angle to the right, radial pre-abduction
is so large that no pitch angle movement in this "pull up" direction

is possible. In the right side of the upper picture of the figure the EMG curves for ulnar abduction i.e. in the "push down" direction is illustrated. The 45° and 60° roll angle permit relatively large "pull up" commands although the curves do not rise as high as those for radial abduction on the left side of that picture. The reason for this is that ulnar abduction of the left hand, which is not illustrated, reaches a limit at these pitch angles before the right hand, thereby preventing further ulnar abduction of the right hand. Of course, release of the control by the left hand would have permitted further movement.

EMG values for ulnar and radial abduction of the right hand is shown in the lower picture of Fig. 13 for left roll at various roll angles. As can be seen on the left side of the lower picture for left roll, right hand "pull up" pitch commands or radial abduction movement is so severely limited by radial pre-abduction of the left hand in all roll angle positions except 0° that further movements are not possible. The range of ulnar abduction or "push-down" commands illustrated on the right side of that picture is slightly reduced by ulnar pre-abduction thereby allowing considerable movement before the ulnar abduction limits are reached.

Discussion of the EMG-Measurements

In designing a range for this control device the following points are the most important to consider.
1. It should permit the largest possible pitch angle in both directions for each of the largest possible roll angles for subjects with 5th percentile wrist movement ranges.
2. Only lower levels of EMG activity should occur most of the time during control operations. Moderate EMG activity levels should occur very briefly and no high EMG activity at all.
These requirements can be satisfied for the subject tested with roll and pitch angle ranges of approximately $\pm 30^{\circ}$ each at which no more than 0.5 units of EMG activity are reached.

REFERENCES

Andersson, G., Broman, H., Magnussen, R., Petersén, I. and Örtengren, R., 1973, Vocational Electromyography: Investigation of Power Spectrum Response to be Repeated Test Loadings during a Day of Work at the Assembly Line, Research Laboratory of Medical Electronics, Mo. 6: 73, Chalmers University of Technology, Göteborg, Sweden.

Basmajian, J.V., 1967, "Muscles Alive", 2nd ed., Williams & Wilkins, Baltimore, Md.

Basmajian, J.V., 1979, Control and Training of Individual Motor Units, in: "Mind/Body Integration; Essential Readings in Biofeedback", E. Peper, S. Ancoli, M. Quinn (eds.), Plenum Publishing Corp.,

New York, 371-375.

Broman, H., Magnusson, R., Petersén, I. and Ötengren, R., 1973, Vocational Electromyography, in: "New Developments in Electromyography and Clinical Neurophysiology", J.E. Desmedt (ed.), Vol. 1, Karger, Basel, 656-664.

Buchthal, F., 1959, The functional Organization of the Motor Unit, Amer. J. Physical Med., 38, 125-128.

Budzynski, T., Stoyva, J. and Adler, C., 1970, Feedback induced relaxation: Application to tension headache, Journal of Behavior Therapy and Experimental Psychiatry, 1, 205.

Gärtner, K.-P. and Rau, G., 1978, Biomechanical analysis applied in designing the operating range of hand controls, in: "Biomechanics VI-B", E. Asmussen and K. Jørgensen (eds.), University Park Press, Baltimore, 213-219.

Grieve, D.W. and Cavanagh, P.R., 1973, The quantitative analysis of phasis electromyograms, in: "New Developments in Electromyography and Clinical Neurophysiology", J.E. Desmedt (ed.), Vol. 2, Karger, Basel, 489-496.

Kadefors, R., Kaiser, E. and Petersén, I., 1968, Dynamic Spectrum analysis of myopotentials with special reference to muscle fatigue, Electromyography, 8, 39-74.

Kogi, K. and Hakamada, T., 1962, Frequency Analysis of the surface electromyogram in muscle fatigue, J. of Science of Labour, 38 (9), 519-528.

Moore, J.C., Fabrication of suction-cup electrodes for electromyography, Electroencephalog. & Clin. Neurophysiol., 20, 405-406.

O'Donnell, R.D., 1979, Contributions of psychophysiological techniques to aircraft design and other operational problems, AGARDograph, No. 244, Technical Editing and Reproduction Ltd., London.

Radl, G.W., 1979, Methodische Untersuchung zur Ermittlung von Erholungszeiten beim Schweißen, Institut für Unfallforschung des TÜV Rheinland, TÜV-Nr. 013 008, Köln.

Rau, G., 1973, Ein verbessertes Meßsystem zur quantitativen Oberflächen-Myographie (An improved measuringsystem for quantification of surface electromyography), Anthropotechnische Mitteilung Nr. 2/73, Forschungsinstitut für Anthropotechnik, Werthhoven, West-Germany.

Rau, G. and Radl, G.W., 1979, Problems in the application of surface EMG as an indicator for muscle strain during electrical welding, to be published in: "Biomechanics VII", A. Morecki and K. Fidelus (eds.), Technical University of Warsaw, Poland.

Rau, G. and Vredenbregt, J., 1970, The Electromyogram and the Force during Static Muscular Contractions, IPO Ann. Progr. Rep. 5, 174-178.

Rohmert, W., 1967, "Untersuchungen über Muskelermüdung und Arbeitsgestaltung", Beuth-Vertrieb, Berlin.

Strong, P., 1979, Muscle Action and the Sensory System, in: "Mind/Body Integration; Essential Readings in Biofeedback", E. Peper, S. Ancoli and M. Quinn (eds.), Plenum Publishing Corp., New

York, 367-370.
Tichauer, E.R., 1978, "The Biomechanical Basis of Ergonomics", John
    Wiley & Sons, New York.
Vredenbregt, J., 1969, Small size Electromyograph, IPO Ann. Progr.
    Rep. 4, 161-162.

THE SMALLEST MANNED SYSTEM: CLOTHING

Wouter A. Lotens

INTRODUCTION

In daily life one would not think of clothing as a system in
the first place. Still, even in civil life, where fashion aspects of
clothing count heavily, the choice of clothing shows rather systematic trends. Most people have a summer and a winter wardrobe, enabling
them to choose the clothes that fit the environmental conditions. It
is recognized that the air temperature is the main factor in this
choice, although other environmental and even ergonomical factors can
play a role. In this way clothing can be considered as a systematic
tool to maintain thermal equilibrium. This may be called a behavioural aspect of thermoregulation.

However important the choice of clothing, it is merely one of
the behavioural aspects. In addition there are activity level and the
continuous look for shelter. Examples are sleeping during the heat of
the day and staying in the shade as much as possible.

It is easily understood that, when there is no free choice of
clothing, as is the case with protective clothing against military or
industrial environments, severe problems can rise in keeping the body
within reasonable temperature limits. In that case deliberate control
of workrate or worktime has to take over the function of behavioural
thermoregulation. If that measure fails to achieve the necessary control, the body heat control mechanisms (shivering, bloodflow, sweating) have to take care of the discrepancy, leading to loss of comfort
when the imbalance is small, developing to discomfort and decrease of
performance for larger imbalance and eventually to illness and complete incapacity.

Quantitative description of this process demands specific knowledge of several physical, biophysical and biomedical stages in this
rather complicated man-clothing-task system. In the next three sections an outline will be given of this system, methods will be explained for experimental approach and finally an example will be
given of application of the theory to a specific clothing item.

THE CLOTHING-MAN-TASK SYSTEM

A comprehensive outline of the clothing-man-task system is given
in Fig. 1, showing the three stages and their interconnections. The
state of the art in each of these stages will be discussed now in
more detail.

Although there is a long history in textile research the quantitative description of the clothing stage is not as sophisticated as
might be expected. Of the parameters that govern heat flow through a
fabric (i.e. insulation, wind penetration, vapour resistance, solar
absorption) usually only two are taken into account, insulation and
vapour resistance (Givoni and Goldman (1972), Nishi and Gagge (1970)).
The others have been documented and sometimes been implemented but a
versatile description has not yet shown up. Even more complicated is
the description of a complete clothing ensemble rather than a piece
of fabric. The effect of special constructions for the purpose of
ventilation is reported to be disappointing (Shivers et al., 1977) and
strong interactions have been found between clothing and subjects,
maybe due to the fit (Lotens, 1979). Mecheels and Umbeck (1977) designed a special coverall with controllable ventilation, that gives
encouraging results when tested on a heated manikin, as has become increasingly popular in recent years. It may well be, however, that the
results would be less convincing with real subjects. Another problem
in calculating the heat flow through clothing is the influence of the
task on clothing (see Fig. 1). Pumping effects and ventilation due to
motion do change the insulation of clothing to a great extent. The
same holds for the other feedback line in Fig. 1, the influence of
sweat, wetting the clothing either by capillary action or by conden-

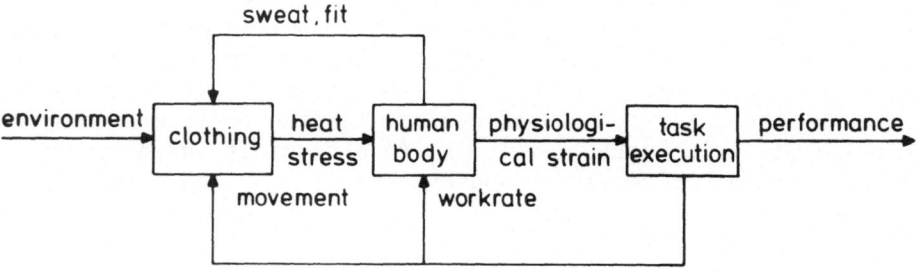

Fig. 1. Three stages in the clothing-man-task system, relating envi-
        ronment to performance.

sation of vapour. This effect is most pronounced, of course, in the cold.

The next stage in Fig. 1 deals with man. The output of the clothing stage is the heat exchange with the skin, serving as an input to the body. Another input is the work rate. If the work rate is such, that the heat exchange through the clothing is insufficient, the body will gain heat. This heat storage will be discussed more extensively in the next section. As a result of the heat storage, physiological parameters like central temperature, cardiac output, body heat content, etc. will rise and the level of those parameters is called the strain. (It should be stressed here that strain relates to the thermal condition of the man (accumulated heat, body temperature), whereas stress relates to the thermal inputs (heat flow, workrate)). A quantitative description of the human thermoregulation, therefore, demands an output in terms of physiological strain as a result of input in terms of stress. Stolwijk (1971) gave an elegant description in his mathematical model of human thermoregulation by dividing the process in two steps, the passive body response, following the physical laws of heat exchange, and an active controller, giving the shivering, blood flow and sweating as a function of skin and internal temperatures. This is a very general and promising approach, still lacking, however, the necessary accuracy of the control functions. Another weakness is that the model does not take into account all kinds of individual variation, just being valid for an 'average man'.

Givoni and Goldman (1972, 1973) took another approach. They integrated the first two stages of Fig. 1 into one set of regression equations, fitting a large amount of data. The result is a description of physiological strain when environment, clothing parameters and workrate are known. The descriptions are reliable and make the important distinction between acclimatized and non-acclimatized man. A disadvantage is that the model does not give any information about other parameters than those that have been explicitly integrated.

In a contest between the two models, over a large range of environmental and clothing conditions, the Givoni and Goldman model shows better agreement with experimental strain values, especially for heavy strain. For certain applications, however, this regression model fails (for example impermeable clothing). Nevertheless it has proved to be very useful and an example of its application will be given in the last section of this paper.

The last stage in Fig. 1 is the task execution. The performance may be conceived of as a result of the physiological strain and the specific demands of the task. As strain increases, performance may decrease and eventually either the endpoint of voluntary tolerance will be reached or the point of physiological disfunctioning, completely disabling the wearer. According to this last view illness is physiological strain beyond the limits of tolerance. One of the main problems to be solved is the establishment of those limits.

Mainly on the basis of available records of heat casualties (heat stroke, heat exhaustion) in Israel (Shibolet et al., 1969) and the U.S. (Schickele, 1947; Yaglou and Minard, 1957) we have made an analysis of risk of heat illness in terms of the strain parameters core temperature (T_c) and accumulated heat storage (Sto) (Lotens, 1978). In this analysis distinction has been made between various tasks and various durations (Table I).

The tasks have been separated in working, standing and lying down. The main difference between these tasks is the load on the circulatory system. At raised levels of body temperature vasodilation causes blood storage in the large vessels of the lower body parts, resulting in insufficient recirculation to the heart. This may be overcome either by muscle action, working as an additional pump (work), or by compensation of gravity (lying down).

Distinction between various durations of the strain has been made because of the common experience that the resistance against heat is decreasing with time. The production of sweat is directly related to the equilibrium that can be maintained. After two hours, however, the sweat production decreases, resulting in lower maintainable equilibrium temperatures. For this reason tolerance limits are lower for longer exposures.

In conditions of high thermal load, the heat has not even time to reach the core within the tolerance time. In these conditions body

Table I. Strain limits for core temperature (T_c, $^{\circ}C$) and accumulated heat storage (Sto, J/g) in order to prevent heat casualties down to a risk level of ca. 20/10,000 man-days.

Duration		Task		
		Working	Standing	Lying down
less than 1 hr	T_c	39.1	38.0	39.1
	Sto	8.4	8.4	10.5
between 1 and 2 hr	T_c	39.1	–	39.1
	Sto	6.3	–	–
over 2 hr	T_c	38.8	–	–
	Sto	–	–	–

- means no limit established

temperature evidently is a bad measure. An analysis of accumulated body heat (Sto) shows that for short tolerance time (relatively hot skin) Sto is only slightly different from that for long tolerance time (relatively hot score). This means that Sto is a suitable strain indicator for a variety of situations whereas T_c is only suitable for long tolerance times.

The decrease in performance with increasing physiological strain, below the level of illness, is still an almost open research area. For mental as well as physical tasks a problem is that performance is strongly related to the motivation and the literature is often very unspecific in this respect. The results are, in consequence, confusing.

EXPERIMENTAL METHODS

For the actual evaluation of a certain clothing ensemble the usual method is to reconstruct the heat equation and establish the clothing parameters. In its most general form the heat equation may be written as:

$$M = S + D + E + Wext \qquad (W)$$

where: M   = rate of metabolic energy production
         S   = rate of heat storage in the body
         D   = dry heat exchange
         E   = evaporative heat exchange
         Wext = rate of external work performed.

The terms D and E are dependent on the clothing parameters I_{cl} (insulation) and I_m (water vapour permeability) according to the following relations:

$$D = \frac{0}{I_{cl}} (T_{skin} - T_{air}) \qquad (1)$$

$$E = \frac{2.2 \; 0' \; I_m}{I_{cl}} (P_{skin} - P_{air}) \qquad (2)$$

where: 0   = surface area of the man (m^2)
         0' = wet surface of the man (m^2), equal to 0 for fully wet
               skin
         I_{cl} = insulation ($m^2 \; {}^{\circ}C/W$)
         I_m = vapour permeability (no dimension) (I_m = 1.0 means equal
               to that of air)
         T   = temperature (${}^{\circ}C$)
         P   = vapour pressure (mm Hg)
         2.2 = conversion factor relating mm Hg to ${}^{\circ}C$ for a wet surface.

When D and E are known, the clothing parameters can be calculated from (1) and (2), serving as an input for predictive models in order to calculate the range of environmental conditions a particular clothing ensemble is suited for, i.e. the psychrometric range. As criterion for tolerance Sto (the integrated S) or T_c may be used, as has been pointed out in the preceeding section.

The quantities of the heat equation may be measured in an experimental set-up by means of various sensors, while the subject is performing work on a treadmill or a bicycle-ergometer, placed in a climatic chamber.

M is calculated from the oxygen uptake as may be measured either by an open system (expiratory air is caught together with an excess of fresh air) or by a closed system (expiratory air is measured directly). Many techniques exist for measuring flow and O_2- and CO_2-concentrations in the laboratory as well as in field conditions.

Sto, the accumulated heat storage, may easily be calculated from the rise in central and skin temperature, using Burton's weighting factors or others.

D is hard to measure directly. Heat flow disks or infrared thermometry may be used but the interpretation of the results is often difficult. A more practical approach is to calculate D from the heat equation when the other quantities are known. A condition is that D is large enough to be obtained with sufficient accuracy, being subject to the combined error of the other terms.

E can rather easily be calculated from the subject's weight loss. To determine that, the subject's task has to be performed on a very stable and accurate balance which allows continuous recording of his weight. An alternative is to interrupt his task for a weight check. Care should be taken of the dripping of sweat, changing the weight without contributing to E.

Wext is dependent on the task. The highest efficiency is about 20%, i.e. for cycling. For walking the external work is virtually zero, except for potential energy gain at inclined surfaces or losses at soft road tops. This does not mean that the efficiency of the muscles is less during walking than during cycling but internal friction converts part of the power of the muscle into heat, thereby decreasing Wext.

P_{skin} has to be known to calculate I_m from E. The vapour pressure near the skin may be measured with electronic humidity sensors of very small dimensions (see Fig. 3).

The total number of sensors that must be applied to the subject amounts to a large number, 20 or more is not unusual. Most of these

sensors need only a slow sampling rate, because temperature, humidity
and heat flow change only slowly. Repiratory parameters and heart
beat (often used as a sensitive control on the condition of the subject), however, demand a relatively high sampling rate. For convenient
experimenting, therefore, a data logger is needed with a lot of slow
channels and a few fast ones. At our laboratory this has been accomplished with a purpose built apparatus that samples the data, pream-
plifies to a standard level, converts the analog signals to digital
information, modulates this information to a pulse coded signal and
eventually transmits it to a receiver (Fig. 2, top). A special feature is the flexible choice of slow and fast channels between the limits of 256 channels at a sampling rate of 1 Hz and one channel only at
a sampling rate of 256 Hz. An usual configuration exists of 64 channels of 1 Hz (32 temperatures, 16 rel. humidities, 16 heat flows), 2
channels of 32 Hz (oxygen concentration and flow), 1 channel of 64 Hz
(cardiogram) and 1 spare channel of 64 Hz.

All hardware is combined in a single box that can be either put
down near the subject in laboratory experiments (thereby avoiding unnecessary cables) or carried by the subject as a backpack in field
tests. The electronics is of a high standard as to operational temperature and power consumption. Fig. 3 shows the apparatus, together
with some sensors.

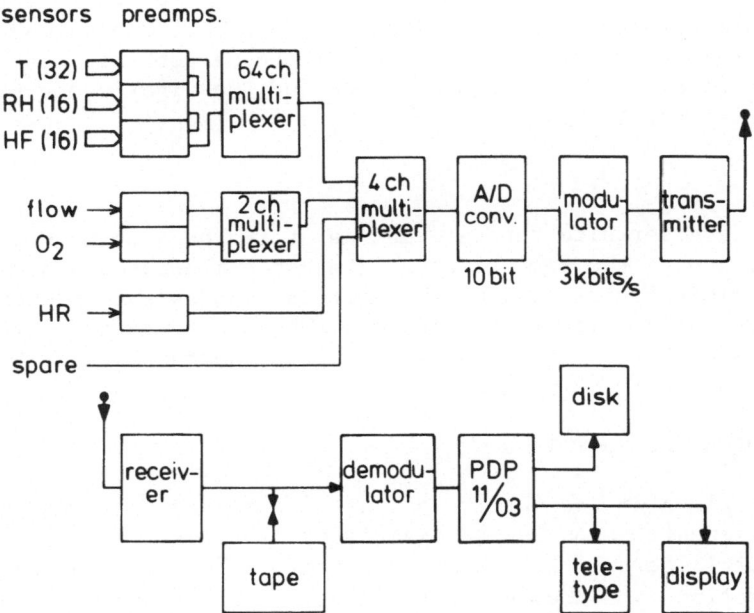

Fig. 2. Schematic diagram of data sampling and processing at the
        transmitter side (top) and the receiver side (bottom).

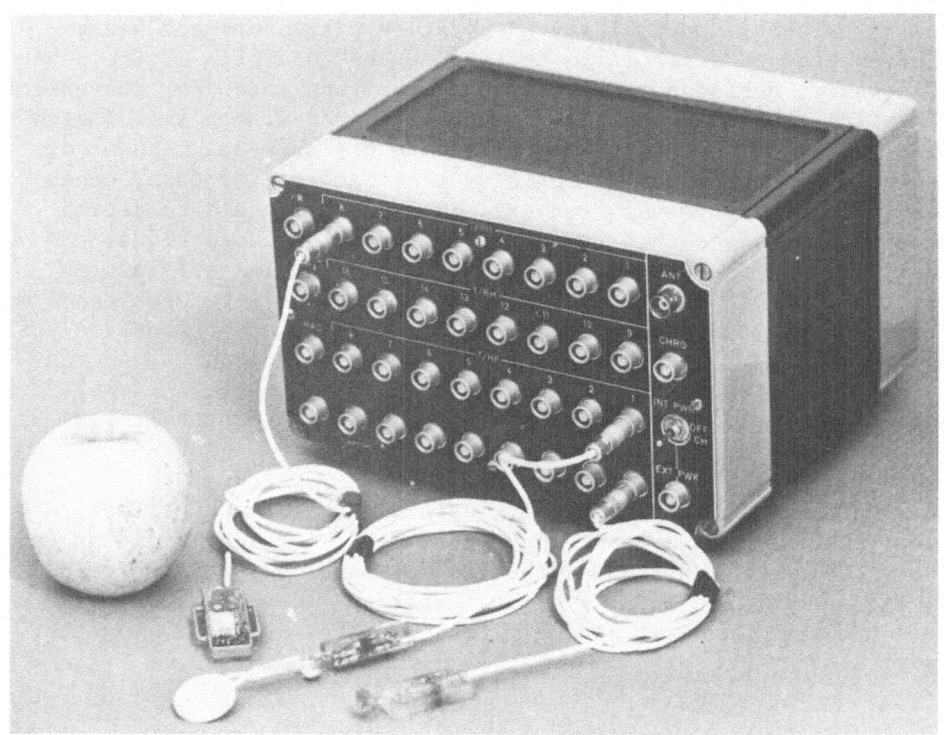

Fig. 3. Portable apparatus, containing the top half of Fig. 2, to-
         gether with some sensors. The left sensor is a combined hu-
         midity/temperature type, the middle one a combined heat flow/
         skin conductance/temperature type and the right one a single
         conductance type.

    At the receiver side the data may be stored on a commercial au-
dio recorder and at the same time or later on demodulated and interpreted by a minicomputer, giving the average results each minute on
the typewriter or on electronic displays. The minute data are filed
and stored on disk for later analysis (Fig. 2, bottom).

EVALUATION OF A NBC-SUIT

    After explaining the interrelations of a manned clothing system
and the experimental methods, an example of clothing evaluation will
be worked out (liberally adapted from Lotens, 1979). The garment under consideration is a chemical protective, water vapour permeable
suit that is usually worn in combination with a gasmask. Because of
the protective demands the construction of the fabric is rather heavy,
resulting in high insulation and low permeability. The garment is

worn over an undergarment and completed with butylrubber boots and
gloves. Eight subjects, who had been medically examined in advance,
participated in an experiment in the climatic room. Before entering
they were provided with sensors and clothing. Inside, they performed
60 W of external power on a bicycle-ergometer during 1 hour, under
environmental conditions of 30.4°C air temperature and 53% relative
humidity. Wind speed was 0.25 m/sec and there was no additional radiation. The bicycle ergometer was mounted on a balance for continuous weight recording. The experimental set-up is shown in Fig. 4.

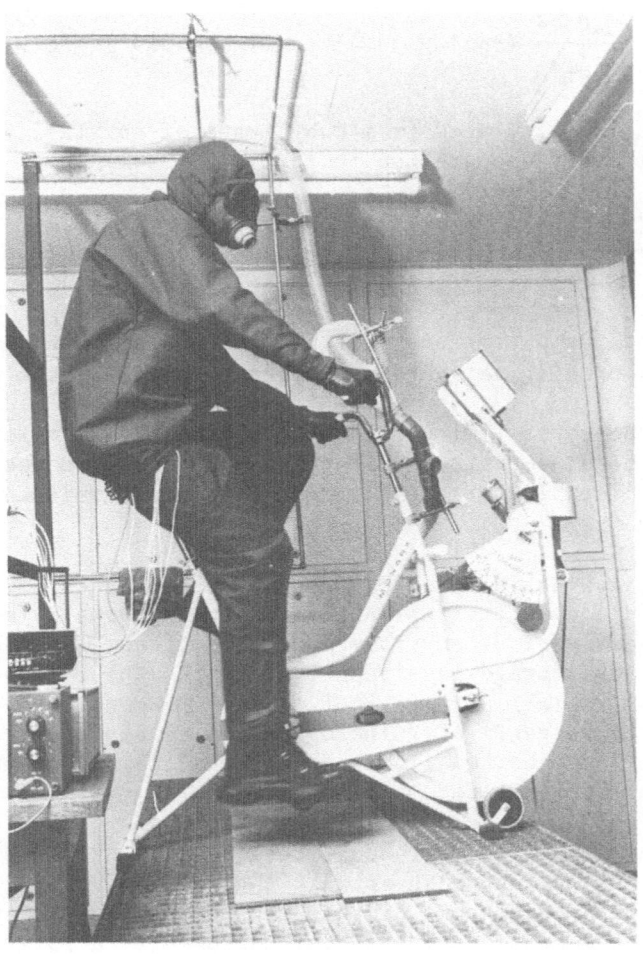

Fig. 4. Experimental set-up, showing a subject in a NBC-suit, working
        on the bicycle ergometer. The tube for O_2-measurement is dis-
        connected.

From the measurements the heat equation, after 60 min of cycling, was reconstructed as:

$$M = S + E + D + Wext$$

$$400 = 153 + 146 + 41 + 60$$

Skin and rectal temperature rose from 33.0 to 36.1°C and from 37.0 to 38.7°C respectively. The skin was fully wet, so the sweat evaporation may be estimated as maximal.

I_{cl} may now be calculated from D by means of eq. (1), applying a small correction (2 W) on D for respiratory dry heat loss.

$$39 = \frac{2.0 \text{ m}^2}{I_{cl}} (36.1 - 30.4) \qquad I_{cl} = .29 \text{ m}^2 \ {}^\circ C/W$$

Analogously I_m is calculated from E by means of eq. (2). The correction for respiratory evaporation is 26 W.

$$120 = \frac{2.0 \text{ m}^2 \times I_m \times 2.2}{.29} (44 - 16) \qquad I_m = .28 \text{ (n.d.)}$$

These clothing parameters will be used now as inputs for the Givoni-Goldman predictive system in order to evaluate the clothing in two ways:
- To predict the upper limit of the psychrometric range for a moderately heavy task, performed during 1 and 2 hours, respectively.
- To predict tolerable work/rest cycles for an intermittently performed heavy task.

The first evaluation has been done by computer calculation of the accumulated body heat (Sto) during a simulated walk (metabolic rate 350 W) under various environmental conditions. Those conditions that cause a heat storage less than 8.4 J/g in 1 hour resp. 6.3 J/g in 2 hours (see Table I) are part of the psychrometric ranges for 1 and 2 hours work respectively. The upper limits of those ranges are drawn in Fig. 5. The borken lines are for unacclimitized men and the solid lines for acclimitized men.

Fig. 5 shows a large difference between 1 and 2 hours walking. The reason is that during a short walk a great deal of the heat can be taken up by the body, thus diminishing the necessity for cooling. For longer periods body heating is only allowed at a lower rate, the cooling becoming increasingly important.

Because of the low permeability of the clothing, the low sweat production of unacclimatized men is already large enough to reach

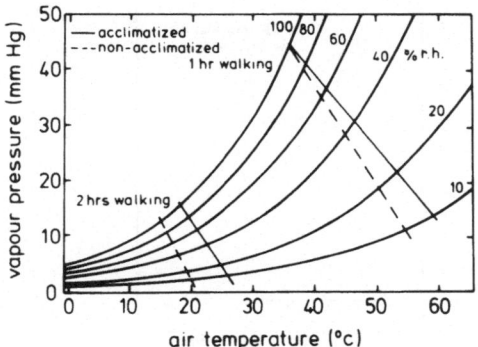

Fig. 5. Psychrometric diagram, showing lines of equal relative humidity in an air temperature - vapour pressure plot. The
straight lines represent the upper limits of the psychrometric range for 1 resp. 2 hrs working at a rate of 350 W in the
NBC-garment. Broken lines are for unacclimatized and solid
lines for acclimatized men.

maximum evaporation. So the larger sweat production in acclimatized
men is not very helpful any more except for the larger sweating skin
area. For this reason there is only a slight difference between acclimatized and unacclimatized men.

The second evaluation is concerned with another way to prevent
overheating, i.e. doing the work intermittently. Again computer calculations have been carried out on the accumulated heat storage, this
time for a heavy task (550 W) during 3 hours. The work time percentage has been varied, keeping the cycle time (work time + rest time)
constant at 1 hour. Fig. 6 gives the maximum tolerable work time percentages for a range of environmental temperatures. The relative humidity is supposed to be the same for all conditions, i.e. 60%.

Even at moderate temperatures (ca. 15°C) a work/rest scheme of
15/45 min turns out to be the limit for this heavy task and this type
of clothing.

These kinds of evaluation make it possible to support wearers of
special clothing with rules of thumb which they can use to protect
their safety. The same procedure is applicable to larger manned systems that have an indoor climate. The model looks then like an onion.
The kernel is the man, his clothing is the first peel and each additional system containing the preceding one may be handled like another peel. The description of the larger peels (for example mobile
shelters, vehicles, buildings, etc.) is often easier (because of
their physical character) than the description of the smallest manned
system: clothing.

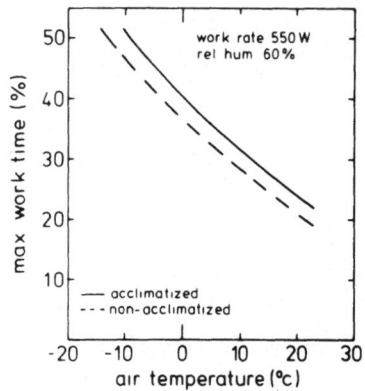

Fig. 6. The maximum tolerable work time percentages as a function of
        air temperature for an intermittently performed task at a
        work rate of 550 W.

REFERENCES

Givoni, B. and Goldman, R.F., 1972, Predicting rectal temperature
    response to work, environment and clothing, J. Appl. Physiol.,
    32, 812-822.
Givoni, B. and Goldman, R.F., 1973, Predicting heart rate response
    to work, environment and clothing, J. Appl. Physiol., 34, 201-
    204.
Givoni, B. and Goldman, R.F., 1973, Predicting effects of heat ac-
    climatization on heart rate and rectal temperature, J. Appl.
    Physiol., 35, 875-879.
Lotens, W.A. and Smienk, E.J.M., 1979, A comparative test of A2-
    clothing II: Thermal aspects (in Dutch), Institute for Percep-
    tion TNO, Report No. 1979-21, Soesterberg, The Netherlands.
Lotens, W.A., 1978, Criteria for maximum tolerable heat stress – a
    proposal (in Dutch), Institute for Perception TNO, Report No.
    1978-13, Soesterberg, The Netherlands.
Mecheels, J.H. and Umbach, K.H., 1977, The psychrometric range of
    clothing systems, in: "Clothing Comfort", N.R.S. Hollies and
    R.F. Goldman (eds.), Ann Arbor Science Publishers Inc., Ann
    Arbor.
Nishi, Y. and Gagge, A.P., 1970, Moisture permeation of clothing –
    A factor governing thermal equilibrium and comfort, ASHRAE
    Trans, 76, part I, 137-145.
Schickele, E., 1947, Environment and fatal heat stroke, The Mil.
    Surgeon, 100, 235-256.
Shibolet, S., Coll, R., Gilat, T. and Sokar, E., 1968, Heat stroke:
    its clinical picture and mechanisms in 36 cases, Quart. J. Med.,
    36, 525-548.

Shivers, J.L., Ych, K., Fourt, L. and Spivak, S.M., 1977, The effect of design and degree of closure on microclimate air exchange in light weight cloth coats, in: "Clothing Comfort", N.R.S. Hollies and R.F. Goldman (eds.), Ann Arbor Science Publishers Inc., Ann Arbor.

Stolwijk, J.A.J., 1971, A mathematical model of physiological temperature regulation in man, NASA Contractor report CR-1855.

Yaglou, P. and Minard, D., 1957, Control of heat casualties at military training centers, Arch. Industrial Health, 16, 302-316.

ANALYSIS OF HUMAN MOVEMENTS FOR WORKPLACE DESIGN

Klaus-Peter Holzhausen

INTRODUCTION: WHY SHOULD HUMAN MOVEMENTS BE ANALYZED?

The design trend in modern systems is towards greater use of automation and computers. This has led to an increase in complexity along with increases in system performance. More functions and modes are now demanded from our systems, the consequence of which are greater rather than lesser task demands on systems' personnel. Some of this task load has been reduced through functional and physical integration of controls and displays, but the overall effect in many cases is for increased operator demands.

Much of these demands involve a large number of interactive processes with the system, such as pressing buttons, moving levers, making keyboard entries or performing continuous control tasks as in manually flying aircraft.

These control activities can be made easier and more efficient by reducing the number and size of movements and decreasing the probability of errors. This is accomplished by ergonomists through:

- proper function allocation,
- use of functional and physical integration (both of controls and displays),
- optimum design of control elements,
- establishing natural control-display relationships,
- optimizing the arrangement and layout of controls and displays,
- determining the proper console shape and dimensions.

In taking these analytic and design approaches to reduce workload, the ergonomist's tasks are complicated by a number of con-

337

straints of various kinds. These include physical space restrictions, limits to human reach and manipulative ability, task sequence requirements and time restrictions of tasks to be performed. Also there may be conflicts between various human engineering layout principles. For example, grouping controls according to function may sometimes conflict with locations according to frequency and criticality. Layouts to satisfy sequence of task requirements may conflict with principles of placement of controls to balance task loads for two-handed operation.

To resolve these conflicts, making optimum trade-offs or to evaluate alternative designs or layout, experienced and knowledgeable ergonomists need certain kinds of qualitative and quantitative anthropometric data to supplement all of the other function, task, and system data which must be used.

The usual anthropometric tables of structural and functional body dimensions and the reach and range of movement data are very useful but insufficient because they are basically static. In trying to obtain dynamic data the design ergonomist can simulate some motions in his office or in his imagination but obviously can only do so in a limited and qualitative way.

The need is for techniques which are both quantitative and dynamic in order to obtain the kind of anthropometric data required to best solve workplace design problems.

Techniques for measuring, analyzing and evaluating human movements during the performance of tasks as a workplace are critically important aids to design and evaluation because they permit us to monitor movements, record errors and understand the complexity and efficiency of a series of movements. Furthermore, we need such techniques because static tables of anthropometric data do not describe the great variety of movements which occur due to the variety of workplace layouts and environments, biomechanical properties of different operators, and different tasks and operator techniques which may be encountered.

An example of the variability in human operators only with respect to major physical dimensions is shown in Fig. 1, taken from a survey of U.S. Flying Personnel. It can be clearly seen that these four people who are at the 5th percentile in height vary a great deal with respect to other important dimensions. For example, subject A with his 0,5 percentile arm/thumb reach and 20th percentile sitting height cannot reach aircraft controls properly but has no problem to see out of the window, whereas subject D can reach all controls but for him the windows are too high (McDaniel, 1980). From these measures can be expected how different required seat positions reach envelopes and line of sights will be in consoles to be designed.

Subject	Height (Stature)	Sitting Height	Thumb-Tip Reach	Buttock-Knee Length	Knee Height Sitting
A	5	20	0.5	3	6
B	5	38	2	16	2
C	5	0.5	24	62	10
D	5	0.05	74	48	55

Fig. 1. Design related dimensions on four U.S. Flyers having 5th percentile stature (McDaniel, 1980).

Techniques for measuring and analyzing human movements may involve the use of real subjects working with existing workplaces, prototype consoles, or with passive or active mockups. Data collected in this way may later be useful in modeling human movements. Another way of analyzing human movements is through the use of simulated subjects or simulated limbs on a computer display using computerized human movement models and computer generated workplace elements.

In detail analysis of movement can help to answer the following questions:

- which is the time consumption of switch actuation or other console operation?
- which is the movement trajectory of the hand/finger selected from different starting points towards a control?
- are dangerous areas (moving parts etc.) close to the path of movement?
- does the trajectory change during repetitive movements as an early signal of fatigue?
- does the required movement path come close to the extreme reach limits (known from reach envelope studies)?
- does the speed and acceleration function with respect to time indicate an improper switch or key design?
- do movements themselves contain a reproducible communicative quality so that these movements may be evaluated as machine inputs or control signals?
- are there major reproducible control movements that lend themselves for new and superior control design?

PROBLEM: WHICH MOVEMENT SHOULD BE ANALYZED?

A clear distinction should be made between different types of movements. The types are defined in Fig. 2.

Free movements can be reach movements that are performed by the human operator to relocate his foot/leg or arm/hand to reach a control such as a console switch, pedal, or lever, etc. Free movements can also be evaluated with respect to their communicative content

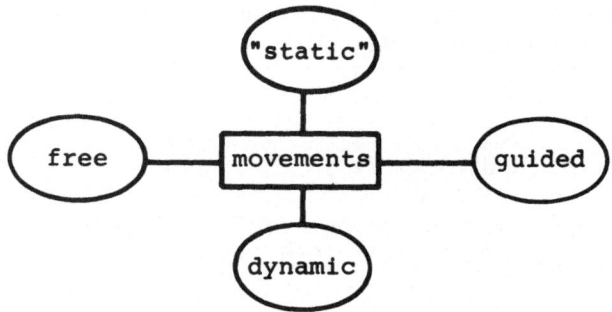

Fig. 2. Types of movement.

such as signalling information to others or expressing system inputs in a way of remote system control. There are no path restrictions on such movements but a dependency on the body's force and
mass and its distribution and the movement experience and pretraining of the operator. The human body is basically an open-chain system of body links rotating around joints. The end members of these
links, the hands and feet, can occupy a limitless number of positions in space as a result of the cumulative ranges of these joints
(Dempster, 1955).

Guided movements can also be described as restricted control
movements which are performed by an operator to actuate a control
element such as turning a steering wheel or pushing a lever. These
movements are restricted by the control's range of movement limitation and task demands. The type of movements made can be as well
control information as in the case of pursuit or compensatory tracking or force input as in industrial plants. The distinction between
"static" and dynamic movements refers to aspects of movement interpretation rather than to different types of movements. It indicates
that there are different points of view of how to evaluate data of
the same movement. "Static" movement data are anthropometric data,
position information of movement for anthropometric workplace layout whereas dynamic movement data describe trajectories as functions
of time whereby travelling time, speed, accelerations and other aspects of dynamic processes are evaluated.

METHODS FOR THE MEASUREMENT OF HUMAN MOVEMENTS

Measurement of human movements can be done by a variety of different technical approaches. These can be subdivided into a number
of classes. They are mechanical, photographic, opto-electronic or
fully electronic.

The later mentioned techniques all deal with movements of the
human arm/hand system with the operator seated in a cockpit or con-

sole chair. This type of workplace is most frequent in consoles for monitoring and control and command purposes. The operator's work under evaluation does not require physical power but consists of monitoring, decision making and switch actuation as well as guidance and tracking operations using the full spectrum of control devices like joystick, roll ball, light pen, Touch Input Device on screen displays etc.

A number of qualities are necessary or desirable in any measuring method:

- Measuring or recording movements should cause little or not degradation of the operator's performance.
- Measurement must be applicable to the workplace of interest with no restrictions because of climate, lighting conditions, etc.
- Measurement apparatus must not require too much installation or set-up effort, especially in existing consoles.
- Measurement should be performed automatically with a small amount of manual labour, nor for the data collection neither the data processing.
- A display device should be on hand for the viewing or recorded movements.
- Algorithms for analysis of the data should be available.

Mechanical Methods

Mechanical methods have historically been associated with anthropometry. It is obvious that acquisition of anthropometric data is still and will be in future the basis of any evaluational process and any modeling approach which depends on

- a broad range of anthropometric data and
- a good knowledge of movement patterns.

Mechanical methods have mostly been used to measure reach envelopes of operators sitting in a cockpit or at a workplace. The measurement of such envelopes in mockups is superior to traditional one-dimensional anthropometric measurements because it takes postural changes during reach movements into account. Therefore design errors based on erroneous assumptions, such as, e.g., the shoulder remaining at rest for certain reach movements, can be avoided. Fig. 3 shows a construction used for reach measurements in the U.S. Air Force. The apparatus allows free movement of the arm/hand system and gives a readout of the achieved reach width by positioning a measuring rod towards the hand or finger. A different device for measuring arm reach envelopes is shown in Fig. 4, which is a very simple setup using moveable rods that are pushed backwards by the operator performing his movement. Measurement or recording using the predescribed methods degrade operator performance because there is no true console environment possible. Rather than measuring under

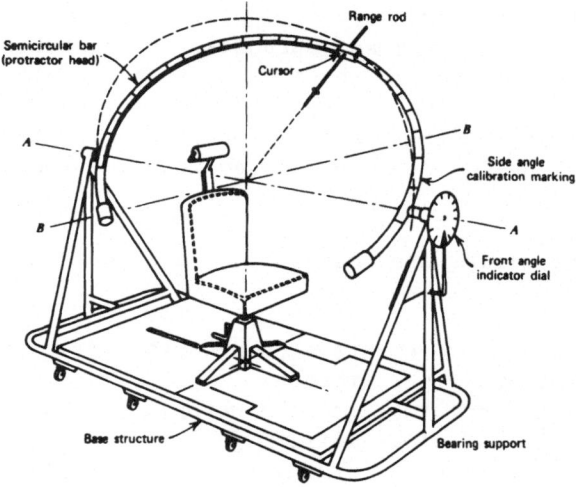

Fig. 3. The Frankenstein Anthropometric Measuring Apparatus (Wright, 1963).

working conditions the operator is performing under conditions more similar to basic research tools. The movement measuring apparatus requires a great deal of installation. One line recording of movement is possible with the Frankenstein apparatus but not in the case of simple arm reach measuring device, interaction between the experimental procedure and the experimenter is not available.

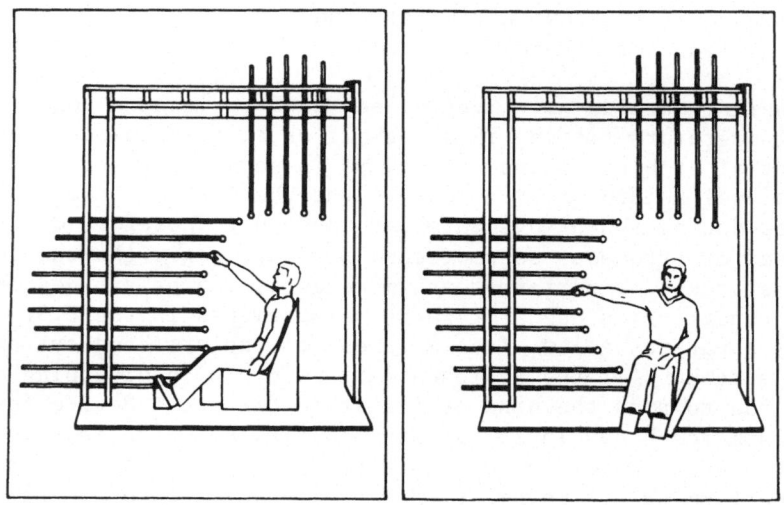

Fig. 4. Arm reach measuring device (Dempsey, 1953).

Photographic Methods

The oldest technical methods for the measurement of human movements are photographic. As early as in the 19th century Braune (1895)
photographed walking persons wearing black trousers with vertical
lines on them in time intervals of exposures on the same picture to
show the position of the human leg as a function of time in a normal walk.

Other experimenters recorded repetitive movements, for example,
rotational plane movements or cycling of the hand by mounting a light
bulb on the hand and photographing the movement in a dark room with
the camera shutter open. Because of these rotational movements this
simple method was first called cyclography. One of the disadvantages
of the method is that the light track on the film has no time reference.

Today this basic method is used in an improved way. To relate
the light tracks to the worksituation a single flash exposure of the
workplace is added to the same picture. Thus, movements can be monitored under normal daylight conditions with IR-diodes and film, instead of lamps and daylight film.

A further refinement is the gliding cyclogram. A camera is e-
quipped with a slot shutter moving at a constant defined speed over
the film plane thus producing a plot of the displacement versus time.
Further refinements of this method led to chronocyclography. By using
a high speed shutter or a rotating disk with a number of holes close
to its edge in combination with a camera, light tracks are interrupted so that a time reference is visible. Fig. 5 shows paths of a
jumping movement in sports. Wide time line distribution indicates high
speed while at the beginning and end of the movement the speed was
lower.

Comprehensive analysis of the kinematics of human movement with
no loss of information is achieved with cinematography (Adrian, 1973).
35 mm film gives a sufficiently high resolution. Filming techniques
have also been used for high speed whole body movements (Kallin et
al., 1979).

All these optical methods have the disadvantage in common that
it is difficult or impossible to use computer aided data processing.
Cyclographic data are evaluated in a qualitative way with no quantitative criterion to be applied though it gives a good overview on
the reproducibility of paths of repetitive movements. Chronocyclo-
graphic data lend themselves to a certain extent to speed evaluation
though the data processing must be done manually and overlapping of
light tracks will render misleading results. Cinematographic films
can be evaluated frame by frame, if body marks were fixed at the subject's body. Digitizing of those landmark positions can be performed

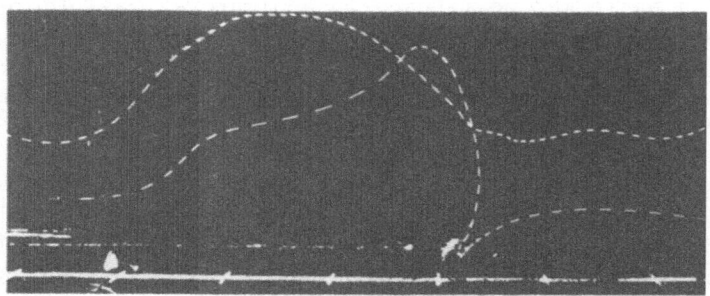

Fig. 5. Trajectories of hip and ankle joint during jump
       (Gutewort, 1968).

by computer controlled digitizing equipment which is a sufficiently
accurate but tedious and time consuming way.

Electronic Methods

     The simplest electronic method is the rubber band goniometry
(Neukomm, 1974). The subject's movements are measured with respect
to that plane in which a potentiometer can be actuated. It is mounted at a distance and connected to the human operator with a special
rubber band. In the proper setup, with selected arm/hand movements
and an appripriate position and mounting attitude of the potentiometer, angular displacement is a measure of the movement carried out.
This method, plus goniometers fixed to joints, render analog signals
ready for computer processing. A similar approach uses acceleration
transducers fixed to the extremities, the signals of which can be
processed to obtain movement trajectories (Nigg, 1977). For the same
purpose compact strain-gauge electrical transducers are used (O'Brian
and Paradise, 1976).

A newer technique is applicable to fine movements in a range of 50 cubic centimeters. The subject moves using a hand held device which produces small electrical sparks recorded by 3 linear orthogonal microphones the positions of which limit the workspace. The precision of the generated 3D trajectories is in the millimeter range (Moritz, 1977).

An analogue to the film recording of movements is the TV recording. The instantly available videotape contains all details of the recorded work situation. Semiautomatic data processing is possible if certain preparations are made. Body landmarks must be sufficiently high in contrast with respect to their surroundings. Moreover camera position should be selected so that landmarks do not cross each other or disappear when the moving extremity, e.g., the arm is turned. Most TV analyses are evaluated in a qualitative way, e.g., by counting movements to certain panel areas or counting actuations of single switches; yet, there are commercially available systems for quantitative analysis.

Besides TV, other optoelectronic methods are promising. Some of them use analog semiconductor sensors built into a camera to track Infrared Light Emitting Diodes which have to be attached to the subject's body. One of these systems is the Selspot-system (Lindholm and Öberg, 1975; and Hamamatsu, 1979)

Up to 30 diodes can be scanned and retrieved even if they are sometimes shaded during the movement. Scanning rates are up to 10 kHz in the case of only one diode. Compared with optical methods and

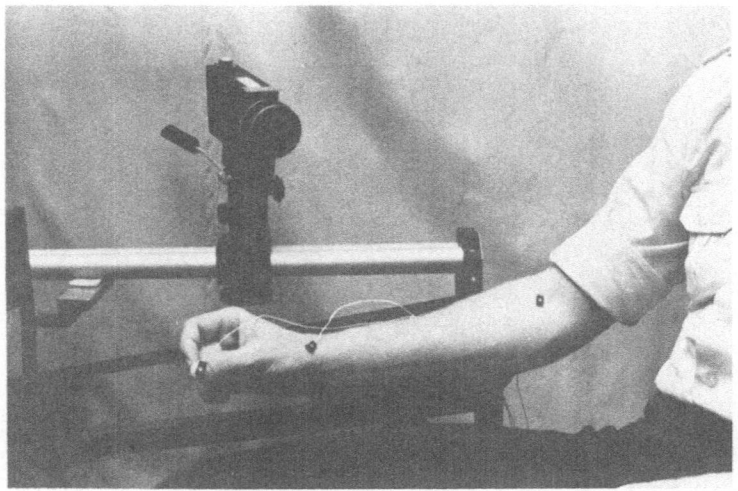

Fig. 6. Selspot-system sensor camera with sensors fixed to a subject's forearm (Holzhausen, 1978).

TV here is a higher degree of independence from lighting conditions, the output data are in digital form (Fig. 6).

Methods such as the Selspot method avoids most of the disadvantages of mechanical and photographical methods. Its disadvantage is the dependency on sensors on the skin that for a certain period of time influence the operator's movements and tend to degrade performance.

Stereo Measurement Techniques

Human movements are not performed in one plane. Consequently, a number of methods have been developed to measure and record three dimensional trajectories. The first approach is stereophotography (Hertzberg, 1956). Two photos from different points of view are taken with the shutter open and periodic flashlight as described for chronocyclography. Developed photographs are then mounted into a stereocomparator. Manually a stereoscopic measuring mark can be moved, the observer looking into the binocular device has the impression of moving the mark in three dimensions following the recorded three dimensional trajectory. By manually repositioning the mark from point to point, the respective x, y, z-coordinates of the mark position can be read out for further processing (Bullock and Harley, 1972; and Gutewort, 1968).

Using two mirrors, one camera can record two orthogonal work situation views (Deivanayagam et al., 1974).

Fig. 7. Optoelectronic measurement of pilot's movements in a fixed base flight simulator (Holzhausen, 1978).

Stereo-cinematography uses two film cameras instead of two pho-
tocameras. The data reduction is performed in a similar way, body
marks are required on the subjects (Van Gheluwe, 1977).

Optoelectronic systems with digital data processing capabilities
perform measuring and data recording in multiplexed time so that on-
line computer data recording from more than one camera is easily
possible. Fig. 7 shows an example of a setup using two optoelectronic
cameras mounted in the fixed base flight simulator of the Research
Institute for Human Engineering (FAT) in Wachtberg-Werthhoven to
test its applicability to cockpit eye movement analyses.

MODELING OF HUMAN MOVEMENTS

Data acquisition using the previously described methods plus a
great variety of anthropometric surveys lead to the development of
extensive data bases. These data bases give a good overview of measures already performed and may to a certain extend avoid the need
for repeating the acquisition of data in cases that have already been
explored. Thus numerous approaches have been used as to combine database information into mathematical models to simulate human movements.

Modeling of human movement is included in this paper because of
the fact that
- models are a consequence of measurement and analysis work,
- of the impact that available computer models have on further analytic work.

The complexity of those models and their intended use varies
widely. Also, the technical terms for their description are sometimes quite misleading.

Modeling of human movement is a versatile design tool. For example:
- It can be a sorting aid for data and data bases achieved through
  analyses.
- A model can be used as a systems analysis tool in a total workplace simulation.
- Modeling is used to suggest designs for later validation in a measurement experiment.

The range of available models is from static to dynamic and active to passive models. It should be made clear that static in this
respect always means static at the level that a problem is looked at.

A further criterion to differentiate modeling approaches is
whether the models are more mechanical or more biological. Though
many researchers in this field use different terms there are basical-

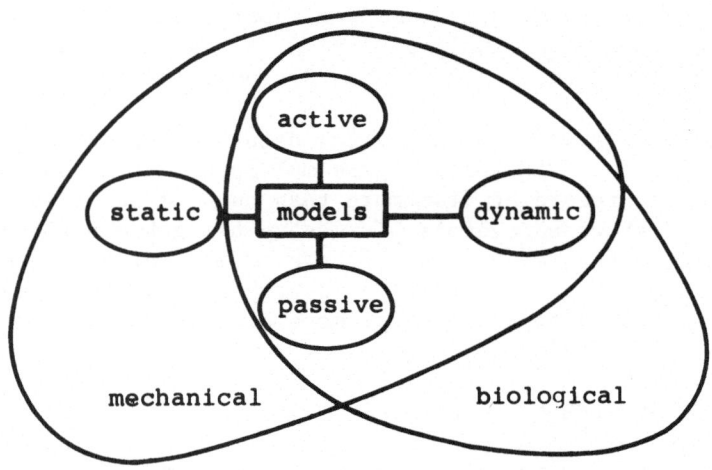

Fig. 8. Different aspects of modeling of human movements.

ly these two different approaches to modeling. Mechanical engineers and construction engineers model the human body as a link and joint system that has its analogue in mechanical designs such as bridges and other framework adding mass distribution and measured capabilities of acceleration for mathematical descriptions. Bioengineers and anatomists try to model the human body by a more biophysiological approach. They model the dynamics of muscle groups, influence of tissues etc. so that such models mainly adapt to partial simulations of the body and lend themselves to basic research. Ergonomic layout work for consoles and control devices does not generally require such a microscopic approach.

Static Models

Static modeling is used for cockpit or workplace layout where dynamic qualities are not required. These models can serve to reproduce reach envelopes, joint positions, limb attitudes etc. There is a wide variety of these models which differ considerably with respect to their complexity, ease of application, computer user guidance, and a more or less intense computer graphics support.

The first approach to computer man modelling were the so-called stickman or linkman models. These are models with a stick type representation of links as rods with no enfleshment. The models use anthropometric data bases to simulate link positions, check required joint angles to achieve static body postures and configurations.

One of the older models was the completely static version of BOEMAN (Ryan, 1971) model and the Crewstation Geometry Evaluator (CGE) (Katz, 1972). These systems, both developed for the Boeing Com-

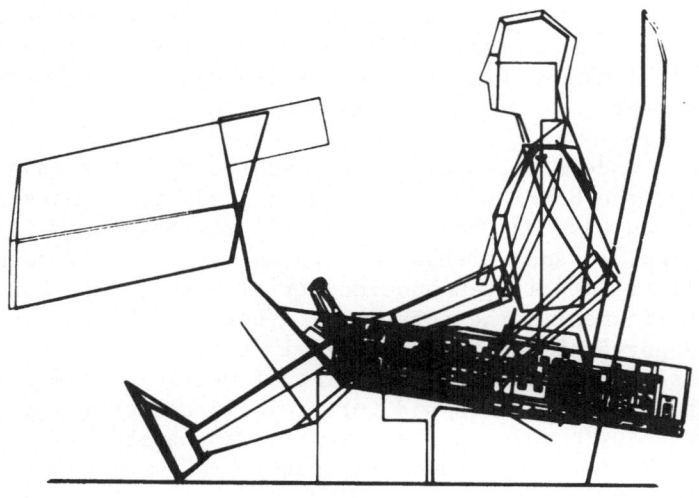

Fig. 9. BOEMAN model.

pany, were batch oriented and not interactive. BOEMAN produces a projection of a 3D man on a screen as shown in Fig. 9.

Simpler to use and less sophisticated are the Computerized Accomodated Percentage Evaluation (CAPE) model (Bittner, 1975) and the Crewstation Assessment of Reach (CAR) (Edwards et al., 1976) model.

The COMBIMAN (Kroemer, 1972) model is a complex model with a higher degree of interactivity and the efficient use of a computer display. Like the BOEMAN, an operator seated in an aircraft cockpit

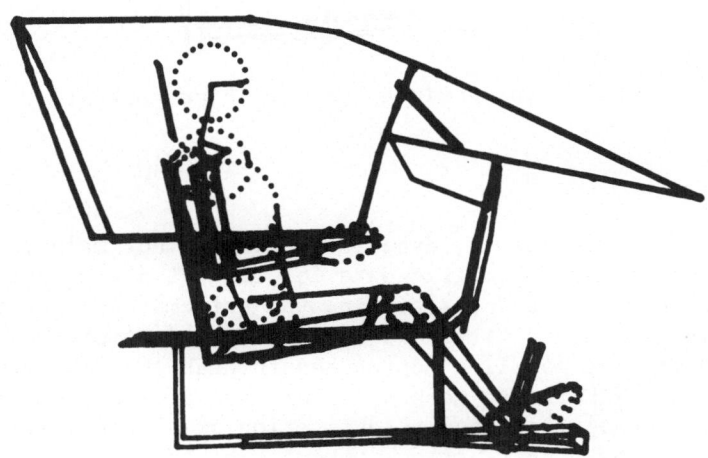

Fig. 10. COMBIMAN link-man in a cockpit (McDaniel, 1976).

is simulated whereby the model user can specify different percentile
measures for each link whereby BOEMAN only supports a simulation of
fixed percentile linkmen. Fig. 10 shows a linkman in a simulated
cockpit environment.

Unlike most models which simulate a seated operator in a cockpit environment, the SAMMIE (Bonney et al., 1979) model uses a simulated man that can have any body posture. It was developed for computer aided workplace and worktask design and is used for cockpit
layouts as well as for kitchen and industrial workplaces. Fig. 11
shows a SAMMIE linkman in the cockpit of a heavy machine. The SAM-
MIE command language allows a comparatively simple computer aided
design of workplace partitions consisting of multiple elements, where
the design ergonomist can move and rotate whole structural parts of
a layout against surrounding parts. Thus it is possible to construct
a car, open its hoods, move its doors, etc.

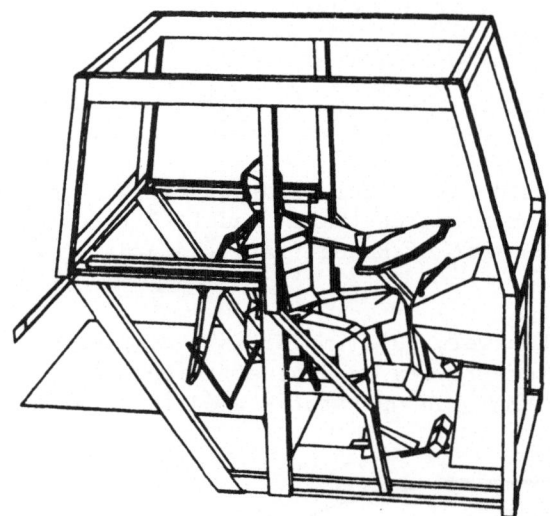

Fig. 11. SAMMIE linkman in a heavy machine cockpit.

Dynamic Models

Contrary to static models, dynamic models are not mainly intended to reproduce anthropometric data for cockpit design but to
take the human body's mass distribution, the extremities' centers of
gravity, muscle forces and tissue influences into account. There is
a variety of different modeling approaches through the literature,
with models being either more mechanical or more biological. Biological models are mostly confined to activation of single muscle
groups, thus modeling finger action, movement of the human hand or
the human arm.

Mechanical models describe human movement as partial or whole body movement with respect to dynamic properties such as acceleration and speed as well as power, energy consumption etc. The Hanavan total body model (Hanavan, 1964) describes the whole body in segments with the parameters being optimized according to the data base used. The model can then be applied for simulation of different manual work situation such as lifting tasks, for gait analysis or the human walk.

Models simulating the spine, neck, and head complex or models for the spine alone (Belytschko et al., 1976) have practical significance for the ejection type problems of modern aircrafts' ejection

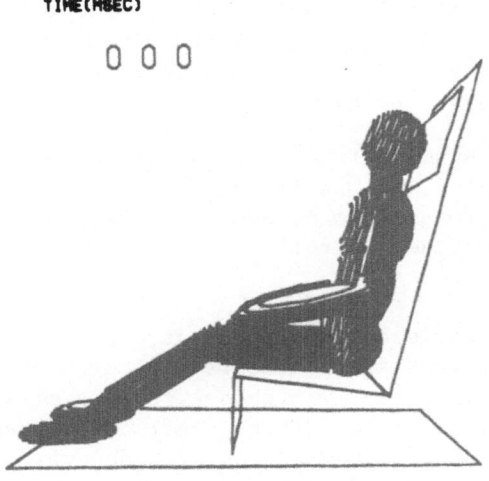

Fig. 12. Ejection from an aircraft, Phase 1.

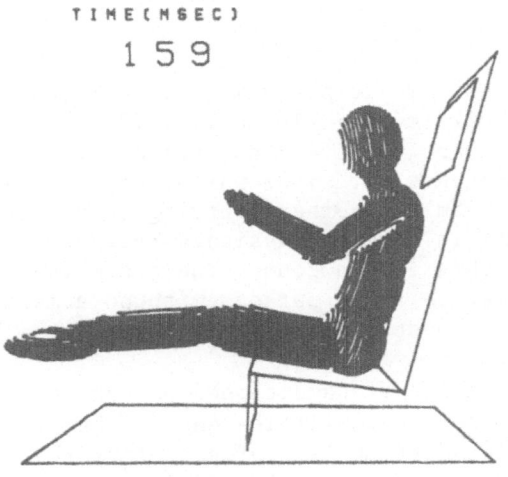

Fig. 13. Ejection from an aircraft, Phase 2.

seats. Figs. 12 and 13 show the simulated body motion of an aircraft pilot during the ejection from the aircraft (von Gierke, 1977), a process, where muscle forces from the human pilot can be neglected against the g-forces imposed on him. The objective here is to avoid injuries and construct a safe cockpit environment.

APPLICATION OF MOVEMENT ANALYSIS TO MANNED SYSTEM DESIGN

Evaluation of Free Movement for Signalling

Besides those movements used to operate a switch board, joystick or machine tool, there is a control potential in human motion such as arm movement without direct connection to a control device. If human control movements can be measured and analyzed by indirect methods, then those movement data could perhaps be used for control purposes in certain cases.

Frequently a machine operator with a limited view is aided by another remotely located man who stands close to the actual workplace signalling the required manoeuvers to the machine operator. With this new approach the controller would be at the work site signalling to the machine.

A research project is being planned at the FAT to find the degree of reproducibility of arm movements. If movement recording and analysis indicate a sufficiently high degree of similarity of a limited number of steering signals (e.g., left hand movement for machine movements to the left, moving the hand downwards for lowering a crane hook etc.) it may make sense in a future system to use such operator hand signals for control purposes. Width of movements and their speeds could possibly command the rate of machine movements.

Design of Robots and Prostheses

Studies of movements while performing manual labour in ordinary work situations were carried out (Tutungi, 1973) by photographic means to find the movement range and degrees of freedom required in the human arm. Measurements were made while the subject was moving a chair from one position to another, putting things on a desk etc. Subsequently an artificial arm was designed (Brudermann, 1978), taking the movement analysis into account. The prototype prosthetic arm is constructed in order to allow persons with no arms to regain a certain degree of limb function.

For the rehabilitation of handicapped persons and for the necessary tests in how far their capabilities enable them to carry out mechanical work in the production line of a factory the so called Available Motions Inventory (Dryden et al., 1980) apparatus was developed. The device is a console type vertical arrangement of keyboards plus a console, equipped with different knobs, cranks, switches etc. The handicapped person is tested here in how far he can cover switch actuation distances, which mechanical power he can exert and which movement patterns are more likely to be carried out than others. Some handicapped people could more easily perform linear movements than radial ones.

Safety Aspects

Even though the system performance may be excellent, a system may have safety defects which are not obvious. For example, there may be no valid data to evaluate whether or not an operator works in dangerous proximity to moving parts or under what circumstances a serious accident may occur. Movement measurement can, even for fast movements, show the obstacle clearance in every movement of the production sequence and how the movements may vary.

This was applied to the problem of egress clearance from an instrument panel when the safety of F-18 aircraft was evaluated (Winkler and Johnson, 1980). There was some danger that the feet of pilots would hit the lower front edge of the instrument panel during ejection from the aircraft. To avoid injuries, a model ejection was calculated and measurement of trajectories of feet in an ejection simulator was accomplished. In this case high speed filming was the measuring method used. Model and experiment coincided in an assumed danger for the pilot which led to an interesting mechanical solution as follows.

Fig. 14 shows in the right half a simulated model ejection indicating contact between the feet and the cockpit, whereas, as shown in the left half figure, a small spring biased toeguide in the pilot's footroom imparted an impulse to the feet during ejection in such a way that the trajectory of the feet is changed towards flatter angles with the result that contact could be avoided. The model result was in a subsequent measuring run in the simulator found suitable to reproduce the experimentally found data.

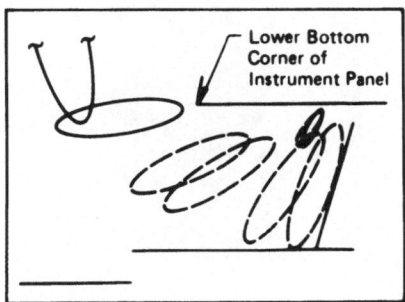 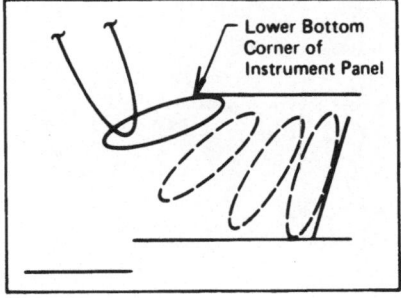

Fig. 14. Pilot ejection with and without a toeguide in the footroom.

Evaluation of existing designs

If an interaction between designer and ergonomist during the redesign phase of a workstation is required, movement measuring and analysis can contribute to enhance the overall system effectiveness. The redesign of a lathe may be an example for such a procedure (Solf, 1973).

 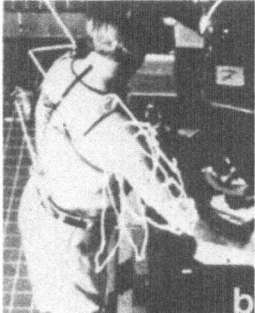

Figs.15a-b. Qualitative movement analysis of a worker at a lathe before its redesign.

The Figs. 15a-d show the qualitative movement pattern of a worker at such a machine using cyclographic measurement. Lamps are not fixed directly at the operator's body but at the end of light rods to amplify the visible motion paths for the camera. Figs. 15a-b show the workers at the original lathe from two viewing directions. Figs. 15c-d show the worker at the redesigned lathe from the same viewing angle. It is obvious that the constructional changes in the machine result in a reduction of required movements during work.

Figs. 15c-d. Qualitative movement analysis of a worker at a lathe after its redesign.

Development of Consoles

Operator movement trajectories, actuation speeds and accelerations were measured in our lab to optimize the arrangement of consoles as well as the selection of the best type of switch input device in the mockup shown in Fig. 16a. The mockup allows for installation of different numeric switches in three planes and in different positions with respect to a seated operator. Experiments are carried out by moving the operator's right hand from a resting position on the armrest to one of the switches. The switch selection is done at random through the computer program. After keying numeric data, the operator is instructed to return his hand to the rest position where he actuates a key switch to start a new keying procedure. This experiment is carried out for longer time periods, monitored with the optoelectric Selspot system which shows influences of switch positions on actuation speed and ease of operation and indicate less fatiguing arrangements of switches in order to reduce operator workload. Fig. 16b shows the graphic representation of an operator's pointing finger trajectory carrying out one single control movement. The vertical lines indicate motion speed as data sampling is at a fixed rate. Also they identify the ground track to see the movement path in the 3D coordinate system more clearly.

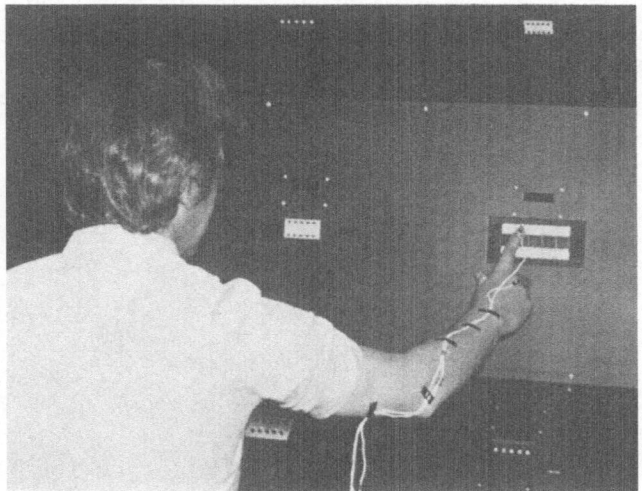

Fig. 16a. Flexible switchboard mockup for selection of ergonomic
           switch positions.

Development of control devices

Two different examples will illustrate the impact that analysis of performed movements may have on the design of control devices.

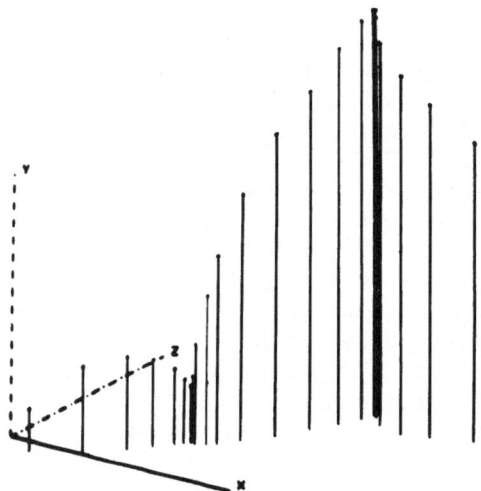

Fig. 16b. Trajectory and speed of the operator's right pointing
          finger.

       The combined use of measurements of guided movement on the twin
grip control yoke shown in Fig. 17 and the recording of correlated
physiological measure of electromyographical activity lead to an e-
valuation of recommended deflection paths in two rotational axes.
The control device was being designed for operation of a vehicle
where smoothness and ease of operation over a long time as well as
safety against erroneous inputs were important design criteria.

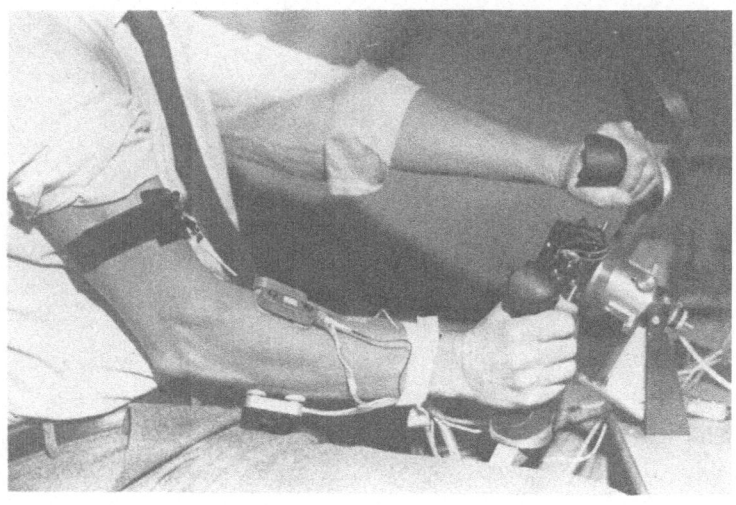

Fig. 17. Measurement of movement on a twin grip control yoke (Gärt-
         ner, 1977).

The second example shows the quantification of required operator movements during free movements to operate a hole punch (Baum,
1978). Cyclographic measurement of movements as shown in Figs. 18a
and 18b indicate progress after adding a paper guide to the tool.
The necessary movements are obviously reduced thus relieving the
subject from tedious paper alignment in the machine and facilitating
the task.

Another application of human movement measurement is referred
to in the case study in the next section, where a two-dimensional
hand joystick is being designed taking free and preferred operator
movements as a basis for design.

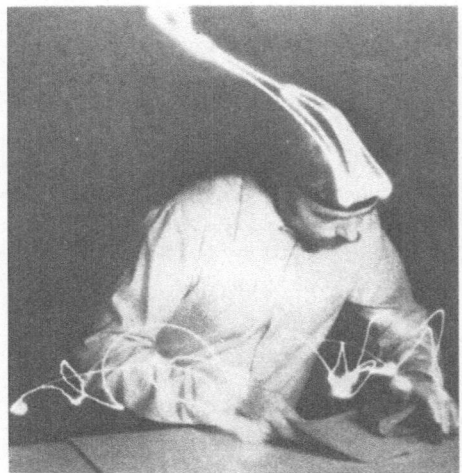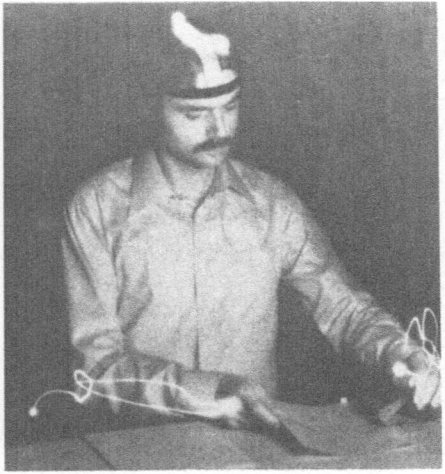

Figs. 18a-b. Reduction of head and hand movements through ergonomic
            redesign of a tool (Baum, 1978).

CASE STUDY: EVALUATION OF OPTIMAL PATHS OF HUMAN MOVEMENTS TO DESIGN
           AND LAY OUT A HAND CONTROL DEVICE ON THE BASIS OF OPERA-
           TOR-PREFERRED MOVEMENTS

Measurement System

The measuring equipment used is an off-the-shelve Selspot-system. This system digitizes position data of up to 30 Infrared Light
Emitting Diodes (IR-LED's) that are fixed on the subject's body as
markers for the measurement of limb positions as a function of time.
For 3D movement analyses two cameras are set up using especially developed calibration procedures. Adjacent electronics measure the
light spot positions in each of the two measuring planes in time multiplex with a rate of up to 300 Hz per diode channel. The data are

stored in a digital computer for later processing. After the end of
a single movement or some observations, software supported stereo re-
combination process takes place. As a result, data files of three-
dimensional position data with respect to an object space coordinate
system are available for further evaluation. This further evaluation
may be an overview of recorded trajectories or parts of motion path.
For this purpose a "quick look" graphic display was developed which
shows movements from any selected viewpoints including a ground
track. Further evaluation is carried out in the computer where data
filtering and calculation of movement parameters like velocity and
acceleration take place.

A Pilot Study

Physical characteristics of control devices should be designed
according to ergonomic principles. This is especially true for tracking controls which are used in such tasks as aircraft guidance, vehicle operation etc. Such controls are usually joysticks, the stick
length, deflection range, required force, and control-display gain
of which are parameters of construction.

After developing a prototype these parameters are usually determined by evaluating its performance. This is relatively easy for adjustable parameters such as force and control-display gain. More difficulties arise if stick length and deflection ranges are concerned,
as these properties influence the whole cockpit design. In case of
marginal results with the first design a second prototype will be
built and evaluated. To optimize parameters may thus require a number of prototypes. This applies to the design of deflection paths for
example. Movement paths are dependent upon the mechanical design of
the joystick. For mechanical simplicity straight angular deflections
around the stick pivot are mostly used with straight left for a left
turn, straight down for pitch down, etc. In a side control equipped
aircraft, hand movements rather than arm movements are applied for
this purpose.

With a new method, optimal deflection ranges and paths of movements for such a device can be determined in an easier way: biomechanical aspects of the human arm/hand system suggest that non-planar
and sometimes nonlinear paths of the operator's hand may be more natural and convenient, compared with those which have to be performed
using the control device designs of today.

A more satisfying approach may be the measurement of operator-
preferred spatial movements and an appropriate control device constructed according to the trajectories found to be best suited.

The measuring technique used is the described optoelectronics
system. Subjects place their forearms on an armrest, as shown in
Fig. 19, holding an unattached handle which does not limit operator

movements. The handle itself contains a number of IR-LED's as markers, so that an accurate 3D trajectory of handle position and attitudes as a function of time during operation can be deducted. The operator's forearm position on the armrest and his hand grip on the handle are very similar to those with control stick operation in an aircraft.

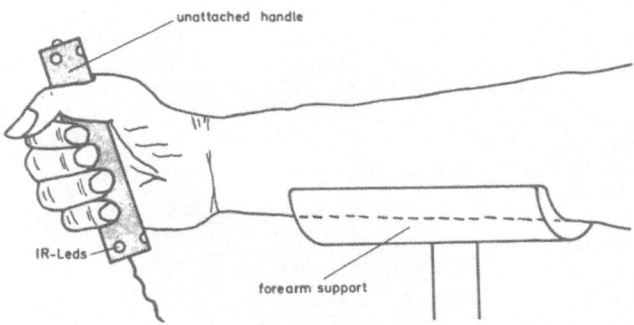

Fig. 19. Control device simulation using a "free control stick" (Holzhausen, 1979).

Subsequently, a randomly moving target on a screen has to be manually "tracked" or followed in a pursuit-type tracking task with no visual feedback. As the target moved in two directions, the handle was used as a zero order two-dimensional hand control stick. Instructions to subjects encourage them to perform movements that they find suitable, natural and not fatiguing. Because there is no visual feedback about the movements performed, a later evaluation can describe the functional correlation between commanded action and operator response on the basis of free operator-preferred movements.

The recorded paths of preferred three-dimensional operator movements are clearly influenced by the fact that an arm rest was used. For this reason, appropriate movements for vertical target travel are fairly symmetrical with ulnary and radial hand abductions of rather limited range.

Horizontal targer travel is followed by wider movements combining hand flexion as well as forearm rotation. Fig. 20 shows typical movements for horizontal tracking as taken from the quick look display. The lines describe the time history of the grip axis with respect to its position and attitude. The movement paths in the x,z ground plane and in the vertical x,y plane are described. Paths of movement show fairly good repeatability without excessive variation.

Recorded paths of preferred three-dimensional human movements, as shown in this example, may be useful in improving the design of new controls and layouts for various applications with actuation

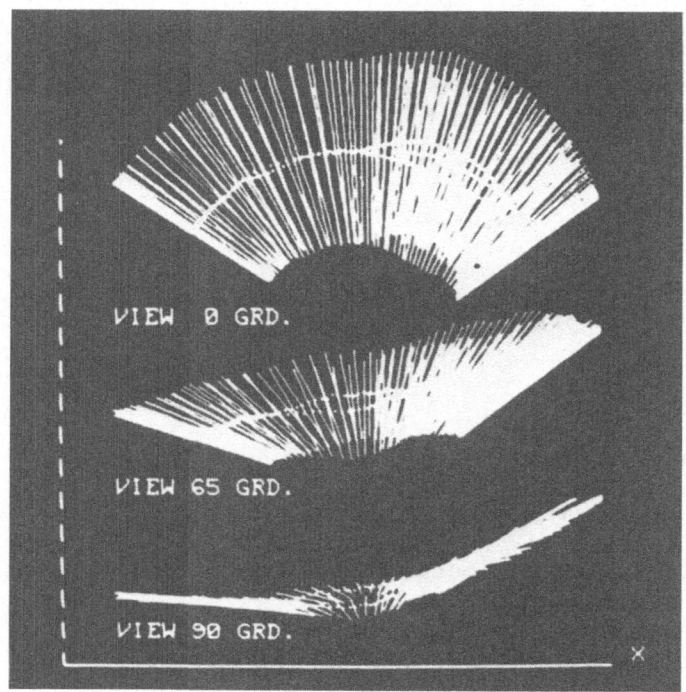

Fig. 20. Operator-preferred movements using joystick type control
simulator (Holzhausen, 1979).

ranges, deflection paths, and control gains optimized using this
stricktly operator-centered approach.

REFERENCES

Adrian, M., 1973, Cinematographic, Electromyographic, and Electrogo-
    niometric Techniques for Analyzing Movements, "Exercise and
    Sport Sciences Review", 1, J.H. Wilmore (ed.), Academic Press,
    New York and London.
Baum, E., 1978, Motografie. Grundzüge einer arbeitswissenschaftlich-
    en Methode zur Bewegingsaufzeichnung, Zeitschrift für Arbeits-
    wissenschaft, Heft 2.
Belytschko, T., Schwer, L. and Schultz, A., 1976, A Model for Analy-
    tic Investigation of Three-Dimensional Head-Spine Dynamics.
    AMRL-TR-76-10, Air Force Medical Research Lab., Wright-Patter-
    son Air Force Base, Dayton, Ohio.
Bittner, A., 1975, Computerized Accomodated Percentage Evaluation
    (CAPE) Model for Cockpit Analysis and Other Exclusion Studies,
    TP-75-49 Pacific Missile Test Center, Pt. Mugu, California.
Bonney, M., Blunsdon, C.A., Case, K. and Porter, J.M., 1979, Man-

Machine Interaction in Work Systems, Int. J. Product Research,
17 (6).

Braune, C.W. and Fischer, O., 1895, Der Gang des Menschen I, Abh.
Math. Phys. CL. Kön. Sächs. Ges. Wissensch. 21.

Brudermann, U., 1978, "Entwicklung und Anpassung eines vollständigen
Ansteuerungssystems für fremdenergetisch angetriebene Ganzarm-
prothesen", VDI-Verlag, Düsseldorf.

Bullock, M.I. and Harley, I.A., 1972, The Measurement of Three-Dimensional Body Movements by the Use of Photogrammetry, Ergonomics,
15 (3), 309-322.

Deivanayagam, S., Ayoub, M.M. and Kennedy, K.W., 1974, Paths of Movement of Body Members in Aircraft Cockpits, 18th Annual Meeting
of the Human Factors Society, Huntsville.

Dempsey, C.A., 1953, Development of a Workspace Measuring Device,
WADC, TR 53-53, Wright-Patterson Air Force Base, Dayton, Ohio.

Dempster, W.T., 1955, The Anthropometry of Body Action, Annals of
the New York Academy of Science, 63, 559.

Dryden, R.D., Leslie, J.H. and Norris, R.H., 1980, Anthropometric
and Biomechanical Data Acquisition and Application to Rehabilitation Engineering, NATO Symposium on Anthropometry and Biomechanics, Cambridge, England.

Edwards, R., Osgood, A., Renshaw, K. and Chen, H., 1976, Crewstation
Assessment of Reach (CAR) Model, Users guide, Report D-180-1932-
-1-1, Boeing Aerospace, Seattle, Washington.

Gärtner, K.-P., 1977, Biomechanical Analysis Applied in Designing
the Operating Range of Hand Consoles, VIth Int. Congress of Biomechanics, Copenhagen.

Gierke, H.E. von, 1977, Information received through personal con-
tacts, Air Force Medical Research Lab., Wright Patterson Air
Force Base, Dayton, Ohio.

Gutewort, W., 1968, Die digitale Erfassung kinematischer Parameter
der menschlichen Bewegung, Biomechanics I, 1st Int. Seminar,
Karger, Basel/New York.

Hamamatsu, 1979, Product Brochure.

Hanavan, E.P., 1964, A Mathematical Model of the Human Body, AMRL-
-TR-64-102, Aerospace Medical Research Laboratory, Wright-
Patterson Air Force Base, Dayton, Ohio.

Hertzberg, H.T.W., Emanuel, I. and Alexander, M., 1956, The Anthropometry of Working Positions: A Preliminary Study, WADC/TR 54-
520, Wright-Patterson Air Force Base, Dayton, Ohio.

Holzhausen, K.-P., 1978, A Contribution for Improved Optoelectronic
Measurement and Immediate Inspection of Spatial Human Movements,
in: Conference Digest, 1st Int. Conf. on Mechanics in Medicine
and Biology, Aachen, Witzstrock Publishing House, Baden-Baden.

Holzhausen, K.-P., 1979, An Operator-Centered Approach to Control De-
vice Design by Optoelectronic Measurements of Operator-Preferred
Movements, 7th Congress of the International Ergonomics Ass.,
Warsaw.

Kallin, K., Stricker, J. and Ferretti, E., 1979, Entwicklung einer
Methode für das Quantifizieren der Leistung beim Salto vorwärts

    und Illustration an zwei Beispielen, ETH, Zürich.
Katz, R., 1972, Crew Station Design and Evaluation Methods, User
    guide, BCS Report 40003, Boeing Computer Services, Seattle,
    Washington.
Kroemer, K.H.E., 1972, COMBIMAN-Computerized Biomechanical Man-Model,
    AMRL-TR-72-16, Aerospace Medical Research Laboratory, Wright-
    Patterson Air Force Base, Dayton, Ohio.
Lindholm, L.-E. and Öberg, K.E.T., 1975, An opto-electronic instru-
    ment for remote on-line movement monitoring, in: "Biotelemetry
    II", P.A. Neukomm (ed.), Basel.
McDaniel, J.W., 1976, Computerized Biomechanical Man-Model, AMRL-TR-
    76-30, Aerospace Medical Research Laboratory, Wright-Patterson
    Air Force Base, Dayton, Ohio.
McDaniel, J.W., 1980, Biomechanical Computer Modeling for the Design
    and Evaluation of Work Stations, NATO Symposium on Anthropome-
    try and Biomechanics, Cambridge, England.
Moritz, W.E., 1977, A System for Locating Points, Lines, and Planes
    in Space, IEEE Transactions on Instrumentation and Measurements,
    26 (1).
Neukomm, P.A., 1974, The Rubber Band Goniometry, Biotelemetry, 1,
    12-20.
Nigg, B.M., 1977, "Biomechanik", Juris Druck und Verlag, Zürich.
O'Brain, C. and Paradise, M.G.A., 1976, The Development of a Portable
    Non-Invasive System for Analysing Human Movement, Proceedings
    of the 6th Congress of the IEA, Univ. of Maryland, College Park.
Ryan, P., 1971, Cockpit Geometry Evaluation. Phase II - Final Report
    Vol. I - Program Description and Summary, JANAIR Report 7000201,
    Boeing Company, Seattle, Washington.
Solf, J.J., 1973, Ergonomische Untersuchung und Umgestaltung einer
    Drehmaschine, Refa-Institut für Arbeitswissenschaft, Stuttgart,
    Technischer Verlag Günter Grossman GmbH, Stuttgart-Vaihingen.
Tutungi, G., 1973, Die Beweglichkeit des menschlichen Armes - Bewe-
    gungsanalyse und Möglichkeiten eines vereinfachten kinematischen
    Ersatzsystems, Diss., T.U., Hannover.
Van Gheluwe, G., 1977, Computerized Three-Dimensional Cinematography
    for any Arbitrary Camera Setup, Biomechanics VI, Copenhagen,
    University Park Press, Baltimore.
Winkler, E.R. and Johnson, E.L., 1980, Application of Dynamic and
    Anthropometric Computer Modeling in Design of an Aircraft Ejec-
    tion/Escape System, NATO Symposium on Anthropometry and Bio-
    mechanics, Cambridge, England.
Wright, I.B., 1963, Applications of a System of Functional Anthropo-
    metry in Pressure Suit Design, Journal of the Brit. Interplane-
    tary Soc., 19, 1963-64.

MEASUREMENTS OF SPINE LOCATION DURING LIFTING

Dennis L. Price

INTRODUCTION

The measurement and recording of the position of the spine during work could lead to information which will contribute to the reduction in injuries of the spine. Seventy nine percent of manual
handling injuries are injuries to the lower back (Snook, 1978). In
the United States, there are 400,000 manual materials handling back
injuries per year (Ayoub et al., 1979). The prevalence of low back
injuries compared with other segments along the spine could be a result of the displacement of the spine from the normal, physiological
null position of the upright relaxed worker. Therefore, considerable
attention ought to be directed to the role of spinal displacement
during manual materials handling.

The role of the angle of the spine from vertical and its effects
on lumbar loads has been investigated by Ortengren et al. (1978).
They used an intradiscal pressure transducer to obtain direct measurements at L^3. Their results showed a high positive correlation of angle of flexion with pressure. However, from their study, only limited
estimates of the amount of displacement of the spine from the physiological null position can be made because the angular measurements
were taken only at T^{10}. This displacement could be important, because
it might result in a considerable reduction in the area of loading
at the disc. It appears of essence to gain measurements of the location of more than one point of the spine during manual materials
handling.

The measurement of the location of the spine has been attempted in research and clinical settings. The research efforts have been developed along lines of those techniques generally developed for human motion analysis. These are described by Ayoub (1972) and include photographic, stereogrammetric and electro-mechanical techniques. When these techniques are applied to spinal location, they either give limited information on specific relative angles along the spine or are not conducive to long term monitoring of spinal location in the dynamic, industrial environment. For example, Tichauer's lordosimetry pantograph (Tichauer et al., 1973) is not easily adaptable for continuous monitoring onsite.

The clinical measurements involve several methods: the use of the Dunham spondylometer (Dunham, 1949), skin distraction (Moll et al., 1972), the inclinometer or pendulum goniometer (Loebl, 1967), radiography and various photographic methods (e.g. Davis et al., 1965). The relative merits of certain of these techniques are discussed by Reynolds (1975). However, these are designed for clinical applications and center around posture and range of movement measurements in a relative static environment. The application to the dynamic industrial setting is largely untried.

It would appear that goniometers and inclinometers or pendular potentiometers could be adapted to the industrial setting to establish the relationship of spinal displacement to high risk back injury occupations, such as manual materials handling and lorry driving (Davis, 1979).

The goal of the effort described in this paper is to develop a very simple device which can be self-contained, portable and worn on the job with as little discomfort and interference as possible. This device would monitor and record spinal displacement. Direct angular measurements are preferred because the device might eventually be used to instruct in lifting without displacing the spine. The populace probably appreciates the meaning of measured angles compared with some measure in other units which correlates with angles. The method of development has been trial and error.

This report is in two parts. The first part describes four prototype spine location transducers examined at Virginia Tech. These were selected or developed because of their simplicity. None is considered fully adequate.

The second part of this report describes the procedure and results of experimentation which served a dual purpose. The experiment involved data to evaluate one of the prototype devices and to test hypotheses about lifting which are related to the importance of developing an adequate transducer for determining spinal location during on the job activity.

PART I

The object of the trial and error development effort described below was to obtain a prototype device which would measure the angle of the spine, with respect to the vertical, at two locations on the spine. The two locations selected, the sacrum and T^{10}, should provide a measure of the displacement of the spine as it most affects the lower back.

Gampots

The first transducer examined was a gravity actuated mercury potentiometer (GAMPOT) invented by a Florida safety engineer, Mr. Henry Hall. Mr. Hall's device consisted of a drop of mercury between two conductive rings encapsulated in plastic. The location of the drop of mercury along the rings determined the magnitude of the output from the potentiometer. Therefore, rotating the potentiometer, when it was properly oriented to the verticle, would result in an output which would vary with respect to the angle from the vertical. This occurs because the mercury always seeks the bottom of the rings.

The accuracy of the GAMPOTS was checked by placing them on a dividing head and measuring the output against the angle from the vertical. Two things were discovered: over time the accuracy of the GAMPOTS deteriorated, and the mercury often would hang inside the rings until a sizeable angular rotation displaced it toward the bottom. Therefore it was not accurately sensitive to small angular rotations.

The GAMPOTS were rebuilt. Pure laboratory mercury was used on the idea that the deterioration over time might be due to contamined mercury that was incompatible with materials in the potentiometer. Also, the surfaces of the rings were found to be rough and poorly aligned. These were machined to a finer smoothness and carefully aligned. The result was a potentiometer which was less accurate than before. Therefore, GAMPOTS were abandoned as an adequate transducer.

Hip-Mounted Goniometer

The next device examined consisted of a rotational potentiometer attached to hip pads strapped around the pelvis and under the crotch. The potentiometer was located over the sacrum and was driven by an arm that was strapped to the T^{10} area of the spine. The arm was slotted to permit it to slide along the potentiometer since the distance between the T^{10} and the sacrum changes with flexion and extension. Two disadvantages to this device existed: it was too cumbersome and awkward, and it could not be effectively stabilized and moved about on the hips during certain operator actions. It was decided not to pursue this device further.

Rod and Cable Goniometer

A rod was firmly attached to hip pads. It extended to a location opposite T^{10}. A pulley was located in the tip of the rod and a cord extended from a reel on a rotational potentiometer mounted to the pads over the sacrum to an attachment point on T^{10}. This alternative was quickly dropped because the extending rod would be an interference on many jobs. It, of course, also had the instability shown from the previous hip pad device.

Pendulated Potentiometer

A single linear potentiometer was mounted over the sacrum and another over the T^{10} and driven by a lead pendulum. The potentiometer selected was a low torque device that was rotated freely by the pendulum. This device was tested on the dividing head and was reasonably accurate in its output. The principal disadvantage of this device is that various body motions other than the angular change of the spine can rotate the pendulum. It was decided to validate this device while performing research on related aspects of manual materials handling. It was felt that reasonably valid measures might occur if frequencies above 2 Hz were filtered, and the output was smoothed by sampling at a frequency of seven samples per second. If the transducer proved effective, it could then be fed into a belt-mounted min ature recorder for gathering data on the job. This research is described in Part II.

PART II

HYPOTHESES

Research was performed to test the following hypotheses:
1. The thoracic and sacral angles measured by conditioned signals from the pendulated potentiometer during the lifting of boxes of various sizes will correlate significantly with the actual angles measured from the series of photographs taken of the lifts.
2. Operators will bend their backs (i.e. displace the lower part of the spine) during lifting of boxes of various sizes.
3. Regardless of the style of lifting, operators will acquire a similar back bent position part way through the lift.
4. The size of the box will affect the maximum difference of angle between the T^{10} and the sacrum so that as box size increases so does the maximum angle of spinal displacement.
5. The height of the lift will affect the maximum angle of spinal displacement so that a floor to knuckle height lift will have less displacement thatn a floor to shoulder height lift.

METHOD

Subjects

Twenty four subjects, paid volunteers, were elected to participate in the lifting experiment after passing a medical examination
of their present health and past health history. Ten volunteers, 8
males and 2 females, were selected to participate as judges of photographs.

All subjects in the lifting part of the experiment were between
the ages of 18 and 30 years. The judges were between the ages of 20
and 40 years.

Apparatus

The following apparatus was used:
1. A backdrop grid consisting of 8.9 cm squares and shown in Fig. 1;
2. Three rectangular wooden boxes, 36 x 43 cm, 49 x 57 cm, and 75 x
   106 cm, each similar to those designed by Snook (1978), including
   a false bottom and a switch which could be activated on the moment
   the lift was initiated;
3. Two low torque one turn pendulated potentiometers mounted on a
   holder which provided a target for photographic measurements of
   the angle of the back;
4. A Butterworth filter which passed signals from the potentiometers
   which were less than 2 Hz only;
5. A Sanborn model 150 four channel recorder;
6. An electric forklift for lowering the box after a lift; and
7. A Grass C4P kymograph camera which takes 35 mm photographs at a
   rate of 7 per second.

The shutter activation was recorded on the Sanborn stripchart
along with the event mark of lift initiation from the box handle
switch, and the continuous outputs from the potentiometers.

Experimental Design

The experimental design is shown in Fig. 2. All subjects were
given all experimental conditions, randomly assigned. Subjects were
given the option to decline any lift.

Procedure

The subjects were instructed in the task and their rights as
subjects. Before each task the amount of weights (welding rods) in
the false bottom of the appropriate box for that task were changed.
The subject was told to add weight to or take weight from the upper
visible section of the box to obtain a weight which was a "comfortable" lift for that type lift and size box.

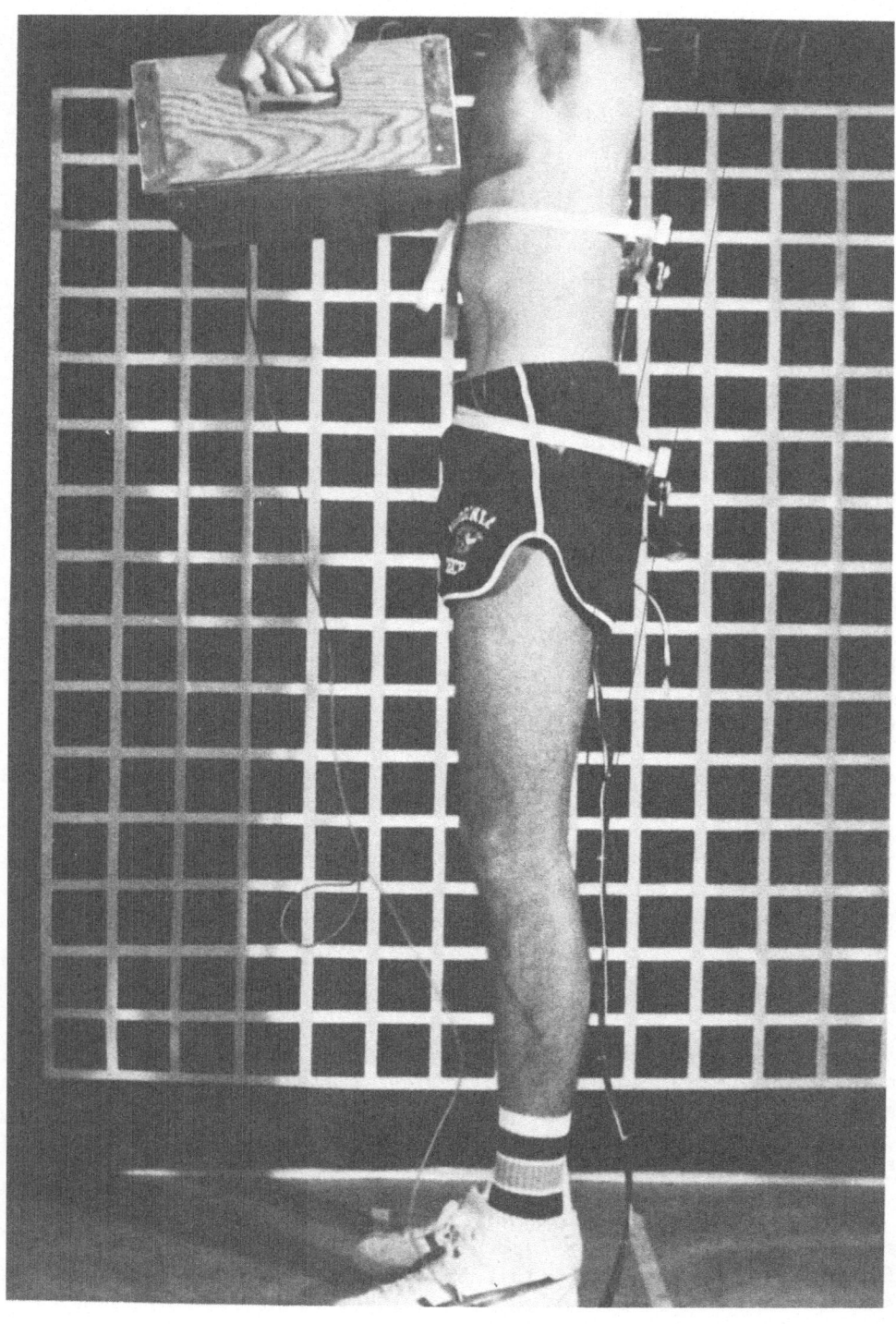

Fig. 1. Subject Lifting Box.

```
L
I
F   Floor to
T   Knuckle    S₁,S₂,S₃,...,S₂₄    S₁,S₂,S₃,...,S₂₄    S₁,S₂,S₃,...,S₂₄

T   Floor to
Y   Shoulder   S₁,S₂,S₃,...,S₂₄    S₁,S₂,S₃,...,S₂₄    S₁,S₂,S₃,...,S₂₄
P                    Small               Medium              Large
E
                             BOX SIZE
```

Floor to Knuckle: $S_1, S_2, S_3, \ldots, S_{24}$ $S_1, S_2, S_3, \ldots, S_{24}$ $S_1, S_2, S_3, \ldots, S_{24}$

Floor to Shoulder: $S_1, S_2, S_3, \ldots, S_{24}$ $S_1, S_2, S_3, \ldots, S_{24}$ $S_1, S_2, S_3, \ldots, S_{24}$

Fig. 2. Lift Experiment.

The purpose of the "comfortable" lift was to provide a genuine lift task without having the weight of the lift determine the displacement of the spine. The subject was told that we were not interested in determining how much he could lift; therefore, he should lift only what was a comfortable weight. He was told that he may lift the boxes any way he chooses. No instruction in how to lift was given except to lift to the task height, knuckle height or shoulder height.

The subject stood in position in front of the grid. The potentiometers were placed in position over T^{10} and the sacrum and their output baseline established. He then performed the task. He lifted the box four times for each task. After each lift the box was placed on the forklift and lowered. The experimenter returned the box to the original position if the lift was to be repeated. Photo data were gathered on the third lift of the series only. Potentiometer recordings were taken for all lifts.

After the photographs were taken and the lifting part of the research completed, the photographs of 18 subjects were printed. The lift sequences, using the floor to knuckle height small box task, were assembled and laid out on tables in one room. The judges were given a standard photograph of a person (not a subject) in mid lift for that task. The judges were to select a picture from each sequence which approximately matched the standard.

Scoring

The angular position of the potentiometer output could be read directly from the stripchart recording. The angles from the photographs were measured from the prints or from developed negatives viewed through a microfilm reader. The stripchart plots were compared with the plots of angles measured from the photos to determine the amount of lag induced by the potentiometer signal conditioning. The lag was then adjusted by matching points on the chart trace with the appropriate photograph, by using the shutter event marks. The adjustment was approximately one seventh of a second. The judging was scored by tallying the selection frequency by photo frame for each lift sequence.

RESULTS

Accuracy of the Pendulated Potentiometers

The Pearson product-moment correlations of the thoracic and sacral angles measured from the photos with the respective potentiometer outputs ranged from moderately low ($r = 0.27$) to moderately high ($r = 0.76$). The correlations are shown in Table I. All correlations are significant. The correlations are depressed, as expected, by the pendular swing from forces other than gravity that occur during the tasks. The correlations for the differences between the thoracic and sacral angles are lowest because the photos show that at times during a lift the sacral pendulum can be displaced from vertical, in the opposite direction of the thorax pendulum displacement. This difference is important as a measure of the bending of the back, i.e. the displacement of the spine.

Table I. Correlations of measured angles of back with potentiometer output.

LOCATION	TASK[*]						
	FLK	FKM	FKS	FSL	FSM	FSS	all data
Thorax (T)	.47	.62	.47	.76	.65	.60	.62
Sacrum (S)	.60	.46	.59	.27	.31	.52	.43
T-S	.33	.19	.16	.47	.30	.28	.35

[*]FKL = Floor to Knuckleheight lift Large box
FKM = Floor to Knuckleheight lift Medium box
FSS = Floor to Shoulderheight lift Small box
etc.

Displacement of the Spine During Lifting

The mean maximum difference between the thoracic angle and the sacral angle was 65°. The values ranged from 35° to 102° difference and the standard deviation of the differences was 15.6^{o}. These values were obtained from measurements taken from the photographs.

In the writer's judgment these figures represent severe displacements of the spine during the lifting of "comfortable" loads. These displacements were not related to the weight lifted. The weights lifted for the floor to shoulder height task averaged 21.3

kg with a standard deviation of 1.9 kg. Floor to knuckle height averaged 21.0 kg with a standard deviation of 1.7 kg. The correlation of these weights with the maximum thoracic-sacral difference of angles during the lift was r = 0.07.

Displacement of the Spine and Lift Start Position

The judges results in selecting a similar bent back position from enlarged photographs (20.3 cm x 25.4 cm) of floor to knuckle height small box of lifting sequences of 18 subjects were highly consistent. Of the 8 subjects whose starting knee angle was "closed" the highest frequency frame was selected by an average of 65.7% of the judges. When this consensus was accumulated with the second most popular frame selected, it accounted for an average of 95% of the selections. These results and the related similar results for the other starting positions are shown in Table II. No judge determined that there was no similar frame for a lift sequence.

Table II. Similar Frame Selection (cumulative average % selecting frame).

Starting Knee Angle[*]	Closed	Semi Open	Open
Most popular frame	63.75	75	73.75
Second most popular frame	95.00	100	93.75
Number of subjects	8	2	8

[*]determined by experimenter from our photographs

The Effect of Box Size on Maximum Displacement of the Spine

A two way analysis of variance was calculated using the SAS GLM procedure. These results are shown in Table III.

The size of the box had a significant effect on the bending of the back during the lift. As one might expect from the box sizes, a Newman-Keuls comparison showed the large box to be significantly different when compared with the medium and small boxes. These boxes were not significantly different from one another on this dependent variable. The greater bending occurred with the large box (68° compared with 63° for each smaller size).

The box x lift task interaction occurred because there was no effect of box size on maximum thoracic-sacral angle difference for the floor to shoulder height task; however, for the floor to knuckle height lift, the maximum thoracic-sacral angle difference increased with increasing box size. This interaction is the only indication of

Table III. Box Size by Lift Task ANOVA.

Source of Variance	df	Sum of Squares	F	p<
Box size (B)	2	853.125	3.60	.05
Lift task (L)	1	55.007	<1	n.s. *
Subjects (S)	23	22974.271		
B x L	2	1005.847	4.87	.05
B x S	46	5454.542	1.20	n.s.
L x S	23	1825.160	<1	n.s.
B x L x S	46	4754.486		

*n.s. = not significant

an effect on spinal displacement related to height of lift. That is, there was no significant lift task main effect. The graphs of the main effect box size and the interaction are shown in Figs. 3 and 4.

CONCLUSIONS

    It is recommended that:
1. The relationship of pendular potentiometric output to back injury prone jobs should be investigated, only after additional transducers or signal conditioning are investigated in an effort to obtain improvements.
2. Operators bend their backs severely during manual lifting, regardless of how they initially address the task, and regardless of general box size.

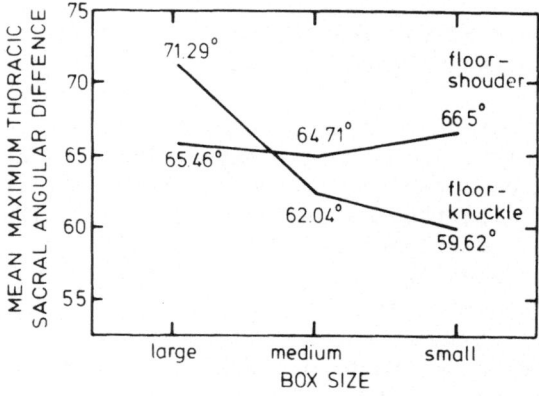

Fig. 3. Box x Job Interaction - Using Group Means.N at each plotpoint = 24.

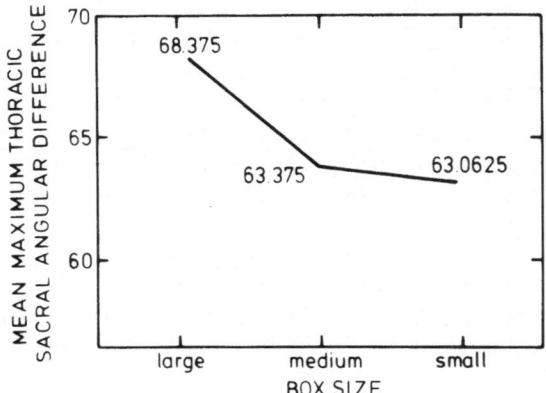

Fig. 4. Main Effect Box Size. Box Main Effect - N = 48.

3. Very large boxes will result in extreme back bending angles.
4. Floor to knuckle height lifts result in increasing back bending
   during lifts as size of box increases. This does not occur with
   floor to shoulder height lifts.
5. The correlations of pendular potentiometric output to actual back
   angles are moderately low to moderately high with the lower correlations being the angular difference between the thorax and the
   sacrum. The pendulated potentiometers might be useful for instruction but need improvement for industrial investigations.

The principal find of this research is that thoracic to sacrum
angular differences of an average of 65° occurs during manual materials handling. These differences may be as large as 100° for some workers.

ACKNOWLEDGEMENT

Funds for this research were provided by the State of Virginia,
Core Research Program for Productivity.

REFERENCES

Ayoub, M.M., 1972, Human movement recording for biomechanical ana-
     lysis, International Journal of Production Research, 10, 35-51.
Ayoub, M.M., Dryden, R.D., McDaniel, J., Knipfer, R. and Dixon, D.,
     1979, Predicting lifting capacity, American Industrial Hygiene
     Association Journal, 40, 1075-1084.
Davis, P.R., Troup, J.D.G. and Bernard, J.H., 1965, Movements of the
     Thoracic and Lumbar Spine When Lifting: A Chronocyclephotographic

Study, Journal of Anatomy, 99, 13.

Davis, P.R., 1979, Reducing the risk of industrial bad backs, Occupational Health and Safety, May/June, 45-47.

Dunham, W.F., 1949, Ankylosing spondylitis - measurement of hip and spine movements, British Journal of Physical Medicine, 12, 126.

Moll, J.M.H., Liyanage, S.P. and Wright, V., 1972, An objective clinical method to measure lateral spinal flexion, Rheumatology and Physical Medicine, 10, 225.

Ortengren, R., Anderson, G. and Nachemson, A., 1977, Lumbar loads in fixed working postures during flexion and rotation, Biomechanics VI-B, Proceedings of the 6th International Congress, Copenhagen, Denmark, University Park Press, Baltimore, Maryland, 1978, 159-166.

Snook, S.H., 1978, The design of manual handling tasks, Ergonomics Society Lecture given at Cranfield Institute of Technology, Bedfordshire, England, April; published in Ergonomics, 21, 963-985.

Tichauer, E.R., Miller, M. and Nathan, I.J., 1973, Lordosimetry: A new technique for the measurement of postural response in materials handling, American Industrial Hygiene Association Journal, 34, 1-13.

EVALUATION OF SYNTHESIZED VOICE APPROACH CALLOUTS (SYNCALL)

Carol A. Simpson

INTRODUCTION TO SPEECH SYNTHESIS

Voice output for computers is an application of synthesized speech with far reaching implications. Talking machines, laboratory approaches to their design, and practical applications of computer speech have been discussed in the literature for years (Scientific American, 1871; Dudley, 1939; Flanagan et al., 1970; Sherwood, 1978; Allen, 1978; and many others). Traditionally, synthesized speech has provided experimental stimuli for the investigation of human speech perception. Speech synthesizers were driven by large computers, and generation of speech was not in real time. Only recently, has it become technologically and economically feasible to build systems commercially that are light weight, low in power consumption, economical in memory requirements, and capable of real time speech generation. Sherwood (1979) gives an excellent overview of this development and reviews the major speech synthesizers currently available. Since his article appeared, the phoneme synthesizers that were then multi-board systems have been reduced to single boards and even to a single Complementary Metal Oxide Semiconductor (CMOS) chip (Votrax 1980). This reduction in size and cost has made mass produced talking machines with unlimited vocabulary feasible. There remains, however, a great deal of research to be done in order to determine the appropriate functions for speech and to develop human factors design principles for its implementation in man-machine systems (Williams and Simpson, 1975; Priest, 1980; Michaelis, 1980).

Types of Synthesized Speech and Applications

There are two basic approaches to the generation of what in lay terminology is called "synthesized" speech. Analog recorded human

speech can be digitized and compressed by any of a number of algorithms. The main algorithms used are based on either the Fourrier
transform or Linear Predictive Coding (LPC), also called autocorrela-
tion. The resulting data is stored in memory and used to reconstruct
the speech via a digital to annlog (D/A) algorithm. Data rates range
from 64,000 bits per second of stored speech for high fidelity speech
to 1500 bits per second for speech that is of moderate intelligibili-
ty (80%) to the naive listener. This digitized/reconstructed speech
is used for fixed vocabulary applications, e.g. talking calculators
for the blind, automatic telephone number referral (Rabiner et al.,
1971), a fixed phrase language translator (Texas Instruments, 1979).
The most widely available commercial application of digitized/recon-
structed speech is probably the Texas Instruments "Speak and Spell"
hand calculator spelling game with a vocabulary of about 200 words
and reconstruction software contained in three positive-type metal
oxide semiconductor (PMOS) chips (Wiggins and Brantingham, 1978).

What has been called "true" synthesized speech is speech generated entirely from algorithms applied to stored constants that describe the different phonemes or speech sounds. Input may be codes
for individual phonemes or a set of values for concurrent acoustic
parameters to generate phoneme strings. No original human recording
is required. Thus an unlimited vocabulary is possible. Data to generate parameter-synthesized speech requires about 900 bits per second
of speech for storage in memory while codes to generate phoneme-syn-
thesized speech requires about 100 bits per second, if a fixed vocabulary is used. If the speech is generated directly from spelled,
ASCII code, the memory needed is on the order of 16K of 8-bit words
for the text-to-phoneme software (for English) plus a small buffer
of no more than 1K. Languages that approach a one-to-one grapheme
to sound correspondence (e.g. Spanish) may require as little as 4K
for text-to-phoneme software. True synthesized speech has been used
for speech prostheses for the speech-disabled (LeBlanc et al., 1980;
Carlson et al., 1980; various articles in Communication Outlook),
reading machines for the blind (Kurzweil, 1980); and is appropriate
for talking computer terminals. It has been implemented for text-to-
phoneme generated speech on large main-frame computers (Ainsworth,
1973; Ellovitz et al., 1976; Allen, 1973; McIlroy, 1974; Rahimi and
Eulenberg, 1974). And, more recently, real time text-to-speech systems for microprocessors have been developed (Colby et al., 1977;
Computalker Consultants and Upper Case, 1980; LeBlanc et al., 1980).
In its phoneme-coded, fixed vocabulary form, phoneme synthesized
speech has been used for speech prostheses (Echo On, 1980), an experimental automatic visual flight rules (VFR) traffic advisory service (Cassell, 1973), an experimental cockpit voice output system
for data link information (Hilborn, 1975) computer assisted instruction for teaching English to Spanish speaking children (Dunklau,
1979), experimental cockpit voice warnings (Simpson, 1976; Hart and
Simpson, 1976; Simpson and Williams, 1980). The best-known commercially available phoneme synthesizers are manufactured by Votrax Di-

vision of Federal Screw Works in Troy Michigan under the trade name "Votrax".

Proposed applications for synthesized speech include warnings and advisories for operators, passengers, and maintenance personnel for various transportation vehicles - aircraft, trains, automobiles (Priest, 1980) and, in the popular electronics literature, talking appliances, voice read-out of digitally broadcast information, and talking computer displays. This list is limited only by human imagination.

Introductory References on Speech Synthesis

Current research on voice synthesis techniques is directed at combining the advantages of digitized/reconstructed speech and synthesized speech. Different languages, dialects, and voice qualities are being investigated. The interested reader will find the latest research reported in the Journal of the Acoustical Society of America, the IEEE Transactions on Communication Technology, the IEEE Transactions on Audio and Electroacoustics, the Reports of the Speech Transmission Laboratory, Department of Speech Communication, Royal Institute of Technology, Stockholm, and the Bell System Technical Journal, among others. For a technical introduction to the field of speech synthesis, the reader is referred to Lehiste (1967), Fant (1973), Flanagan and Rabiner (1973), Atal and Hanauer (1971), Makhoul (1975), Wiggins (1974), Rice (1976a and 1976b), and Sherwood (1979).

Given the availability of synthesized speech displays for man-machine systems, research is needed to study suggested applications for speech and design principles for speech displays (Williams and Simpson, 1975). This paper reports one such study, for which new performance measures were developed.

STUDY RATIONALE

A number of air carrier approach and landing accidents during low or impaired visibility have been associated with the absence of approach callouts. This fact at first suggests that the absence of altitude and deviation callouts may have contributed to inadvertent flight into terrain during these approaches. Such a conclusion was drawn by the United States (U.S.) National Transportation Safety Board (NTSB) in a special study on flight crew coordination procedures in accidents during air carrier instrument landing system approaches (1976). However, due to the absence of data on callout performance for successful approaches made during the same period (1970-1975) no test of this suggested relationship of cause and effect can be performed. However, one can ask whether the current procedure can be improved, especially when unexpected events cause a high workload

in the cockpit. One alternative medium for information transfer
would be a speech synthesizer which could receive raw data from the
onboard central air data computers and automatically make the callouts at appropriate times during the approach.

EXPERIMENTAL APPROACH

The purpose of the present study was to compare a pilot-not-
flying (PNF) approach callout system to a system composed of PNF
callouts augmented by an automatic synthesized voice callout system
(SYNCALL). A full task flight simulation experiment was conducted to
determine if one or the other system transfers altitude and deviation
information more reliably than the other and/or results in better
flight performance by the pilot-flying. This objective was accomplished by a collaborative effort between American Airlines Flight
Academy and NASA Ames Man Vehicle Systems Research Division. Scheduled air carrier operations were observed from the flight deck for
the purpose of familiarization with airline operations and the collection of data on auditory events including callouts, Air Traffic
Control (ATC) voice communications, and other auditory signals. Then,
a controlled experiment was conducted in a DC-10 airline training
simulator, flown during normal recurrent flight training using crews
as assigned but on a strictly voluntary basis. All pilots offered the
opportunity to participate did so. Twenty 3-man crews flew several
types of approaches half with the standard callout procedures and
half with a synthesized voice callout system (SYNCALL). Measurements
were made of flight performance, callout reliability, callout interference with other audio events, and pilot ratings of callouts on
several scales. Reliability of PNF approach callouts made during actual operations was compared to reliability of PNF callouts and SYN-
CALL callouts during the simulator experiment.

FLIGHT DECK DATA COLLECTION

During the flight deck observation, it was found that 15% of
the time during the descent from cruise altitude and the approach
to landing was spent listening to or talking to Air Traffif Control
(ATC). It was also found that certain approach callouts were sometimes not made but that no approaches were flown for which no callouts were made. Omitted callouts fit definite patterns, and their
omission was usually associated with conflicting audio or oral tasks.
See Simpson, 1980, for further details.

SYNTHESIZER CALLOUT SYSTEM (SYNCALL)

Figs. 1a and 1b show the experimental SYNCALL system designed
to test the concept of automatic approach callouts and the approach

callout procedures used by American Airlines at the time of the study. SYNCALL was modeled after American's procedures so as to control for pilot familiarity with the callout set. The system was designed to include all of the types of automatic approach callouts that might be useful. As such it was an experimental tool rather than a prototype system. The synthesizer was programmed to make both normal and deviation callouts. These callouts were the standard callouts then required for all approaches by the participating airline. Callout wording was designed according to principles derived from previous research on synthesized voice message wording (Simpson, 1976). See Simpson (1980) for a full description of callout triggering conditions and wording.

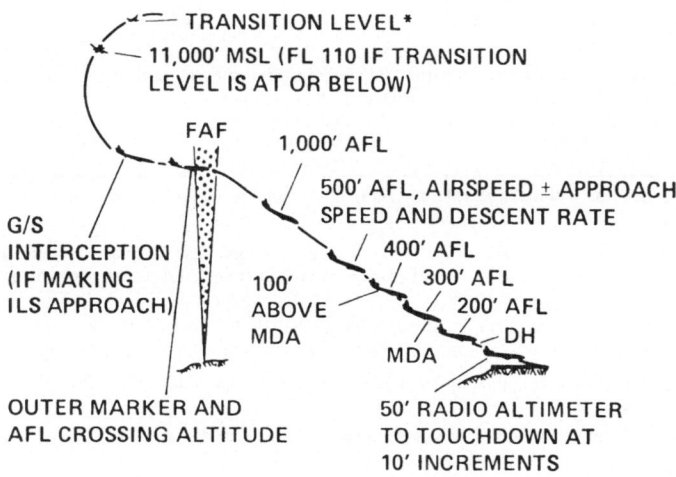

Fig. 1a. Experimental SYNCALL System Design.

NORMAL PROCEDURES

DC-10 Operating Manual

Section 3   Page 63

2-27-76

STANDARD CALLOUTS — ALL DESCENTS, APPROACHES AND LANDINGS

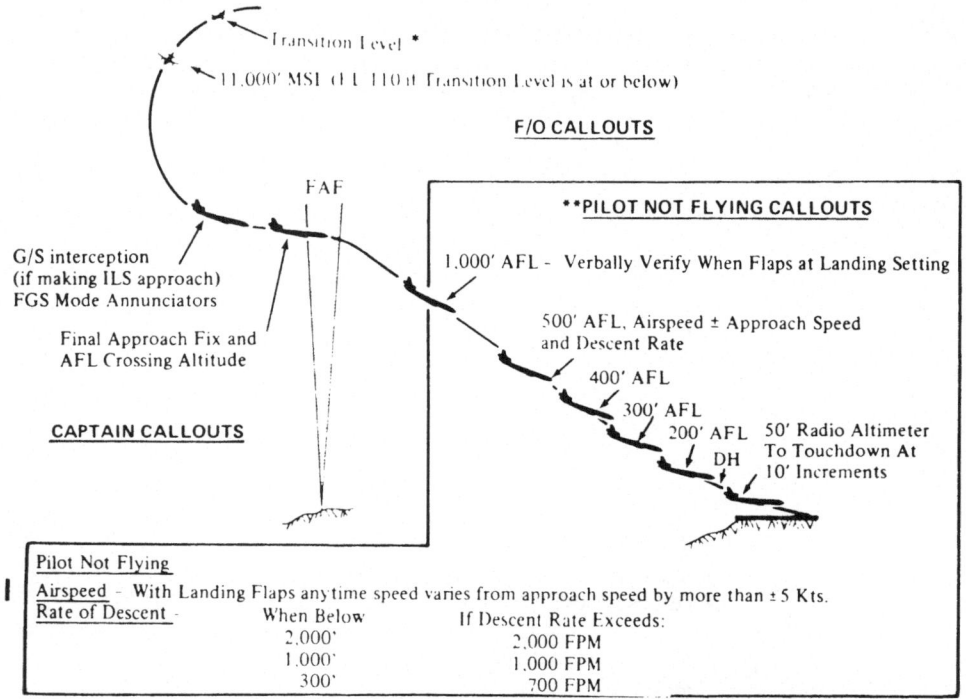

Transition Level *

11,000' MSL (EL 110 if Transition Level is at or below)

F/O CALLOUTS

FAF

**PILOT NOT FLYING CALLOUTS

G/S interception
(if making ILS approach)
FGS Mode Annunciators

1,000' AFL - Verbally Verify When Flaps at Landing Setting

500' AFL, Airspeed ± Approach Speed
and Descent Rate

Final Approach Fix and
AFL Crossing Altitude

400' AFL

300' AFL

CAPTAIN CALLOUTS

200' AFL   50' Radio Altimeter

DH   To Touchdown At
10' Increments

Pilot Not Flying

Airspeed - With Landing Flaps anytime speed varies from approach speed by more than ± 5 Kts.

Rate of Descent -

When Below	If Descent Rate Exceeds:
2,000'	2,000 FPM
1,000'	1,000 FPM
300'	700 FPM

During Cat II Approaches F/O will make all callouts from 500' to landing, including Airspeed or Descent Rate if not within limits.

LOC and GS indication Callout On final, anytime any crewmember observes LOC displacement greater than 1/3 dot and/or G/S displacement greater than 1 dot, other pilot will acknowledge this deviation.

NOTES:

1. Use the Captain's or F/O's altimeter for all altitude callouts, except for the CAT II approach, in which case use the radio altimeter for the 300' callout and remaining callouts below 300' to touchdown.

2. When executing a non-precision approach, the pilot not flying will call out "100' ABOVE MDA," MDA and MAP.

3. Callouts not applicable on non-precision or visual approaches should be disregarded.

* For Transition Level outside U.S. refer to avigation charts.

F/E MONITORING

In addition to his normal monitoring of engine instruments and F/E panel, the F/E will monitor flight instruments, especially altitude and airspeed. He will assist in maintaining a watch for traffic and other factors that could adversely affect safety. He will monitor HSI and ADI navigation indications and call out any discrepancies between the two instruments. On final, call out LOC displacement anytime it is greater than 1/3 dot and/or G/S displacement anytime it is greater than 1 dot.

Fig. 1b. Pilot-Not-Flying (PNF) Callout Procedures. Reprinted with
permission from American Airlines.

DEVELOPMENT OF PERFORMANCE MEASURES

Flight performance measures were designed which are thought to
relate directly to airline operators in a way that allows the system
users (pilots, airline management, manufacturers, government regulatory agencies) to base decisions on system implementation directly
on the data. They are based on the concept "out of operational tolerance". It was assumed that a pilot generally attempts to fly an
aircraft within certain tolerances or "windows" and that as long as
performance remains within the desired window, no correction will
be deemed necessary by the pilot. This is supported by the observation that both government and airline company regulations are written
with allowable tolerances. Also, pilots if asked will say they fly
to tolerances rather than try to keep a needle exactly centered or
positioned: Indirect support is also suggested by the need to incorporate an "indifference threshold" into control models of human operator performance in order to obtain an acceptable fit between model-
predicted performance and observed human operator performance (Curry
et al., 1977). If this assumption is correct, measures of mean deviation from an ideal value for, say, airspeed would have little operational relevance to what a pilot judges to be good performance. It
would seem more relevant to keep track of whether a pilot flies inside or outside the tolerance window for a given parameter (airspeed,
sink rate, localizer deviation, or glideslope deviation) and, when
outside tolerance, to monitor the average and maximum deviations from
the boundary of the tolerance window - not from the ideal value. This
"out of tolerance" concept is embodied in the following performance
measures.

Percent Time out of Tolerance

The percentage of the total duration of the segment that was
flown outside the tolerance window for each of four flight parameters was computed and labeled "percent time out of tolerance".

Maximum Deviation

The measure "maximum deviation" was defined as the deviation
with the largest absolute value, beyond the tolerance window boundary for a given flight parameter. Four parameters were measured for
deviation outside allowable tolerance windows: airspeed, sink rate,
localizer displacement, and glideslope displacement. These were
chosen since they are the parameters which the pilot-not-flying and
the flight engineer are required to monitor during the approach. An
"operational tolerance" for each flight parameter was derived by adding its operating manual ("book") tolerance and its "indicator" tolerance. The resulting values were: 1) Airspeed +/-8 KTS; 2) Sink
Rate between 1000 and 300 ft AFL 1174 fpm; 3) Sink Rate below 300 ft
AFL 874 fpm; 4) Localizer +/-0.56 dot (0.70 deg); 5) Glideslope +/-
1.28 dots (0.32 deg).

MEASURES OF PILOT JUDGMENTS OF SYSTEM PERFORMANCE

     Preference grids, a measure that had been successfully used with
airline pilots in a study of line pilot preferences for design of
cockpit warning systems (Williams and Simpson, 1976) were used to
study the effects of type of approach conditions (Night VFR, Day VFR,
IFR to Weather Minimums, or Abnormal/Emergency) on whether or not
pilots would want a particular callout to be made by the Pilot-not-
Flying and/or by SYNCALL. The last item was a list of possible modifications to the SYNCALL system as configured for the study. Pilots
were asked to rate each modification on a five-point scale of 1)
Highly Desirable, 2) Desirable, 3) No preference, 4) Undesirable, or
5) Unsafe. At the end of the list were blank lines with instructions
to fill in additional modifications the pilots thought they would
want. They were asked to rate their own suggested modifications using
the same 5-point scale. This concluded the main debriefing form. A
second, optional form was used to obtain more information from those
pilots who were interested and willing to spend additional time. This
form consisted of a set of semantic differential scales (Osgood et
al., 1957) chosen for their relevance to cockpit flight information
systems. Nine different types of callouts (e.g. one hundred-foot callouts, airspeed deviation callouts) as spoken by each system were measured with these eleven scales. For purposes of comparison, the current GPWS "Whoop, Whoop. Pull up!" and "Glideslope" voice messages
were added to the set of callouts to be rated on each of the semantic
differential scales. These items were included because they were then
and still are the only examples of electronic speech heard by all air
transport pilots.

PROCEDURE

     Each crew received a taped briefing explaining the scope and
purpose of the experiment, and describing the SYNCALL system as compared to the PNF system. They received copies of the diagrams in
Figs. 1a and 1b, and heard a tape of SYNCALL pronouncing each callout to familiarize them with the electronic voice quality and "ac-
cent". After the 15-minute briefing, the instructor pilot conducted
the normal pre-simulation briefing on the flight maneuvers to be
practiced. This was followed by four hours of recurrent simulator
training in accordance with government and company requirements,
using the appropriate set of experimental conditions for each crew.
Pilots flew the first "set" of approaches with one callout system
(SYNCALL or PNF) and a second set with the other system. Approach
Set 1 consisted of a Localizer Back Course Approach, an Autoland
Approach, and then a 2-Engine or Engine-out Approach. Approach Set
2 consisted of a VOR Raw Data Approach, another Autoland Approach,
and, finally, a Single-Engine Approach. A pre-recorded set of Approach and Tower transmissions, recorded and edited from actual ATC
communications, were played over the cockpit headsets to the pilots.

The pilot instructor, acting as ATC controller, and the pilots used
ATC phraseology via the microphones and headsets, as in the actual
aircraft. The ATC tapes were edited from actual ATC communications
recorded at the airport that was simulated for the study. Flight
performance parameters were sampled every 0.8 sec by the simulation
computer. Data collection began at 3000 feet AFL (1000 meters QNF)
and continuted until either the initiation of a go-around on a missed approach or the touchdown on a landing. The cockpit audio including crew conversation, ATC, approach callouts, and cockpit noise
was recorded on audio tape. After the simulator session, each pilot
completed the briefing form and discussed the experiment with the author. Pilots were offered a copy of the final report, when published.

RESULTS

Flight Performance Results

     Three-way Analysis of Variance (AOV) for the effects of 1) type
of callout system, 2) pilot flying, and 3) order of pilot flying was
performed for each approach. Type of Callout System (SYNCALL or PNF)
was the only factor that had statistically significant effects on
flight performance. Furthermore, significant differences in flight
performance resulted only for the non-precision approaches (Localizer Back Course and VOR) only in the lower altitute segments (III:
500 AFL to MDA: and IV: MDA to Landing/Go-Around) and only for airspeed and sink rate performance - not course tracking performance
(Localizer and Glideslope where applicable). In each case, flight
performance was better with SYNCALL than with PNF callouts.

     Airspeed Deviations During Localizer Back Course Approach. Ta-
ble I shows airspeed performance for Segment IV (MDA to Landing/Go-
Around) for the Localizer Back Course Approach for each callout sys-
tem. There were 19 such approaches flown by either a Captain or a
First Officer for which SYNCALL was used and 17 approaches for which
PNF callouts were used[1]. Percent time out of tolerance for the SYN-
CALL approaches was 13% compared to 34% time out of tolerance for
the PNF approaches. This difference was significant ($F=4.30$, df=
$=1,28$, $p<0.05$). Similarly, fewer approaches were flown out of tole-
rance at all (5) with SYNCALL compared to PNF (12).

     When approaches were categorized as in or out of tolerance at
500 feet and at MDA (Minimum Descent Altitude), there was no significant correlation between callout system and airspeed performance

---

[1] If the crew on a given approach initiated a go-around before reaching MDA (as sometimes happened) then there was no data collected
for that approach for that crew. This is the reason for the unequal N number of approaches for the two callout systems.

Table I. Airspeed performance for localizer back course approach
         during segment IV: MDA to Landing/Go-Around.

Callout System	Mean percent time out of tolerance	Number of approaches out of tolerance
SYNCALL (19 approaches)	13 %	5
Pilot-not-Flying (17 approaches)	34 %	12

at 500 AFL (X(2)=0.07, df=1, p>0.10). But the Chi-Square for the data at MDA was significant (X(2)=5.56, df=1, p<0.02). Fig. 2 graphs
the mean airspeed deviation from Selected Approach Speed at 500 AFL
and at MDA for SYNCALL and PNF approaches. Average airspeed deviation for PNF approaches tends to increase slightly (+6.5 KTS +/-4.9
KTS at 500 AFL increasing to +7.9 KTS +/-4.3 KTS by MDA). In contrast, average airspeed deviation for SYNCALL approaches tends to
decrease (+6.3 KTS +/-3.9 KTS at 500 AFL decreasing to +4.6 KTS
+/-3.2 KTS by MDA). These analyses taken together suggest that airspeed performance improved during the descent when SYNCALL was in
use but did not improve during the same part of the descent when only the PNF callouts were in use.

**MEAN AIRSPEED DEVIATIONS FOR LOCALIZER
BACK COURSE APPROACH**

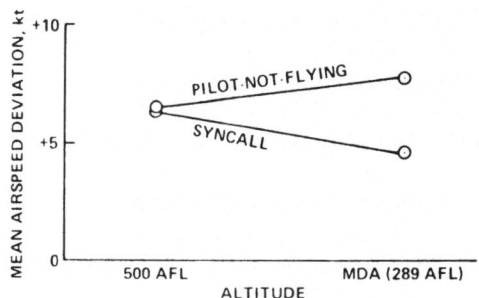

Fig. 2. Mean Airspeed Deviation at 500 Feet AFL and at MDA.

Sink Rate Deviations During the VOR Approach. Table II shows
the sink rate data for Segment III (500 AFL to MDA) for the VOR Raw
Data Approach. Analysis of Variance for effects of callout system,
pilot flying, and order of pilot flying on sink rate performance
for this approach yielded a significant difference (F=5.01; df=1,31;
p<0.05) due to callout system on the maximum deviation beyond op-

Table II. Sink rate performance for VOR Raw Data Approach during
          Segment III: 500 feet AFL to MDA (435 Feet).

Callout system	Number of approaches out of tolerance	Mean maximum deviation
SYNCALL (19 approaches)	0	--
Pilot-not-Flying (19 approaches)	4	1412 FRM +/- 167 (for 4 approaches)

erational tolerance. The effect of callout system on percent time out
of tolerance approached significance (F=4.06; df=1,31; p=0.053). When
SYNCALL was in use, there were no instances for the 19 approaches
flown with SYNCALL of sink rate exceeding the operational tolerance
of 1174 FPM. Of the 19 approaches flown with the PNF callouts 4 had
sink rate deviations in excess of tolerance. The mean maximum deviation of these four "out-of-tolerance" approaches was 1412 FPM +/-167
FPM or 238 FPM beyond the operating tolerance of 1174 FPM. Finally,
for those 4 PNF approaches that were out of tolerance the mean percent time out of tolerance was 37.3%.

1-Engine Approach and Landing Airspeed Performance. The 1-Engine
Approach and Landing is flown by Captains only and thus had to be
analyzed separately. The unique approach speed procedures used for
this approach resulted in SYNCALL making inappropriate excessive airspeed deviation callouts between 600 and 100 ft AFL (see Simpson,
1980, for details). To determine if this adversely affected performance, airspeed was measured at 100 feet AFL, the point at which procedures call for the speed to be slower than the speed to which SYN-
CALL was referenced for deviation callouts. Table III shows mean Indicated Airspeed (IAS) deviation at 100 AFL for the two systems.
This difference was not significant (T=1.01; df=17; p>0.10). However,
the difference between the variances in airspeed performance was significant by two-tailed test for differences between independent variances (F=4.69; df=9,8; p<0.05). Thus the captains' airspeed at 100

Table III. Airspeed performance for 1-Engine Approach and Landing -
           Captain only Airspeed measured at 100 Feet AFL.

Callout system	Mean Airspeed Deviation	Standard Deviation
SYNCALL (10 approaches)	+ 9.4 KTS	+/- 7.8 KTS
Pilot-not-Flying (9 approaches)	+ 6.5 KTS	+/- 3.6 KTS

feet during the Single-Engine Approach and Landing was more variable with SYNCALL than with PNF callouts. This is certainly an adverse effect on airspeed performance from an operational standpoint.

SYNCALL and Pilot-Not-Flying Callout Performance

The cockpit audio recordings for ten of the twenty crews were analyzed to obtain callout reliability (percent of required callouts that were actually made) for each callout system. A tabulation was made of other concurrent audio events that occurred in the cockpit when SYNCALL callouts would have been made for the approaches that used PNF callouts.

Reliability of normal callouts. Reliability of PNF and of SYN-CALL for making normal approach callouts in the simulator was compared. Of the 746 callouts required during the 53 PNF approaches, 507, or 72%, were made by PNF. SYNCALL made 609, or 87%, of its required 685 normal callouts for its 53 approaches. It should be noted that there were no approaches for which no callouts at all were made, for either system. The difference in number of callouts made by SYN-CALL and by PNF was significant ($X(2)=91.24$; $df=1$; $p<0.002$). The reliability of PNF callouts in the simulator was also compared to the reliability of PNF callouts in the actual aircraft on the line. This comparison could only be performed for normal callouts and could not be done for matched sets of approaches. In an attempt to make the line data and the simulator data more comparable, the percentage of normal callouts made under the three conditions: 1) PNF on the line, 2) PNF in the simulator during the experiment, and 3) SYNCALL in the simulator during the experiment were re-calculated, deleting from considerations all the MDA and DH callouts, since these were rarely required on the line due to the good weather (4 of the 64 line approaches were flown in weather close to the published minimums). Table IV shows the resulting percentages with the previous percentages that did include MDA and DH callouts, for comparison. No statistical analysis of this three-way comparative data was performed since the validity of such a comparison is doubtful. Note that the 83% reliability of PNF normal callouts on the line would appear to be very

Table IV. Reliability of normal approach callouts made by Pilot-not-Flying on the line and in the simulator and by SYNCALL in the simulator.

	MDA & DH callouts included	excluding MDA & DH callouts
On the line	79 %	83 %
PNF in the simulator	72 %	74 %
SYNCALL in the simulator	87 %	85 %

close to the 85% reliability of SYNCALL normal callouts in the simulator.

    Reliability of deviation callouts. The reliability of the de-
viation callouts made by SYNCALL and by PNF is shown in Table V.
When PNF deviation callouts were used, the PNF made 39 out of 89, or
49%, of the required deviation callouts. This is compared to 75 out
of 83, or 90%, reliability for deviation callouts made by SYNCALL.
The difference in numbers of callouts made by the two systems was
significant (X(2)=41.63; df=1; p<0.002).

Table V. Reliability of deviation callouts by SYNCALL and by Pilot-
         not-Flying in the simulator.

Callout system	was callout made?		total	percent
	yes	no		
SYNCALL	75	8	83	90 %
Pilot-not-Flying (SYNCALL OFF)	39	50	89	49 %

Data are totals from 53 approaches for each callout system.

    Other audio information that overlapped SYNCALL callouts. 23%
of the SYNCALL callouts would have occurred during crew checklists
and other callouts. In contrast, the Pilot-not-Flying when making
Approach Callouts was observed to sequence these with other crew
checklist calls so as to minimize overlap. Like SYNCALL, however,
the PNF callouts often overlapped with Other ATC.

Pilot Judgment Data

    The pilot judgment results are reported in detail in Simpson,
1980. Nearly all (82%) of the pilots who flew SYNCALL wanted SYNCALL
to make callouts during approaches when weather was down to IFR minimums. Only 55% of the pilots wanted SYNCALL callouts during Day
VFR conditions. Thus, type of approach conditions affects pilots'
judgment of SYNCALL usefulness. Many pilots noted independently that
the electronic voice quality of the synthesizer made it very dis-
tinctive and left no doubt regarding the source of the voice.

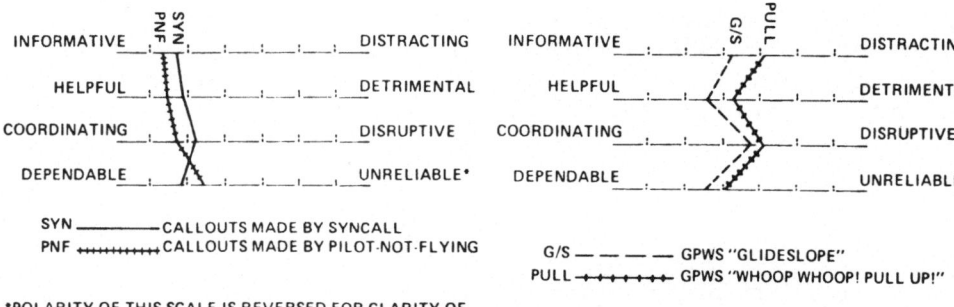

Fig. 3. Semantic differential ratings of SYNCALL and PNF.

Fig. 3a shows the mean ratings for all callouts on four of the semantic differential scales for each of the callout systems (SYNCALL and PNF). These data come from the 28 of the 40 pilots or 70% who volunteered to complete this optional form on their own time. Differences between ratings for SYNCALL and PNF callouts were significant for all four scales (Informative-Distracting: t=-3.23; p<0.05; Helpful-Detrimental: t=-5.12; p<0.002; Dependable-Unreliable: t=7.98; p<0.002; Disruptive-Coordinating: t=-6.42; p<0.002; df=8 for each scale). While both systems were judged as informative, helpful, dependable, and coordinating, callouts given by PNF were judged higher on these scales than the same callouts by SYNCALL. On the other hand, callouts by SYNCALL were judged as more dependable than the same callouts by PNF. For purposes of comparison to an existing system that produces voice messages on the approach to landing, the pilots ratings of the GPWS, "Whoop, Whoop, Pull Up!" and "Glideslope" messages are given in Fig. 3b.

Finally, of the 31 modifications to SYNCALL suggested by the 40 pilots, 5 received a median "highly desirable" rating from the 13 pilots. These were 1) Obtain FAA and company agreement not to use the cockpit voice recordings of SYNCALL for disciplinary actions; 2) On/Off switch; 3) Eliminate nuisance callouts; 4) Provide test switch; 5) Manual volume control (for the entire list, see Simpson, 1980).

DISCUSSION

Implications of Improved Flight Performance

Airspeed and sink rate performance with SYNCALL making the callouts was better during non-precision approaches than when only the
current PNF callouts were used. Airspeed (as an input to estimated
ground speed) and sink rate constitute the raw data used by a pilot
to fly a desired glide-path during a non-precision approach. A non-
precision approach is considered more difficult to fly just because
of the lack of direct glide-path information as compared to the easier precision approach which provides glide-path information displayed directly to the pilot. Thus the use of SYNCALL was associated
with improved flight performance for a type of approach that is more
difficult to fly. Non-precision approaches result in a large percentage of the air carrier approach and landing accidents and incidents,
53% according to the NTSB study (1976). If a system which improves
flight performance for this type of approach in particular made a
sufficiently large contribution to lowering the over-all approach
and landing accident rate it might pay for its installation. It could
also be that the higher attention workload that is usually but not
exclusively associated with non-precision approaches is the underlying factor whose effect could be mitigated by SYNCALL. If this is
correct, then SYNCALL would be expected to be helpful during any approach with high attention workload.

Implications of Differential Callout Reliability

The data on PNF and SYNCALL callout reliability suggest that the
function of SYNCALL should be to alert the crew to deviations from
desired flight parameters but not to make routine callouts. Altitude
callout reliability by PNF was comparable to that of SYNCALL. Also
23% of the SYNCALL callouts overlapped crew checklist callouts. In
contrast, the PNF minimized this overlap by sequencing altitude callouts between checklist callouts whenever possible. Thus one cannot
attribute improved flight performance to any greater reliability of
SYNCALL over PNF for routine altitude callouts. And it is possible
that SYNCALL altitude callouts actually interferred with pilots'
comprehension of checklist callouts. Pilots did, in fact, rate SYN-
CALL less "coordinating" than PNF. All this argues against automatic
altitude callouts by SYNCALL. The case for deviation callouts by
SYNCALL is very different. They were substantially more reliable
when made by SYNCALL than by PNF. Also, because they occurred in-
frequently compared to altitude callouts, their interference with
crew checklist callouts would be minimal and thus could be an acceptable price in increased audio workload to pay to the benefit of
timely and reliable alerts to flight performance deviations. Occasionally, however, an altitude callout can function as a warning of
an unnoticed deviation. One can speculate that certain airline accidents might not have occurred, had automatic altitude deviation

callouts been made (cf. 6 and 7). If a reliable algorithm to detect unnoticed altitude deviations during the approach could be developed, this could be a valuable component of a SYNCALL system. One algorithm for this is proposed in Simpson, 1980. It would trigger SYNCALL altitute deviation callouts only when sink rate was extremely small or extremely large. Such an algorithm assumes that when sink rate is within a reasonable tolerance window, the pilots know they are descending and are aware of the rate. Consistent with a SYNCALL system that calls out deviations only would be the incorporation of the Ground Proximity Warning System (GPWS) modes into SYNCALL. If this reasoning is correct, a SYNCALL system that automatically called out deviations in airspeed, sink rate, glideslope, localizer, altitude and terrain closure, only when deviations actually existed, could be a valuable addition to airline cockpit systems.

CONCLUSION

Improvements in flight performance by airline pilots were obtained for approaches that have high manual and visual workload for the pilots. By extrapolation, similar improvements might be obtained for any high workload approach. The detrimental effect on performance associated with inappropriate deviation callouts requires that such callouts be designed out of the system. Because of comparable reliability for normal altitude callouts by PNF and SYNCALL, a SYNCALL system should not make routine altitude callouts. Since SYNCALL was more reliable than PNF for deviation callouts, the system should make deviation callouts. For consistency, it is recommended that the modes of the Ground Proximity Warning System (GPWS) be incorporated into a SYNCALL system. Pilots generally favored the concept of SYNCALL and judged it more reliable than PNF callouts. They suggested modifications before it would be appropriate for operational use. Clearly the concept of SYNCALL should be given further study, building on the findings reported here.

REFERENCES

Allen, J., 1978, An Approach to Reading Machine Design, Human Factors, 20, 287-293.
Atal, B.S. and Hannauer, S.L., 1971, Speech Analysis and Synthesis by Linear Prediction of the Speech Wave, Journal of the Acoustical Society of America, 50, 637-655.
Carlson, R., Galyas, K., Granström, B., Pettersson, M. and Zachrisson, G., 1980, Speech Synthesis for the Non Vocal in Training and Communication, Paper presented at the London Seminar on the International Project on Communication Aids for the Speech-Impaired, London, April 30. Also available as Speech Transmission Laboratory Report STL-QPST 1/1980, Department of Speech Communication, Royal Technical Institute, Stockholm, Sweden.

Cassell, R., 1973, The Knoxville Automatic VFR Advisory Service Operational Report, Terminal Branch, Air Traffic Control Systems Division, Systems Research and Development Service, Federal Aviation Administration, Department of Transportation, Washington, D.C. 20591, USA.

Colby, K.N., Christinaz, D. and Santiago, G. Jr., 1977, A personal, Portable and Intelligent Speech Prothesis, Memo of the Algorithmic Laboratory of Higher Mental Functions, Memo ALHMF-9, Department of Psychiatry, University of California, School of Medicine, Los Angeles, CA 90024, USA.

Communication Outlook, A publication of the International Action Group for Communication Enhancement, Artificial Language Laboratory, Computer Science Department, Michigan State University, East Lansing, MI 48824, USA.

Computalker Consultants and Upper Case, 1980, ANGLO2: English Text to Speech System User's Manual, available from Computalker Consultants, 1730 21st Street, Santa Monica, Ca 90404, USA.

Curry, R.E., Kleinman, D.L. and Hoffman, W.C., 1977, A design Procedure for Control/Display Systems, Human Factors, 19 (5), 421-436.

Dudley, H., 1939, The Vocoder, Bell Labs., Record 17, 122-126.

Dunklau, W.M., 1979, Voice Synthesis for Early Elementary Computer-Assisted Instruction, The Best of the Computer Faires, Vol. IV, Conference Proceedings of the 4th West Coast Computer Faire, San Francisco, CA; available from Computer Faire, 333 Swett Road, Woodside CA 94062 USA.

Echo On, Publication of H.C. Electronics, 250 Camino Alto, Mill Valley, CA 94941, USA.

Ellovitz, H.S., Johnson, R.W., McHugh, A. and Shore, J., 1976, Automatic Translation of English Text to Phonetics by Means of Letter to Sound Rules, Naval Research Laboratory Report NRL-7948, Department of the Navy, Office of Naval Research, Arlington, VA 22217, USA.

Fant, G., 1973, "Speech Sounds and Features", MIT Press, Cambridge, MA.

Flanagan, J.L. and Rabiner, L.R. (eds.), 1973, "Speech Synthesis", Dowden, Huchinson & Ross, Inc., Stroudsburg, PA.

Flanagan, J.L., Coker, C.H., Rabiner, L.R., Schafer, R.W. and Umeda, N., 1970, Synthetic Voices for Computers, IEEE Spectrum, 7, 22-45.

Hart, S.G. and Simpson, C.A., 1976, Effects of linguistic redundancy on synthesized cockpit warning message comprehension and concurrent time estimation, Proceedings of the 12th Annual Conference on Manual Control, Moffett Field, CA, National Aeronautics and Space Administration, Ames Research Center, NASA TMX-73,170, 309-321.

Hilborn, R.H., 1975, Human Factors Experiments for Data Link: Final Report, Report No. FAA-RD-75-170, prepared for U.S. Department of Transportation, Federal Aviation Administration, Systems Research and Development Service, Washington, D.C. 20591, USA; available from National Technical Information Service, Springfield, VA 22161, USA.

Kurzweil Reading Machine Updata, No. 1, 1980, Kurzweil Computer Products, 33 Cambridge Parkway, Cambridge, Mass. 02142, USA.

LeBlanc, M.A., Simpson, C.A., Williams, D.H. and Lingel, C.D., Progress Report: Research and Development of a Versatile Portable Speech Prosthesis, prepared by Children's Hospital at Stanford under NASA Ames Grant No. NSG-2313 for the National Aeronautics and Space Administration, Ames Research Center; available from the first author, Childrens Hospital at Stanford, 520 Willow Rd., Palo Alto, CA 94304, USA.

LeHiste, I. (ed.), 1967, "Readings in Acoustic Phonetics", MIT Press, Cambridge, MA.

Makhoul, J., 1975, Linear Prediction: A Tutorial Review, Proceedings of the IEEE, 63, 561-580.

Michaelis, P.R., 1980, An Ergonomist's Introduction to Synthesized Speech, paper presented at the International Conference on Ergonomics and Transportation, Sept. 8-12, sponsored by the Ergonomics Society; Conference Chairman: D. Osborne, Psychology Dept., University College of Swansea, Singleton Park, SWANSEA SA2 8 PP, West Glamorgan, Wales, UK.

National Transportation Safety Board, 1972, Aircraft Accident Report: Eastern Airlines Inc., L-1011, N310EA, Miami, Florida, report no. NTSB-AAR-73-14.

National Transportation Safety Board, 1976, Special Study: Flight-crew Coordination Procedures in Air Carrier Instrument Landing System Approach Accidents, report no. NTSB-AAS-76-5, Washington, D.C.

National Transportation Safety Board, 1978, Aircraft Accident Report: National Airlines Inc., Boeing 727-235, N4744NA, Escambia Bay, Pensacola, Florida, report no. NTSB-AAR-78-13.

Osgood, C.E., Suci, G.J. and Tannenbaum, P.H., 1957, "The Measurement of Meaning", University of Illinois Press, Urbana.

Priest, J., 1980, Synthesized Speech: Ergonomic Implications for the Transportation Industry, paper presented at the International Conference on Ergonomics and Transportation, Sept. 8-12, sponsored by the Ergonomics Society; Conference Chairman: D. Osborne, Psychology Dept., University College of Swansea, Singleton Park, SWANSEA SA28PP, West Glamorgan, Wales, UK.

Rabiner, L.R., Schafer, R.W. and Flanagan, J.L., 1971, Computer Synthesis of Speech by Concantenation of Formant Coded Words, Bell System Technical Journal, 50, 1541-1558.

Rahimi, M.A. and Eulenberg, J.B., 1974, A Computer Terminal with Synthetic Speech Output, Behavior Research Methods and Instrumentation, 6 (2), 255-258.

Rice, L., 1976, Friends, Humans, and Country Robots: Lend me your Ears, Byte Magazine, number 12.

Rice, L., 1976, Hardware and Software for Speech Synthesis, Dr. Dobbs Journal of Computer Calisthenics & Orthodontia, 1, 6-8; available from Hayden Books, 50 Esse St., Rochelle Park, NJ 07662, USA.

Scientific American (Anon.), 1871, Talking Machines, Sci. Amer., 24, 32.

Sherwood, B.A., 1978, Fast Text-to-Speech Algorithms for Esperanto, Spanish, Italian, Russian and English, Int. J. Man-Machine Studies, 10, 669-692.

Sherwood, B.A., 1979, The Computer Speaks, IEEE Spectrum, August, 18-25.

Simpson, C.A., 1976, Effects of Linguistic Redundancy on pilot's comprehension of synthesized speech, Proceedings of the 12th Annual Conference on Manual Control, NASA TMX-73,170, 294-308.

Simpson, C.A., 1980, Synthesized Voice Approach Callouts for Air Transport Operations, Report prepared under NASA Contract Number A64184B by Psycho-Linguistic Research Associates for Man-Vehicle Systems Research Division, Ames Research Center, National Aeronautics and Space Administration, NASA Contractor Report NASA CR-3300.

Simpson, C.A. and Williams, D.H., 1980, Response Time Effects of Alerting Tone and Semantic Context for Synthesized Voice Cockpit Warnings, Human Factors, 22, 319-330.

Texas Instruments, Product Announcement, 1979, 13500 North Central Expressway, Dallas, TX, USA.

Votrax, Product Announcement, 1980, Votrax Division of Federal Screw Works, 500 Stephenson Highway, Troy, MI 48084, USA.

Wiggings, R., 1974, Formation and Solution of the Linear Equations used in Linear Predictive Coding, Project No. 6550 prepared by the MITRE Corporation under Contract No. F19628-73-C-0001 for the Electronic Systems Division, Air Force Systems Command, United States Air Force, L.G. Hanscom Field, Bedford, MA, Report ESD-TR-74-301.

Wiggins, R. and Brantingham, L., 1978, Three-chip System Synthesizes Human Speech, Electronics, August, 109-116.

Williams, D.H. and Simpson, C.A., 1976, A systematic approach to advanced cockpit warning systems for air transport operations: Line pilot preferences, Proceedings of the Aircraft Safety and Operating Problems Conference, NASA Langley Research Center, Hampton, VA, October 18-20, in: NASA Special Publication SP-416, 617-644.

ACKNOWLEDGMENTS

This project could not have happened without the collaborative effort of NASA Ames Research Center and American Airlines Flight Academy. Many Colleagues and friends have their expertise. Dr. Robert C. Houston and Captain J.A. Brown of American together with Dr. Edward M. Huff and Thomas E. Wempe of NASA envisioned the potential benefit of collaboration and are responsible for its coming to pass. The research reported here was supported in part by the National Research Council, by NASA Grant NGL-05-046-002 to San Jose State University, and by NASA Grant NGR-45-003-108 to University of Utah.

THE USE OF VIDEO-TECHNIQUES IN TRAFFIC RESEARCH

Richard van der Horst

INTRODUCTION

At the Institute for Perception TNO, the study of human information processing is a main research area including the study of perception, decision making, and action. Road-user behaviour in particular is studied by using various methods ranging from mere observations in real traffic situations to highly controlled laboratory experiments, sometimes using advanced simulation techniques. With respect to the last type the Institute has a driving simulator at its
disposal, in which subjects, seated in a car, are able to drive
through a scale-model as a substitute of the real world by means of
a TV camera system. The TV pictures are projected on a large screen
in front of the car (Institute for Perception, 1978).

In what follows an objective method for observation and analysis
of the behaviour of road users in real traffic will be discussed. As
an example a study will be reported in which new road design elements
were evaluated in a demonstration project on cycleroutes in the cities
of The Hague and Tilburg.

The questions were how the design elements are functioning,
whether they are leading to the desired traffic behaviour and whether
they have an effect on road safety.

The study was carried out under contract with the Ministry of
Transport. The developed method also seems to offer good possibilities
for the development of a conflict analysis technique in which time-
to-collision (TTC) is the final criterion.

OBSERVATION AND ANALYSIS

The use of observation techniques generally is needed for the unobtrusive study of actual road user behaviour, e.g. in studies of conflicts between road users. The interest in conflicts resulted from the idea that they could be of relevance in studying road safety, being an alternative to accidents as a criterion measure. Especially on single spots they are more frequent than accidents and therefore form a more efficient basis for road safety studies. In the past various conflict-observation methods have been developed, mostly using individual observers. Although they may be highly trained and experienced, large differences between individual observers remain, sometimes due to inadequate definition of conflict, sometimes due to, for example, inaccurate time estimation in case of an interaction between road users.

To reduce observer subjectivity, it was felt necessary to develop an observation technique which enables objective quantification.

In order to measure motion and positional parameters involved in an interaction between road users, it is almost imperative that recordings are made by means of film or video. In a preliminary study both techniques were compared, each with its own characteristic advantages and disadvantages.

Advantages of film compared with video:-
(a) Better resolution
(b) Analysis equipment available
(c) Additional information of colour, e.g. for marking positions.

Disadvantages of film compared with video:-
(a) Higher costs (20x)
(b) Frequent change of filmcassettes necessary (limits practical use)
(c) No direct check on the quality of recordings
(c) Changing contrast afterwards not possible
(e) More expertise needed
(f) Low potential for automation.

In view of the picture quality and facilities for analysis, film seems to be in favour. However, due to costs and practical aspects it was felt necessary to check whether video gives comparable results. Both techniques, film and video were tried out in real traffic situations, i.e.: four locations on the aforementioned demonstration cycleroutes.

Because of facilities for analysis 16-mm colour film was chosen. Normal black-white video was used because it was expected that the resolution of colour video was too low for further analysis.

A quantitative analysis was carried out on both techniques. It was also investigated what kind of aspects could be measured more directly from the recordings by individual observers.

The quantitative analysis of film pictures is a well-known technique, equipment is available in the form of a motion-analyser. No equipment was available for the analysis of video pictures. So, special equipment for analysis had to be developed.

Quantitative analysis

The film- and video-recorded vehicle movements were analysed quantitatively for the description of road-user behaviour in terms of course, course changes, speed and speed changes. Also measures were taken for the interactions between road users, for example time-to-collision (TTC, see p.402). Positions are selected of some points of the vehicle on stills. By means of transformation rules positions in the plane of the picture can be translated to positions in the plane of the street. By differentiating successive positions in time, the speed of the vehicle can be calculated. The selection of one picture from each number of six (one picture/0.24 sec) seemed to be a reasonable compromis between accuracy and analysis time.

At the time of this first study no equipment was available for selecting video pictures directly. To overcome this problem a 16-mm film was made from a TV monitor. This film was analysed with a motion-analyser. This intermediate step was admissible because video mainly determined picture quality.

Transformation from film co-ordinates to street co-ordinates

A point on the film plane (x_F, y_F) has to be transformed to a point on the plane of the street (x_S, y_S) in a unique way (see Fig. 1). Assuming that all points of the street are lying in one flat plane and that no reproduction errors occur by the camera or by the projector or monitor the following transformation rules can be derived (Hallert, 1960):

$$x_S = \frac{C_1\, x_F + C_2\, y_F + C_3}{C_4\, x_F + C_5\, y_F + 1}$$

$$y_S = \frac{C_6\, x_F + C_7\, y_F + C_8}{C_4\, x_F + C_5\, y_F + 1} \tag{1}$$

The coefficients C_1 to C_8 can be calculated from (1) if the co-ordinates of at least four points in both planes are known. Substituting the x_F, y_F, x_S and y_S of four points in (1) gives a system of eight linear equations with C_1 to C_8 as the unknown elements. This

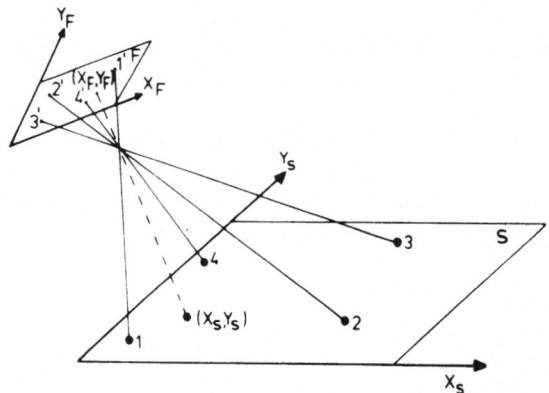

Fig. 1. Schematic representation of the projection of points in the
        plane of the street (plane S) on the plane of the film
        (plane F).

system can be resolved if none of a combination of three points in
both planes are lying on a straight line. This transformation offers
the great practical advantage that nothing has to be known about position and orientation of the camera. All information is included in
the way in which the four points on the street are projected on the
filmplane. On the street only the distances between the points have
to be measured. However, the accuracy of this transformation depends
on the selection of the four reference points. For a check on the
transformation it is advisable to measure some points in addition.

    Software is available for computing the transformation rules,
the speeds, accelerations or decelerations and interaction measures
like time-to-collision curves, minimum passing-times, etc. Also facilities are available for plotting the course of the vehicles involved on a map of the location.

16-mm colour film versus black/white video

    For a comparison of both techniques a number of manoeuvre com-
binations were analysed from 16-mm colour film as well as from black
and white video (by means of the intermediate step of black/white
16-mm film). Both techniques gave comparable results in terms of accuracy and analysis time. However, video was preferred because of
costs and practical aspects: especially the mounting facilities of
a TV camera on the spot are important.

    Yet special equipment had to be developed for the direct anal-
ysis of video pictures.

Video-equipment

Because of the time-consuming character of the quantitative analysis the design of the equipment was based on **a semiautomatic** procedure. A fully automatic procedure is not yet possible because of the complexity of the scenes at an intersection. By using a mini-computer a flexible use is possible.

To find the right video frame automatically a special labelling system was developed. Each frame (50 per sec) is labelled uniquely by mixing and storing digital information at the beginning of each video line (see Fig. 2).

The digital label (24 bits) is repeated four times in each frame. So the information is always available independent of the position of the 'noise bar' (the separation of two successive frames on stills).

In Fig. 3 a blockdiagram of the video-recording equipment is given. The timer mixes the time in the video picture. This is very helpful in selecting the manoeuvres and relating those with other parameters like traffic intensities, etc. A blockdiagram of the video analysis equipment is given in Fig. 4.

Fig. 2. Video still with digital label at the left, electronic cross-hairs, time and noise bar.

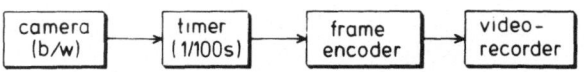

Fig. 3. Blockdiagram of video-recording equipment.

        The central part of the system consists of a small minicomputer
(PDP 11/03 with 28 K memory). An 8-channel digital interface (24
bits per channel) interconnects the computer with the other elements.
A small modification of a standard joystick remote control unit of
the video recorder makes computer control of the video recorder possible (operation control and control of the tape speed). The digital labels are read by the frame decoder continuously and passed at
a command of the computer. Two of the 24 bits are used for numbering
the four labels of each frame separately. So the missing label gives
information about the position of the noise bar. In this way the
noise bar can be shifted down automatically.

        The operator is communicating with the system in two ways:-

(1) By means of a terminal for normal in- or output of the programme
    and

(2) By means of a special keyboard (Fig. 5) consisting a.o. of 16
    push buttons, to which a function may be related in software, for
    example, 'point ready', 'picture ready', 'manoeuvre ready', 'oth-
    er point', etc.

        The operator is indicating a point on the monitor by positioning
two crosshairs continuously by a joystick or step-by-step by four
push buttons. The crosshairs are mixed electronically in the video.
For example, the computer reads the x and y co-ordinates on the command 'point ready' and positions then the crosshairs on predicted x
and y positions of the next point to be measured. The operator has

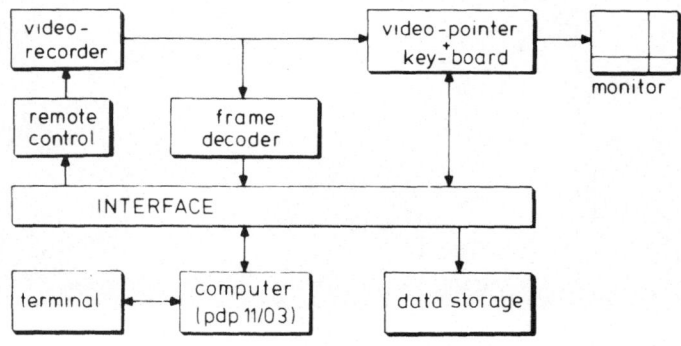

Fig. 4. Blockdiagram of video-analysis equipment.

Fig. 5. Video-analysis equipment: monitor, remote-control of the vi-
        deo-recorder (at the right) and the special keyboard (black)
        with 16 programmable push buttons, joystick and at the left
        4 push buttons for control of the electronic crosshairs.

only to correct these co-ordinates with a few steps. After finishing
a manoeuvre a datafile is submitted to a disc of a PDP 11/40 computer for further data analysis.

BEHAVIOURAL STUDY CONCERNING PRIORITY INTERSECTIONS OF THE DEMONSTRA-
TION CYCLEROUTES

Background

     The increasing number of cars on the road leads to an increas-
ing demand on the available space. Therefore, the government's policy aims at a restricted car use, especially in urban areas and during
peak hours, and to promote the use of the bicycle and/or public transport instead. A safe and highly comfortable system of cycletracks
might promote the use of the bicycle. To stimulate the construction
of cycleroutes in urban areas, the central government had designed

and constructed the two aforementioned demonstration cycleroutes in
The Hague and Tilburg.

These cycleroutes have their own tracks on the road, separated
from motorised traffic, traced through areas with low traffic intensities, crossing other traffic streams as less as possible, giving,
right-of-way to bicyclists at non-signalised intersections, special
priority measures at signalised intersections and, if necessary, even
a viaduct or tunnel over/under heavy traffic streams.

Traffic conflicts as a measure for road safety

As stated before, traffic conflicts could be relevant in studying road safety, being an alternative to accidents as a criterion
measure. In a restricted area and in a restricted time period conflicts are more frequent and therefore may form a more efficient
basis for on the spot studies of road safety. An example of a conflict-observation technique using individual observers is the General Motors Traffic-Conflicts Technique, developed in the United
States (Perkins et al., 1967). Perkins et al. defined a traffic conflict as any potential accident situation leading to the occurrence
of evasive actions; for the latter over twenty criteria were defined
(breaking, weaving, etc.). The great strength of Perkins' method lies
in the very simple way of application. But in later studies observer
subjectivity was felt as a great disadvantage. Therefore, it was considered necessary to develop an observation technique based on objective quantification of conflict measures. According to Hayward (1972)
and Hyden (1977), the time-to-collision measure (TTC) is a promising
one for describing the danger of a conflict situation objectively.
TTC is the time required for two vehicles to collide if they continue at their momentary speed and course. It is a measure continuous
with time. The theoretical shape of a TTC-curve as a function of time
is shown in Fig. 6.

If the vehicles are not on a collision course the value of TTC
is infinite. However, a change in speed of one of the vehicles may
lead to a collision course, implying that TTC will decrease. This
will be linear as long as the speeds and courses of both vehicles
are constant. If neither one would take action, it will result in a
collision (TTC = 0). An evasive action (decelerating, swerving) may
lead to a minimum value for TTC which then increases to infinity a-
gain. It often happens that road users are on a collision course,
but very rarely this will result in a real collision, because drivers
are making constantly speed and heading changes.

Hayward (1972) and Hyden (1977) use a minimum TTC value of less
than 1.5 sec as a critical measure for the risk involved. They define a traffic situation with a minimum TTC < 1.5 sec as being a
serious conflict. The number of conflicts (defined in this way) related to the number of vehicles and a given time period then gives

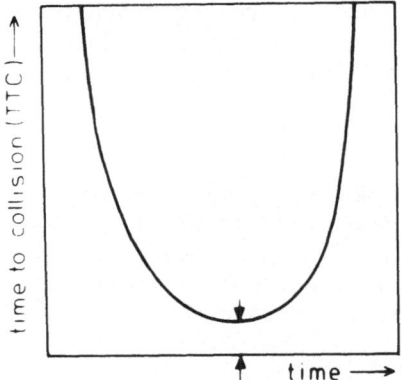

Fig. 6. Theoretical TTC curve as a function of time (Hayward, 1972).

a risk index (RI) for a particular intersection. With this RI intersections can be compared on a relative basis. However, a value of
1.5 sec is chosen rather arbitrarily. The seriousness of a conflict
with a minimum TTC of 0.8 sec may be greater than the one with a
value of 1.5 sec and will depend on the kind of manoeuvre and type
of road users involved; car/car will by different from car/cyclist.
Apart from this, how can TTC curves be obtained? Hayward (op. cit.)
calculated TTC curves of a number of traffic situations by analysing
film pictures quantitatively. Hyden (op. cit.) had individual observers estimate minimum TTC values after an intensive training with
video recordings.

     Although an interaction measure like TTC seems to be a promising
one, a number of questions still remain. Trying out the measure in
practical situations ultimately may lead to a more profound insight.

Recordings and analysis procedures

     At fifteen locations of the cycleroutes (non-signalised inter-
sections) and at four control locations video recordings were made
during six hours on one day. The video recorder was started by hand
when a vehicle arrived and stopped when the manoeuvre had occurred.
In most cases the camera was mounted in an adjacent building, so
that road users were not influenced by the observations.

     For each location the road user behaviour to be recorded was
discussed with municipal authorities. Relevant behaviour was that of
cars crossing the cycletracks (path, speed, place of stopping) as influenced by specific design elements (humps, hobbles, constrictions,
curves, etc.) and, of course, the interaction with bicyclists on the
cycletracks (conflicts). At some intersections also the velocity of
through-traffic was investigated in relation to the design elements.
With the video-analysis equipment a quantitative analysis of the vi-

deo pictures was carried out. Where clear behavioural alternatives could be distinguished, registrations were done by individual observers directly from the video-recordings.

Results

In this paragraph some preliminary results will be given. Important characteristics of crossing car traffic in approaching and negotiating a priority intersection are speed control, stopping (time and place) on or after passing the cycletrack in presence of through-traffic at the main road, the path chosen and interactions with cyclists.

Speed curves of crossing cars. For four locations speed profiles of crossing cars as a function of the distance to the cycletrack are given in Fig. 7. Each point gives the mean value of n vehicles, together with the standard deviation. Although the speed curves differ between locations, it seems that in general the speed control elements (humps for instance) are functioning according to the expectations. The mean speed at the begin of the cycletrack amounts 3.3 m/s with a s.d. of 1.4 m/s.

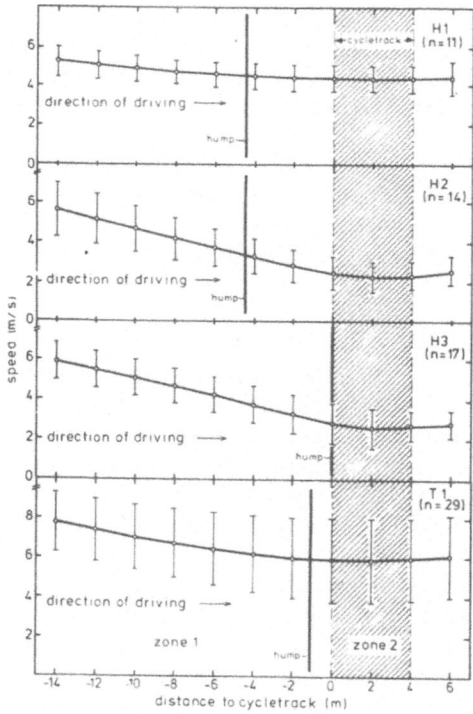

Fig. 7. Speed profiles of cars from the minor street at four locations (H1, H2, H3 and T1).

Speed curves of through-traffic. For some intersections the influence of a raised intersection plane (humps + plateau of brick
pavement) on the speed of through-traffic (cars) was investigated,
the results of which are given in Fig. 8.

A speed reduction of about 4 m/s at the intersection with the
humps could be observed (H7) compared with the intersection without
a hump (H5). Situational maps of both intersections are given in
Figs. 9a and 9b.

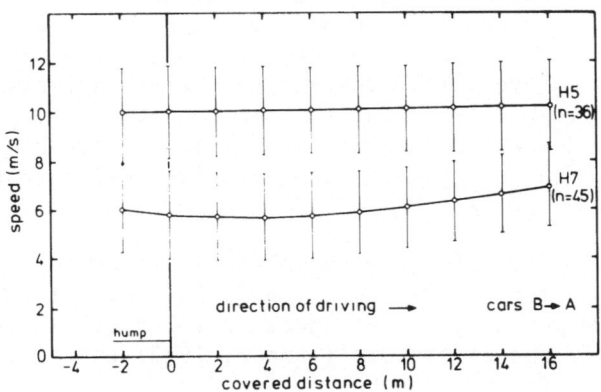

Fig. 8. Speed profiles of through-going cars at the main road; location H7 with humps and plateau, location H5 without.

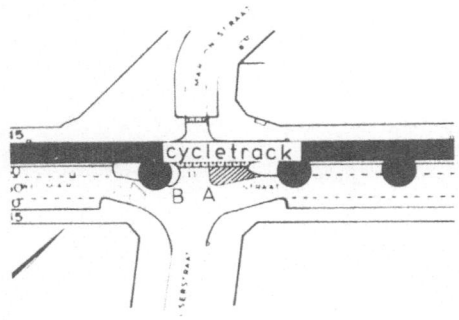
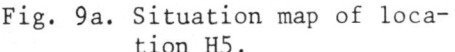
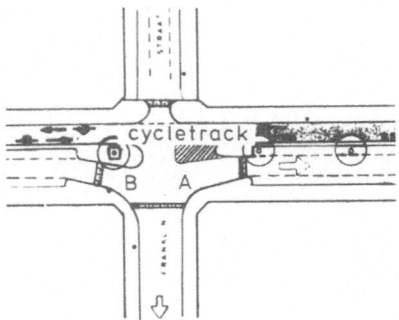

Fig. 9a. Situation map of loca-   Fig. 9b. Situation map of loca-
        tion H5.   tion H7.

Interactions between crossing cars and cyclists at the cycletrack. An important aspect in evaluating intersections is road safety, in this study especially in relation to cyclists. The number of
accidents at a single location cannot be used as a evaluation criterion for reasons given before. Conflicts seem to be an alternative

to accidents as a criterion measure for road safety. In the following TTC will be given as a possible measure for describing interactions between road users.

By means of the video-analysis equipment a large number of manoeuvre combination were analysed. On the basis of the x and y positions of the vehicles at successive moments TTC curves can be calculated with the help of a computer programme. Fig. 10 illustrates the
output of a given manoeuvre combination: a serious conflict between
a crossing motorist and two cyclists on the cycletrack. In a situational map of the intersection the courses of the vehicles are plotted. Each point gives the position of a vehicle at successive time
intervals of 0.24 sec. The plot in the bottom corner gives the TTC
curve of the car and cyclist 1. The car driver did not give right-of-
way to the cyclists. Cyclist 1 had to stop (points close together),
while cyclist 2 rode behind the car. The minimum value of TTC is
0.7 sec. Cyclist 1 had to take a strong evasive action to avoid a
collision.

The number of conflicts (for example defined as the number of
interactions with a minimum TTC less than 1.5 sec), related to an exposure measure E might give a risk index (RI) for two intersecting
traffic streams. In the literature (Tanner, 1953) the exposure E of
two traffic streams i and j has been defined as

$$E = \sqrt{I_i * I_j} \, ,$$

where I_i and I_j are the number of vehicles in stream i and j during
a given period.

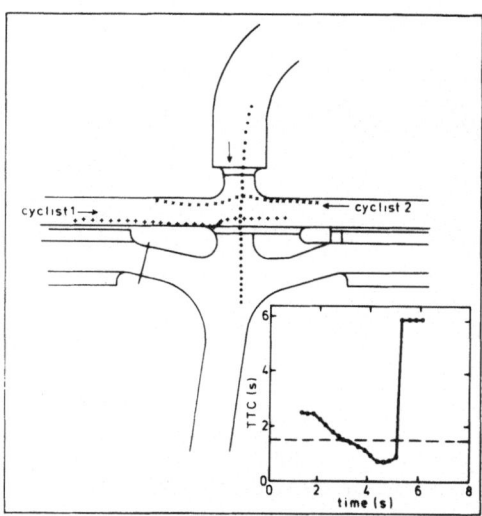

Fig. 10. Example of a serious conflict between a car and cyclist 1.
         Bottom right: TTC curve.

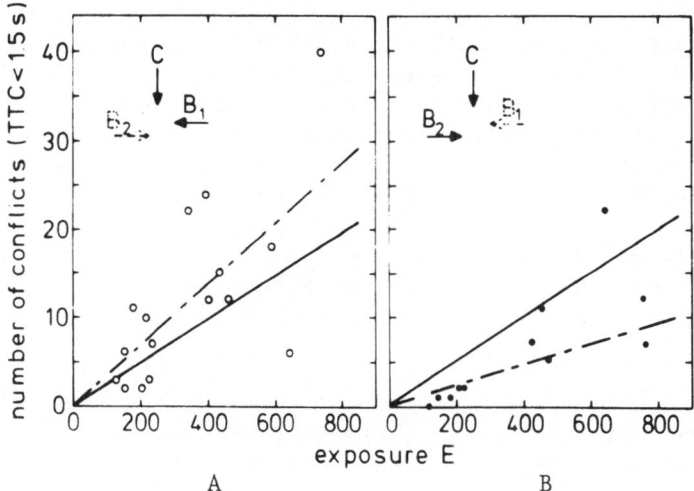

Fig. 11. The number of conflicts (TTC < 1.5 sec) as a function of exposure E (E = $\sqrt{C.B_i}$) for two types of manoeuvre combinations.

In Fig. 11 the number of conflicts is given as a function of
the exposure E for two types of manoeuvre combinations at priority
intersections of the cycleroutes. In Fig. 11a it concerns the conflicts between car drivers from the minor street (C) and cyclists coming from the left (the first bicycle stream B1), while in Fig. 11b between stream C and cyclists coming from the right (the second stream
B2). Each point represents the relevant type of manoeuvre at the intersection. The quotient of the number of conflicts and E gives the
risk index RI. The solid line in both figures represents RI, averaged
over all points (C-B1 and C-B2 combined). Interactions between C and
B1 at an average are scoring above this line, while those between C
and B2 are below. Cyclists B1 are involved more frequently in a serious conflict with cars from C than cyclists from B2. The width of the
cycletrack gives some extra space between a car from C and a cyclist
from B2. The reversed situation holds for cyclists from B2 and cars
coming from the opposite direction (data not presented here). The
mean RI per type of manoeuvre combination in Fig. 11 is given by the
dashed lines. Intersections above this line are relatively more unsafe (in terms of conflicts) than intersections below the line, that
is for the particular type of manoeuvre combination.

FINAL REMARKS

The study described here has not been finished yet. Nevertheless, several aspects in the evaluation of road design elements at
non-signalised intersections have been investigated using the video
equipment as described.

However, in spite of a semi-automatic analysis procedure by the use of a minicomputer the quantitative analysis remains time consuming. The method, in particular seems to offer good possibilities in developing a conflict-analysis technique in which time-to-collision (TTC) is the final criterion. Some aspects have to be worked out further before overall comparison between different solutions at priority intersections with slow traffic can be made.

Other applications of the equipment are possible as well, see for instance Blaauw and Poll (1980) and Riemersma (1979).

REFERENCES

Blaauw, G.J. and Poll, K.J., 1980, Human Factors and Riding Motorcycle: Effects of Type of Motorcycle and Rider's Experience, Proceedings of the International Motorcycle Safety Conference, Washington, D.C.

Hallert, B., 1960, "Photogrammetry, Basic Principles and General Survey", McGraw-Hill Book Company, New York.

Hayward, J.E., 1972, Near-Miss Determination through Use of a Scale of Danger, Report No. TTSC 7115, The Pennsylvania State University, Pennsylvania.

Hyden, Ch., 1977, A Traffic-Conflicts Technique for Examining Urban Intersection Problems, Lund Institute of Technology, Lund, Sweden.

Institute for Perception TNO, 1978, A large-scale Simulation System for the Study of Navigation and Driving Behavior, Documentation sheet 39e, Soesterberg, The Netherlands.

Ministry of Transport, 1979, Interim-report of the study group on design aspects of the demonstration cycleroutes in The Hague and Tilburg, The Hague, The Netherlands (in Dutch).

Perkins, S.R. and Harris, J.T., 1967, Traffic Conflict Characteristics: Accident Potential at Intersections, General Motors Corporation, Warren, Michigan.

Riemersma, J.B.J., 1979, The Perception of Deviations from a straight Course, Report IZF 1979 C-6, Institute for Perception TNO, Soesterberg, The Netherlands.

Tanner, J.C., 1953, Accidents at rural three-way junctions, Journal of the Institution of Highway Engineers, 2 (11), 56-57.

PART IV
SIMULATOR DESIGN AND EVALUATION

EVALUATING SIMULATOR VALIDITY

Jan Moraal

INTRODUCTION

Simulators are means and no ends in themselves, at least not for those who use them for the study of human behaviour or, for training people on the required skills in complex man-machine systems. By definition, simulation is never complete, otherwise it would result in a perfect copy of the real system.

The term fidelity of simulation refers to the resemblance between simulation and reality, or, in other words, "to the degree of accuracy required of a simulated task, element or situation in reproducing the real counterpart" (Mudd, 1968).

Judgments of fidelity (or validity) are often made subjectively by having experts or skilled operators give their opinion whether a simulator 'looks' or 'feels' realistic. When indeed it does, the simulator is considered to have at least "face validity' (Cronbach, 1960). Face validity is clearly of some relevance when one wants people to believe in a simulator and to become motivated to use it seriously. However, of primary concern should be the question whether a simulator does what it is expected to do, which defines the 'functional correspondence' with the simulated system. For instance, a simulator for training purposes is said to have functional correspondence with the real system when transfer-of-training is considerable, so that only a small number of training sessions on the simulated system is necessary. In case of a simulator for research purposes a high functional correspondence means that performance results are congruent to those which were to be found if the experiment had been carried out with the simulated system itself. Functional (or behavioural) correspondence should be defined with inclusion of the human

operator, i.e. the man-machine system. When excluding the human operator the correspondence between simulator and simulated system can
be considered only physically. 'Physical correspondence' refers to
the extent both simulator and simulated system behave similarly so
far only the hardware component, i.e. the machine, of the man-machine
system is involved. For example, the question whether a simulator
matches or approximates the dynamics of a given vehicle represented
by way of equations of motion. Physical and functional correspondence
are not unrelated, although they do not always correlate highly. Together with the notion of face validity, the three are complementary
in defining the general concept of validity.

It is stated here that, when using simulators, validation studies are absolutely necessary (see also Sanders, 1976). If not done,
one will not know which factors may be responsible in case a simulator does not do what it is expected to do. This means that the evaluation of the validity of a simulator should be primarily concentrated on the functional correspondence with its real counterpart.
When this turns out to be low, appropriate measures should be undertaken for improvement. Improving the physical correspondence could
be one method to raise the functional correspondence.

In the following paragraphs three studies will be described in
which simulators were validated, illustrating some particular problems in using validation criteria and also the fact that functional
correspondence indeed can be raised by improving physical correspondence.

THE VALIDITY OF A SHIP NAVIGATION SIMULATOR

The first study concerns a combined system for the simulation
of ship navigation and driving behaviour (Institute for Perception,
1978). In both cases the human operator (helmsman or driver) has a
120° horizontal field of view. The outside view is recorded by TV
cameras (black and white) linked to endoscopes and moving by an x,y
system through a scale model, representing either a waterway or a
highway. Fig. 1 gives an impression of the TV recording-system moving
through the model, in this case of a waterway in the vicinity of Rot-
terdam, which was also the area of the study to be discussed here.

Fig. 2 gives an impression of what one sees from the mock-up of
the navigation bridge, in this case of a 50.000 tons containership
entering a harbour.

To assess the degree of functional correspondence between this
navigation simulator and real ships an empirical validation procedure is necessary. One appropriate way could be to measure performance
on some defined critical variables in the simulator as well as on the
simulated ship. The functional correspondence then would be the cor-

Fig. 1. Overall view of scale model and TV recording system

Fig. 2. View from the mock-up of the navigation bridge.

relation between both sets of measurements. However, experimenting at sea is practically impossible. It would require repeated measurements in order to overcome variations in circumstances and operator behaviour which would lead to extreme costs, apart from the dangers being involved.

Confronted with this situation Truijens and Schuffel (1978; see also Truijens, 1978) used an alternative procedure. The question was to make recommendations by means of a simulator study for the possibilities to manoeuvre through a narrow canal (the situation modelled in Fig. 1), of a specialized tug, pushing a train of barges on the Rhine river, having a total length of 190 m approximately. To do so, a validation study was necessary. Taking measurements onboard the real ship turned out to be very difficult and impractical; the area under study is hardly ever free from other traffic and the installment of measurement equipment on the ship was impossible. However, the captains who were going to participate in the study as subjects make the passage three or four times a week. Therefore, they have a profound knowledge of the nautical characteristics of the local situation. For instance, it is their daily routine to decide whether to use an extra tug or not for passing the bridge, based on knowledge of wind-direction, wind-force and tide. Another decision to make is whether or not to mount a bow rudder on the foremost barge, being of considerable help to the manoeuvring, but requiring much extra work.

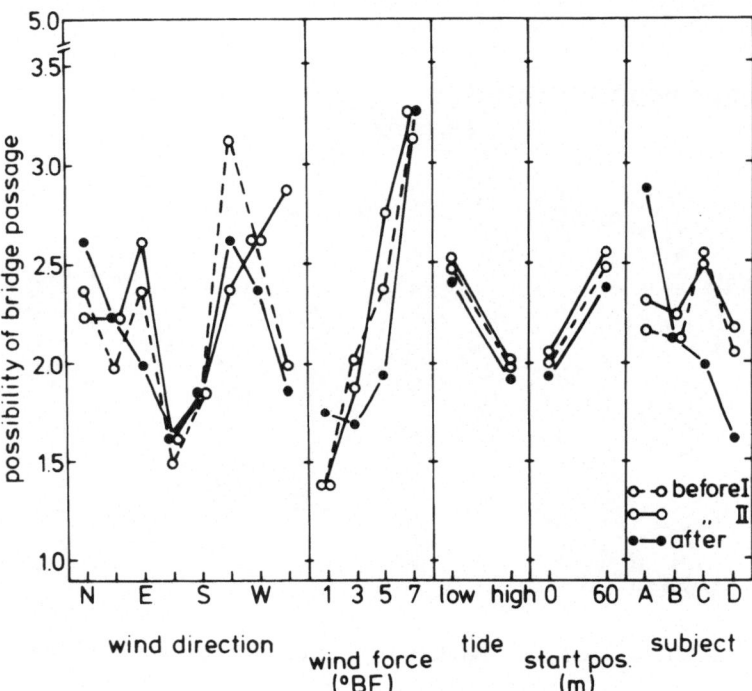

Fig. 3. Predicted ('before') and reported ('after') possibility of the passage of the bridge as a function of the experimental conditions. Ordinal scale from 'perfect passage' (1.0) to 'failing passage' (5.0). Before I: situations not followed by simulator runs; before II: situations followed by simulator runs (Truijens and Schuffel, 1978).

This knowledge of the captains was used as the criterion in the validation study. Before the start of the simulation trials, each of four captains were given a number of descriptions (32) of possible conditions under which the bridge had to be passed. These descriptions included variations in wind-direction and -force, tide and starting position (the middle or 60 m east of the middle of the canal). Then, questions were asked about the possibility of a perfect passage and about the necessity to use a bow rudder. Answers could be given on a 5-points and a 3-points scale respectively. After this, the captains made 16 runs in the simulator, i.e. half of the conditions they had responded to before. Following the runs a questionnaire had to be completed with the same questions as before the experiment, but now requiring a report of what had actually happened.

It was hypothesized, that if the simulator was representative of the real situation, the way the captains were successful in passing the bridge, using a bow rudder or not, would correlate with what they had predicted. The results are shown in Figs. 3 and 4.

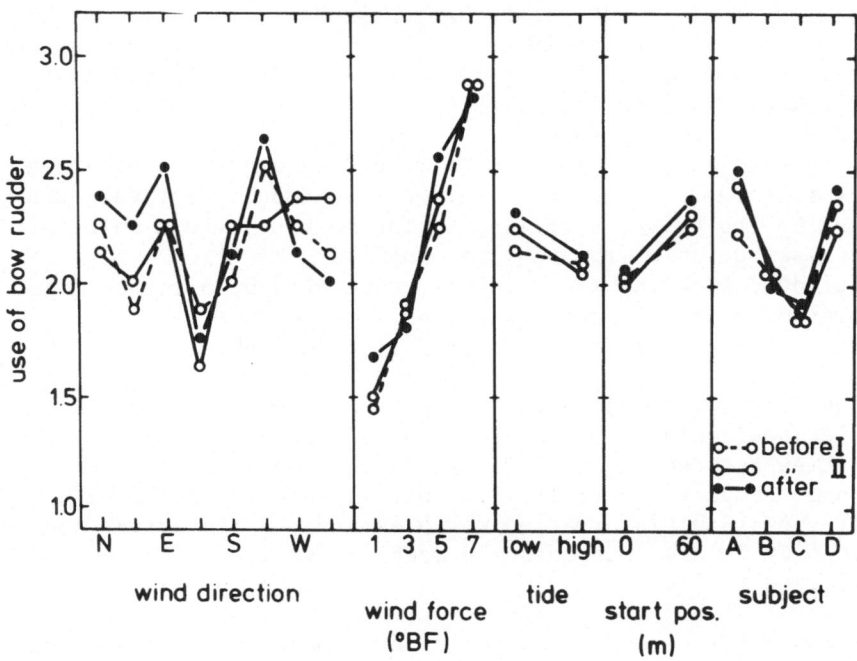

Fig. 4. Predicted ('before') and reported ('after') use of bow rudder as a function of the experimental conditions. Ordinal scale from 'no need for bow rudder' (1.0) to 'bow rudder absolutely necessary' (3.0). Before I: situations not followed by simulator runs; before II: situations followed by simulator runs (Truijens and Schuffel, 1978).

In both figures, the captains' predictions and reports are plot-
ed as a function of the various conditions. None of the differences
between predictions and reports proved to be significant although the
differences between subjects in Fig. 3 came close: subject A was too
optimistic, whereas D was too pessimistic about his performance. The
overall functional correspondence between predictions and reports for
bridge passage as well as for bow rudder use was rather convincing.
Therefore, it was concluded that for this particular purpose the simulator is sufficiently valid.

THE VALIDITY OF A DRIVING SIMULATOR FOR RESEARCH PURPOSES

Although numerous vehicle simulators have been developed, validation studies are very scarce and mostly are restricted to experts'
judgments of fidelity (face validity). As stated already, in case of
a research simulator one should investigate empirically whether man-
machine performance results are congruent to those which were to be
found if the experiment had been carried out with the simulated system itself.

In Blaauw et al. (1978) a validation study is reported of the
driving simulator of the Institute for Perception. The main features
of the driving simulator, apart from those already mentioned are:
- a vehicle mock-up, identical with the institute's instrumented car
  for road user studies (ICARUS)
- computer controlled movements of TV recording system, consisting
  of three translations and one rotation (yaw around vertical axis),
  initiated by the driver's actions in the mock-up, via a mathematical representation of the vehicle dynamics
- calculation of steering-wheel forces, generated by a torque motor
- simulation of sound of engine and wind
- a fixed base, i.e. absence of kinesthetic feedback.

At the time of the validation study the simplified and lineari-
sed vehicle model only allowed lateral accelerations up to 3 m/s^2
and velocities between 70 and 110 km/h in the top gear. A drive
started under computer control to an initial velocity of 72 km/h
whereafter the visual information was presented to the driver and control taken over by him. Due to scale model dimensions straight driving was limited to 50 s.

Functional correspondence between performance of simulator and
simulated system (the instrumented car) was studied with subjects'
driving experience and type of task as the independent variables
(between subjects), and lateral control as the dependent variable.
Because experts are not representative of the normal population of
drivers, normal experienced as well as inexperienced drivers served
as subjects.

Significant differences between both systems were found. In the
simulator, driving experience had a significant effect on the standard deviation of steering-wheel angle, lateral position and yaw rate
($p<.05$), all three being smaller for the experienced drivers, while
no such differences were found when subjects drove the instrumented
vehicle in a real world environment. The effect of type of task also
differentiated between simulator and instrumented vehicle. The tasks
'drive as straight as possible' versus 'just keep in lane' showed no
significant effect on steering-wheel angle and lateral position in
the simulator, while they did in the instrumented vehicle ($p<.01$ and
$p<.05$ respectively). The conclusion then was, that a sufficient functional correspondence between both systems was absent.

Several hypothesis were put forward, in particular the one that
lack of physical correspondence might have been responsible for lack
of functional correspondence. As the first step the dynamics of the
instrumented car were thoroughly investigated. Lateral vehicle dynamics are described primarily by the transfer function between steering-wheel angle and yaw rate. In the simulator this transfer function
combines that of the mathematical vehicle model with the computer
controlled realization of the desired yaw rate of the TV recording
system. It turned out that the dynamics of the simulator showed a
phase lag in the relevant frequency range 0-0.5 Hz, resulting in a
time delay of about 0.3 s. The mathematical model and the servo characteristics of the TV recording system contributed about equally to
this time delay.

So, one of the measures taken was the elimination of the time
delay, which resulted in an almost perfect physical correspondence.
Other modifications were:
- implementation of a moving belt system to enable unlimited straight
  driving;
- a velocity range of 0-120 km/h, instead of being launched at 70
  km/h.

Subsequently, the improved simulator system was investigated in
a second validation study (Blaauw, 1980). Driving experience and task
demands again were taken as the independent variables (between subjects), while lateral vehicle dynamics again were taken as the dependent variables. Task demands were defined with regard to forced lateral control ('drive as straight as possible') versus free lateral
control ('just keep in lane'). The system, simulator versus instrumented car, acted as a within-subjects variable.

The results of the first and second validation studies are presented in Fig. 5. The figure shows the standard deviations (S.D.) in
yaw rate and lateral position as a function of driving experience
and system for both validation studies: the first study in which
there was a time delay of $\Delta\tau$ of 0.3 s in the simulator and the second study in which this time delay had been eliminated. In the first

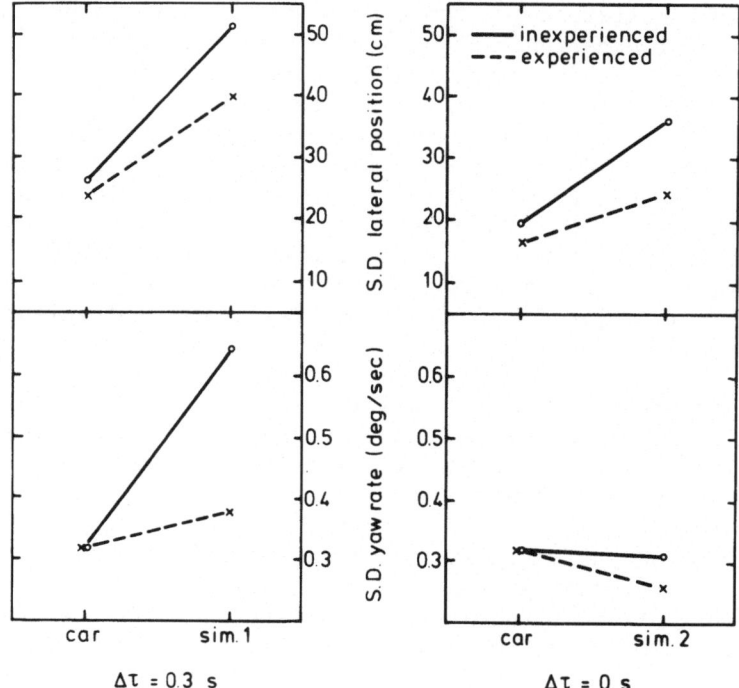

Fig. 5. Standard deviation of yaw rate and lateral position as a
function of driving experience and system. Left: time delay
of simulator dynamics Δτ=0.3 s. Right: equal dynamics for
simulator and car (Blaauw, 1980).

study, the yaw rate values of the simulator are significantly higher
compared with the instrumented vehicle, whereas in the second study
the results are about equal for both systems. The significant interaction driving experience x system in the first study is absent in
the second. These results mean that in the second study, simulator
and instrumented car have equal dynamics.

With regard to the standard deviations in lateral position (task
demands combined) it is clear from Fig. 5 that in both studies the
values are higher for the simulator than for the instrumented car.
However, in the second study the absolute values are significantly
lower for both systems. As the dynamics of simulator and car have become equal, the higher lateral position values in the simulator could
be explained by the absence of kinesthetic information to the driver,
because the simulator is fixed-base. Similar results were explained

in the same way by McRuer and Krendel (1974), McRuer and Klein (1976), McRuer et al. (1977) and McLane and Wierwille (1975).

As was already stated in the introduction, functional correspondence or validity in case of a research simulator means that performance results in the simulator are congruent with those to be found when the experiment is carried out with the simulated system itself. The question then is, whether the data presented in Fig. 5 are of such relevance that one may conclude about the simulator's validity. For yaw rate performance both systems behave rather similar, i.e. show neither a difference in absolute values nor a significant effect of driving experience. With regard to both variables rather precise quantitative predictions can be made from simulator to car.

However, for the lateral position data the situation is different. For the simulator the absolute values remain significantly higher than for the car and also a significant interaction driving experience x system remains. In this case no precise quantitative predictions from simulator to real car can be made. The only thing one possibly could predict is, that inexperienced drivers will show 'higher values' in the real system than experienced drivers, which is a prediction of a qualitative nature.

To test the assumption that qualitative predictions are justified, a third experiment was done. In this experiment (Blaauw, 1980), the same independent variables between subjects were studied as before, namely driving experience and lateral control (forced versus free), and in addition longitudinal control. With regard to the last, subjects were instructed to drive with either a 'very constant velocity of 100 km/h' (forced longitudinal control) or to drive with a 'normal highway velocity', i.e. somewhere between 80 and 120 km/h (free longitudinal control).

So, driving experience, lateral control instructions and longitudinal control instructions together accounted for 2 x 2 x 2 = 8 different conditions per system (simulator or car). Again, the systems acted as the within-subjects variable and the standard deviation of lateral position as the dependent variable.

The effects of driving experience and task demands on the standard deviation of lateral position are shown in Fig. 6. Lateral position values were higher in the simulator than in the car for both experienced and inexperienced drivers, while the difference between both groups was significant only in the simulator. Therefore, these results confirm those already found (see the upper right part of Fig. 5). With regard to task demands, longitudinal control had a significant effect (p<.01). In both systems forced longitudinal control led to higher standard deviations of lateral position for the inexperienced drivers, while the reverse was true for the experienced drivers. No significant effects were found due to variations in lateral control.

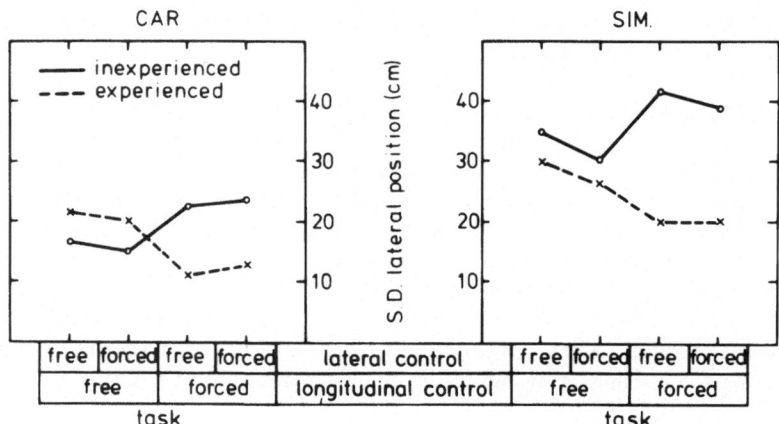

Fig. 6. Standard deviation of lateral position of instrumented car
and simulator as a function of driving experience and task
demands (Blaauw, 1980).

However, even more important than the absolute values of the
results, is the overall impression from Fig. 6 that the trends of
the curves are remarkably similar. This means that, although the absolute values are higher in the simulator, the relative effect of
the experimental conditions is similar in both systems, i.e. subjects change their behaviour in the same way in simulator and real
car according to changes in instructions or task demands.

Based on the results of the final experiment it seems that one
indeed may conclude that the simulator proves to have a sufficient
functional correspondence with its real counterpart to justify qualitative (relative) predictions in sofar lateral position is concerned. When yaw rate is concerned it seems that more quantitative (absolute) predictions are possible.

THE VALIDITY OF A TANK DRIVER TRAINING SIMULATOR

The third study to be discussed concerns the validity of a training simulator. Training simulators are used as aids for teaching
skills that are necessary in order to operate a real system safely
and efficiently. Training simulators are valid to the extent that
they shorten the time needed for candidates to become trained on the
real system. Therefore, validity should be defined primarily with
regard to the amount of transfer-of-training from simulator to simulated system. Performance levels per se should be of secondary importance. In validating training simulators one usually compares the
training results of an experimental group of candidates, those who

both receive simulator and real system training, with the training
results of a control group of candidates, receiving real system training only. The results give the amount of transfer-of-training. In
formula:

$$Tr = \frac{(t - s_t)}{t} \times 100\% \tag{1}$$

in which: Tr = percentage transfer-of-training
          t = training time on real system (control group)
          s_t = additional training time on real system, after simu-
               lator training (experimental group).

If simulator training is perfect (Tr = 100%), no additional real
system training would be necessary, which seldom happens. However,
Tr should reach a minimum value in order to make the simulator cost-
effective.

Recently the Royal Dutch Army put into use a training simulator
for tracked vehicles (tanks). The main principles of this simulator
are the same as those of the navigation simulator of the second paragraph, with the exception that in this case the driver's cabin has
three degrees of freedom of rotation: yaw, pitch and roll (see Fig.
7).

After using the simulator for one year, however, the instructors did not react favorably and some were even quite negatively.
Their main argument was that simulator candidates did not learn the
skills in a reasonable time, or even not at all, compared with candidates who only received real tank training. In other words: Tr
was estimated to approach 0%. If this were true, one or both of the
following factors could have been responsible for this:
- the lack of sufficient validity of the simulator in terms of transfer-of-training;
- inadequacy of instruction or inability of instructors.

In a study by Moraal and Poll (1979) two groups of candidate
tank drivers were trained: one group exclusively on the tank (control group) and one group on the simulator as well as on the tank
(experimental group). So it was possible to calculate Tr according
to formula (1). Obviously then, a criterion problem had to be met.
Driving a tank is a complex task, including lane and speed control,
changing gears, performing manoeuvres, driving through all kinds of
terrain, etc. When taking complex performance output as the criterion, for example the time to drive a course during which all of
the aforementioned tasks have to be performed without making errors,
no problem arises when Tr is high. However, when Tr is moderate or
low, one does not know whether this results from an overall poor
transfer-of-training, or whether it is originated by more specific

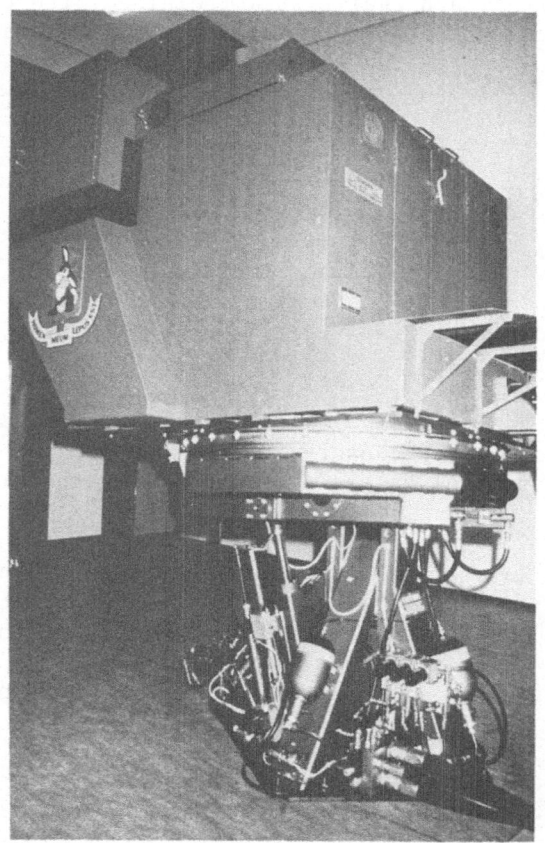

Fig. 7. Driver's cabin of tank driver training simulator. Manufac-
        turer: Link-Miles division of Singer Ltd, U.K.

control actions for which transfer-of-training is insufficient. Therefore, it was concluded necessary to be more precise in defining performance criteria.

    Four specific tasks were considered:
a. Changing gears without steering, i.e. driving a straight road;
b. Steering without changing gears, i.e. driving a slalom in the bottom gear;
c. Steering and changing gears in combination, i.e. driving a stretch
   of road that included two curves which had to be negotiated in the
   bottom gear;
d. Driving through a terrain obstacle, i.e. a bump followed by a hole.

    Both the experimental and control group were trained on the four
tasks. As criterion level for each task was taken the performance of
a normally experienced tank driver, carrying out the tasks without

making any errors. Time to criterion served as the unit of measurement of performance. Another unit of measurement is the number of
kilometers driven in order to reach criterion performance. Both are
included in Table I.

Table I. Mean training time, mean training kilometers and transfer-
of-training (%) of four tasks. The number of simulator- and
tank-candidates was 14 and 15 respectively.

		tasks			
Mean time (min)		a	b	c	d
exp. group on tank	(s_t)	7.9	8.2	14.3	0.7
control group	(t)	70.6	38.8	30.4	13.3
transfer-of-training	(Tr)	88.8%	78.0%	53.0%	94.7%
Mean nr. kilometers		a	b	c	d
exp. group on tank	(s_t)	1.5	0.4	1.9	0.1
control group	(t)	13.8	2.8	4.5	0.9
transfer-of-training	(Tr)	89.1%	85.0%	57.8%	88.9%

From Table I it can be concluded that Tr is considerable which
is extremely in contradiction with the opinion of the instructors.
Therefore one might conclude to inadequacy of instruction or inability of the instructors as being responsible for the poor initial results. However, from the way Tr was calculated a third possible factor emerged, namely total training time. It turned out that total
training time for the experimental group in general is higher, sometimes much higher than for the control group.

The increase in training time can be defined as:

$$Ti = \frac{(s_s + s_t) - t}{t} \times 100\% \qquad (2)$$

in which: Ti = percentage increase in training time
s_s = training time of experimental group on simulator
s_t = id., on tank
t = training time of control group.

Total training times and Ti values for the four tasks a-d are
mentioned in Table II.

From Table II is it clear that in case of task a, an almost ideal situation exists: a high percentage of transfer-of-training and a

Table II. Mean time (min) needed to become trained on four tasks,
          and increase in training time (%) for experimental group
          relative to control group of tank driver candidates.

Category		tasks				
		a	b	c	d	total
control group	(t)	70.6	38.8	30.4	13.3	153.1
exp. group on sim.	(s_s)	63.2	79.3	38.6	21.8	202.9
exp. group on tank	(s_t)	7.9	8.2	14.3	0.7	31.1
total time exp. group	$(s_s + s_t)$	71.1	87.5	52.9	22.5	234.0
% increase in training time for exp. group	(Ti)	0.7%	125%	74%	69%	35%

negligible increase in training time for simulator candidates. Task
b, on the contrary, requires a 125% increase in training time for
the experimental relative to the control group, i.e. in total $2\frac{1}{4}$
times as long. Tasks c and d show intermediate results. That, at
least for some specific aspects of the driving task, the experimental
group was in need of more, sometimes even much more training time,
was neither expected nor anticipated by the users. In fact they
wrote training syllabi scheduled in the same way as used formerly by
the tank candidates, thereby unconsciously underestimating necessary
training time.

When, during the actual training of the candidates, the instructors discovered that candidates were not able to acquire some of the
skills according to the time schedule based on tank training, they
immediately started to develope a negative - 'the thing doesn't work'
- attitude towards the simulator, with the consequence of also demo-
tivating the candidates. The main fault made could have been that the
simulator simply was conceived as a substitute of a tank instead of
a mere training aid, having its very own characteristics which only
can be traced by investigating thoroughly its validity.

CONCLUDING REMARKS

Simulators are aids and no substitutes of real systems; they
have their own characteristics and directions for use. If not conceived that way proper evaluations cannot be made. This was illustrated by the study of the tank driver simulator in the foregoing paragraph.

When using simulators, validation studies are required to assess the degree of functional correspondence with their real counter-
parts. Results of validation studies should form the basis of knowledge for further refinement of the simulator in question. However,

for empirical evaluation of the functional correspondence or validity, it is not always practically possible to do comparable experiments with the real system. In that case one has to look for other
relevant criteria, an example of which was shown in the second paragraph.

In general, it is too unspecific to say that a simulator is just
valid. Functional correspondence or validity should be defined with
specific reference to the purpose of what the simulator is developed
or going to be used for. A training simulator is valid in sofar it
shows transfer-of-training to the real system, while a research simulator is valid when experimental results of simulator and real system are congruent, so that performance predictions can be made from
one to the other.

The type of predictions one can make from the results of a research simulator to performance in its real counterpart can either
be quantitative (absolute) or qualitative (relative). Quantitative
predictions can be made when the parameter in question shows identical absolute values in simulator and real system. Qualitative predictions can be made when, for instance, the effects of different levels of certain independent variables show more or less similar curves in both systems, while the absolute values differ. An illustration was given in the third paragraph.

When functional correspondence is insufficient, improvement of
the degree of physical correspondence might be a way to compensate
it. This was also illustrated in the third paragraph. The example is
also of principal importance. The tendency to decide for the present
that simulators should be such that maximum reality is reached certainly may lead to extreme costs, but at the same time it may lead
to a surplus value which remains unused. Therefore, the necessity to
maximize physical correspondence should evolve from empirical studies. For the yaw rate values of the driving simulator, discussed in
the third paragraph, it was felt necessary to improve the system dynamics, while the lack of kinesthetic information is not felt to be
a serious lack for the purpose it is used for. Also, the highly detailed terrain model of the tank driving simulator of the fourth paragraph seems to be of relative value. A very simple one was developed for the particular driving criteria taken into consideration, resulting in a considerable amount of transfer-of-training.

REFERENCES

Blaauw, G.J., Horst, A.R.A. van der and Godthelp, J., 1978, The ve-
    hicle simulator of the Institute for Perception: a validation
    study of straight road driving (in Dutch), Institute for Percep-
    tion TNO, Progress Report I: IZF 1978-16, Soesterberg, The Neth-
    erlands.

Blaauw, G.J., 1980, Driving experience and task demands in simulator and instrumented car: a validation study, Institute for Perception TNO, Progress Report II: IZF 1980-9, Soesterberg, The Netherlands.

Cronbach, L.J., 1960, "Essentials of Psychological Testing", Harper and Row, New York.

Institute for Perception TNO, 1978, A Large-Scale Simulation System for the Study of Navigation and Driving Behaviour, TNO Documentation sheet 39e, Soesterberg, The Netherlands.

McLane, R.C. and Wierwille, W.W., 1975, The influence of motion and audio cues on driver performance in an automobile simulator, Human Factors, 17 (5), 448-501.

McRuer, D.T. and Krendel, E.S., 1974, Mathematical models of human pilot behaviour, AGARD Meeting AG-188, NATO, Brussels.

McRuer, D.T. and Klein, R.H., 1976, Comparison of human driver dynamics in an automobile on the road with those in simulators having complex and simple visual displays, Paper 173A, presented at the 55th Annual Meeting of the Transportation Research Board, Washington, D.C.

McRuer, D.T., Wade Allen, R., Weir, D.H. and Klein, R.H., 1977, New results in driver steering control models, Human Factors, 19 (4), 381-397.

Moraal, J. and Poll, K.J., 1979, The Link-Miles driver training simulator for tracked vehicles; a validation study (in Dutch), Institute for Perception TNO, report nr. IZF 1979-23, Soesterberg, The Netherlands.

Mudd, S., 1968, Assessment of the fidelity of dynamic flight simulators, Human Factors, 10 (4), 351-358.

Sanders, A.F., 1976, Experimental Methods in Human Engineering, in: "Introduction to Human Engineering", K.F. Kraiss and J. Moraal (eds.), Verlag TÜV Rheinland GmbH, Köln.

Truijens, C.L., 1978, The validity of marine simulators, in: "Proceedings of the First International Conference of Human Factors in the Design and Operation of Ships", D. Anderson et al. (eds.), Gothenburg, 463-476.

Truijens, C.L. and Schuffel, H., 1978, Ergonomic Research "Open Hartelkanaal", Part III: The validity of the simulation of pushing convoy navigation in the Hartel area (in Dutch), Institute for Perception TNO, report nr. IZF 1978-C6, Soesterberg, The Netherlands.

JUSTIFICATION FOR, AND DESIGN OF, AN ECONOMICAL PROGRAMMABLE MULTIPLE FLIGHT SIMULATOR

John G. Kreifeldt, J. Wittenber and G. Macdonald

INTRODUCTION

The fidelity of an aircraft is of continuing importance for human factors aviation research and training. Practical and experimental interests usually collide over the desires to extrapolate results to the real world without encountering simulation penalties so great in attempting to mimic real world conditions as to prohibit meaningful research. The opposite concern, of course, is that a simulation lacking the essential mimicry may jeopardize transfer of training or results back to the real world. The challenge in aircraft simulation based research is to uncover and produce the necessary simulated reality in a practical, budget-limited manner.

As much as simulation of an individual aircraft is important, simulation issues for a system of multiple aircraft for air traffic control studies are of similar concern and there are numerous advantages/disadvantages to be weighed in experimentation via either real world or laboratory manipulation (simulation). Some of these issues, relative advantages and postulated simulation criteria are listed in Table I from Bogdanoff et al. (1960), Sackman (1967) and Parsons (1972).

However, while single aircraft simulation studies tend to stress pilot-aircraft performance, ATC simulation studies stress the organization, interactions, conflicts, communications and data flows created by decision making and program execution usually in an environment of mixed stochastic and deterministic events. Man-machine system simulation is the only systematic way available for studying these complex phenomena (Geisler, 1960). A particularly relevant practical and experimental issue concerns the necessary numerosity and role fi-

Table I. Simulation Concerns

Issues	Real World	Simulation	Simulation Criteria
Cost	high-to-prohibitive	low-to-high	Cost
Variables	difficult-impossible to control	selectably control-lable	
Environment	may be unavailable	may be simulatable	
Danger	high under mishaps	low-to-nonexistant	
System Alternatives	difficult to arrange	can be simulated	
Events	may be too infrequent to study	can be induced	Precision and Reliability
Inputs	difficult to arrange	controllable and available	
Replication	difficult	simple	
Administration	difficult-to-complex	simple-to-difficult	Ease and Flexibility of Inputs
Events	natural	artificial-hypothetical	
Operations	heuristic	algorithmic	
System Extent	open	closed	Fidelity to Original
System Alternatives	operational	symbolic	
Reality	psychologically valid	logically valid	
Time	conditional real time	temporal invariance	Psychological Reality
Experimentation	open	tight control	
Situations	largely probabilistic	largely deterministic	
Results	locally to globally valid	transferability in question	Sufficing Level of Detail

delity of the human factors in the simulation, as opposed say to their behavioral emulation by computers, and the level of fidelity of the simulation equipment-specifically the fidelity of the simulated aircraft. While the complete replacement of pilots and controllers by computer models is manifestly unacceptable in ATC simulation, the appropriate level of human involvement and role versimilitude is not clear. Documented results comparing performances obtained from real world-vs-simulated ATC conditions are quite limited.

The Air Traffic Control Simulation Facility (ATCSF) at the National Aviation Facilities Experimental Center (NAFEC) in Atlantic City, New Jersey appears to have the largest on line complement of human "pilots" and controllers in the USA. Personnel, tasks and equipment are high fidelity simulations for controllers but low fidelity for "pilots". Pilot simulation consists essentially of numerous but non-pilot keyboard or panel operators performing pilot designated functions such as heading, altitude and speed changes of targets in response to controller requests. The "pilots" are present more to effect the numerous commanded changes rapidly and provide voice communications than to serve as actual pilots. Buckley et al. (1978) reiterate the position that if the critical skills are mostly in the areas of decision making and communication rather than precise sensory-motor skill then completeness rather than precise realism is sufficient. As will be seen, however, that statement may be highly questionable.

Our own research interests in ATC revolve about the concept of distributed ATC management based on the assumption that the pilot has a cockpit display of traffic and navigation information (CDTI) via CRT graphics. The basic premise is that a CDTI equipped pilot can, in coordination with a controller, manage a part of his local traffic situation thereby improving important aspects of ATC performance such as controller workload, system safety and capacity as well as providing "assurance" to the pilot of the proper conduct of management particularly in areas of dense traffic. This assumption and its ramifications are experimental issues which must be addressed through simulation. Kreifeldt (1980) discusses possible CDTI uses and results from a number of CDTI studies and it appears that the Federal Aviation Administration (FAA) proposes incorporating CDTI with new definitions of pilot responsibilities and controller roles in the late 80's to early 90's time frame (Klass, 1980). However, CDTI and ATC issues must continue to be studied through simulation.

Simulation studies of CDTI and distributed management appear to require a higher level of fidelity for piloting functions than NAFEC's ATCSF is configured for. As the traffic complexity of the ATC task being studied increases, as for example managing arriving traffic on curved approaches to parallel runways, the continuing ATC performance is stronly influenced by individual piloting performances and the consequent pilot-pilot and pilot-controller interactions.

System behavior strongly affected by air-to-air interactions argues even more strongly for multiple independent pilots in order to study the effects of air-to-air interactions conditioned by the shared information presented by each CDTI. Realistically complex traffic management problems themselves require a multiplicity of aircraft targets.

The traffic in ATC simulations can be generated by using: (1) actual pilots flying simulators; (2) "pseudo pilots" at keyboards or panels (as in the ATCSF) controlling heading, speed, altitude, etc. of targets with simple inertial characteristics; (3) prerecorded piloted flights which, except possibly for speed changes, are real time unmodifiable; (4) preprogrammed computer generated targets which are real time unmodifiable except possibly for "holding" at preselected points on the programmed trajectory; and (5) "live" traffic obtained via a display repeater from an actual controller site and on which simulation traffic can be superimposed. The above methods are listed in order of decreasing realism and/or flexibility of piloting controls. Practically, a mix of piloted simulator(s), pseudo piloted targets and preprogrammed or "canned" traffic would be used to achieve the traffic density needed for the study. However, CDTI is only amemable to piloted simulators and perhaps pseudo-piloted targets. Therefore both the number of independent CDTI's and the level of piloting fidelity (pilot-vs-pseudo pilot) are objects of concern in attempting to extrapolate simulation results to the real world.

SIMULATOR NUMEROSITY AND FIDELITY

Several CDTI-distributed management ATC experiments have been conducted since 1973 in the Man-Vehicle System Research Division of the NASA-Ames Research Center using a specially designed facility consisting of three piloted simulators, one pseudo-pilot responsible for speed controlling multiple targets via keyboard input and an air traffic controller station. Professional pilots and controllers were used as subjects in all experiments. Careful analyses of the data from one experiment studying the use of CDTI in making a curved rather than the usual long straight-in approach to a runway sheds some light on the possible differences in results obtainable using (1) piloted-vs-pseudo piloted targets when both respond only to controller commanded speed changes and (2) multiple (i.e. three), one or no piloted simulators in an ATC study. A full description of the experiment may be found in Kreifeldt et al. (1977). The specific simulation fidelity dependent results will be discussed here in support of requiring multiple manned simulators in ATC simulation with the concomitant need for inexpensive but adequate simulators.

Methodology

The basic task (Fig. 1) required that targets at 2000 feet me-
tered into the area at nominal one minute intervals and randomly assigned to one of the numbered curved approaches (except 18L) descend
along the approach crossing the 1000 ft missed approach point (MAP)
to runway 36L at 60 second intervals on a 6° glide slope and at the
landing speed appropriate for the aircraft type (65 or 80 knots).
There were approximately seven targets approaching 36L at any time.
Three of the targets were fully controllable piloted simulators and
the remainder were pseudo piloted speed controlled targets following preprogrammed trajectories. The pseudo pilot effected the commanded speed changes via a keyboard input. The three piloted simulators had simplified but realistic flight dynamics responding to
throttle, aileron and elevator inputs while the pseudo piloted targets responded to speed changes with simple inertial characteristics.

Two traffic management conditions were studied. In the distributed management condition, each of the three simulator pilots had
full traffic information displayed with the approach routes on his
CRT forming a CDTI picture. Pilots could select such options as the
orientation of the picture and the range displayed. In this condition,
the controller issued to each pilot only flight number of the target
that was to proceed him over the MAP. Each pilot then modified his
flight to achieve the requested sequence using his CDTI. Fig. 2 shows

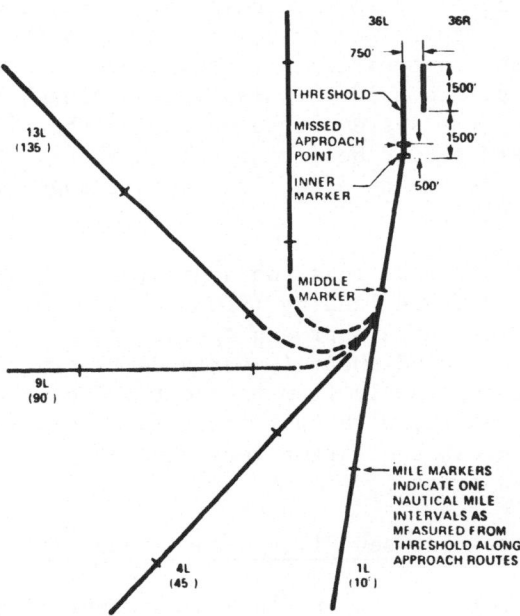

Fig. 1. Curved Approach Scenario.

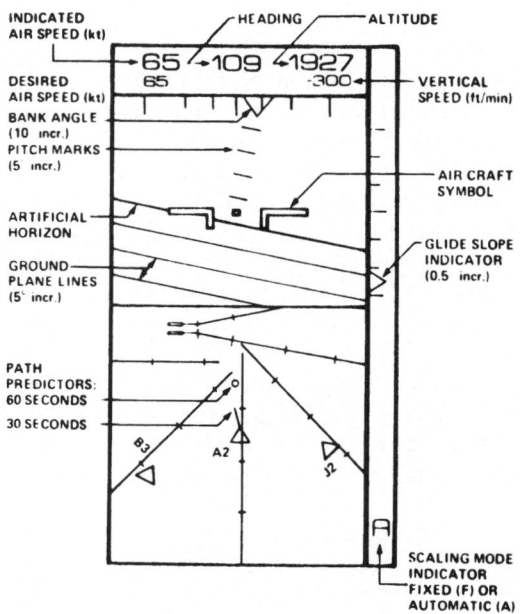

Fig. 2. Pilot CRT Display of EADI and CDTI with Ownship Centered.

a typical CRT display with CDTI. The controller effected speed control of each of the other targets via the pseudo pilot.

In the centralized management condition, pilots saw their own aircraft only on the display and thus, similarly to the pseudo pilot, had to follow the controller initiated speed commands. As in the previous condition, pilots used the displayed approach routes for navigation. Ostensibly, the system performances of the piloted and pseudo piloted targets should be indistinguishable in the centralized condition since they were each effecting speed changes only as requested by the controller.

In both management conditions, an approximate 50-50 mix of two simulated short take off and landing (STOL) speed characteristics was maintained with terminal area/approach speeds of 120/60 knots and 150/80 knots. A 25 knot wind shear linearly decreasing to 15 knots at ground level blew constantly from the north. The pseudo piloted targets flew the approaches with appropriately programmed altitude and speed profiles unless the latter were changed through controller request to the pseudo pilot.

Performance Differences Between Piloted and Pseudo Piloted Targets

In addition to information provided about centralized-vs-CDTI distributed management, careful data analyses also revealed significant differences between piloted and pseudo piloted targets as they

crossed the missed approach point. These differences are particularly relevant in the centralized management condition which was a standard ATC simulation in that all targets responded to controller requested speed changes. However, the three piloted simulators effected
speed changes through direct manual throttle control while the other
targets were speed controlled via the keyboard inputs from the pseudo pilot.

Since spacing is a critical varialble in ATC particularly for
landing, time between targets crossing the MAP (ICT-intercrossing
time) was recorded as one of the performance measures used to compare distributed-vs-centralized management and piloted-vs-pseudo piloted targets. The random mix of piloted and pseudo targets produced
four paired MAP crossing sequences. Using S for piloted (simulator)
and C for the pseudo piloted (computer programmed) targets, the four
pairs were:

pilot active $\begin{cases} \text{SS – simulator precedes simulator} \\ \\ \text{CS – pseudo precedes simulator} \end{cases}$

pseudo pilot active $\begin{cases} \text{SC – simulator precedes pseudo} \\ \\ \text{CC – pseudo precedes pseudo} \end{cases}$

Fig. 3 presents the cumulative ICT values for these four pairs
and Fig. 4 presents their means and standard deviations. Both centralized and distributed management conditions are represented.

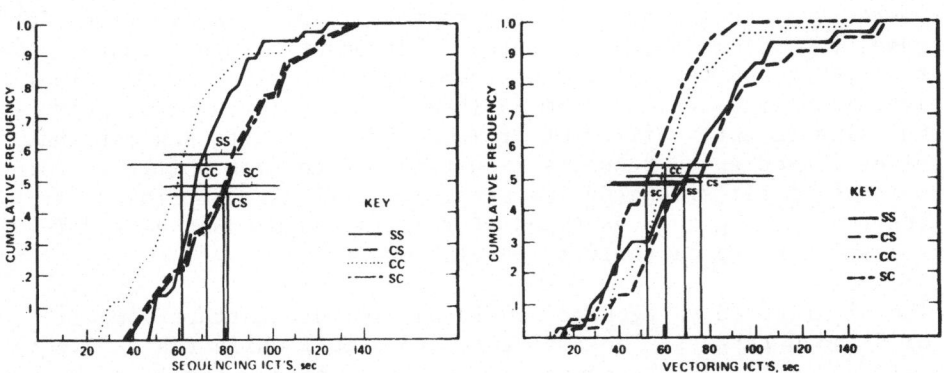

Fig. 3. Cumulative Display of the Intercrossing Times (ICT) for the
four Possible Piloted (S) and Pseudo Piloted (C) Target Pairings Crossing the Missed Approach Point in the Distributed
(Sequencing) and Centralized (Vectoring) Management Conditions.

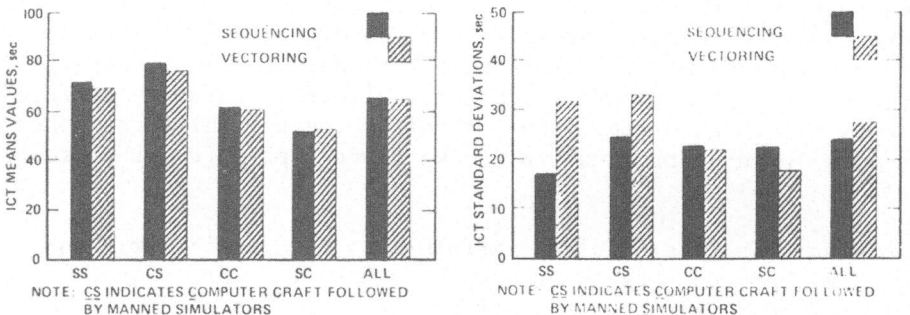

Fig. 4. Means and Standard Deviations of the Intercrossing Times
(ICT) for the Four Possible Piloted (S) and Pseudo Piloted
(C) Target Pairings Crossing the Missed Approach Point in
the Distributed (Sequencing) and Centralized (Vectoring)
Management Conditions.

In interpreting the figures, SS would represent the most realistic and CC the least realistic "manned" target ICT data. The CS and
SC pairings would be somewhat comparable to using a single simulator
among pseudo piloted targets. The differences attributable to piloted-
vs-pseudo piloted control of the targets are apparent both within and
between the two conditions.

Simulator Numerosity

In the centralized management condition, the SS pairings produced
a slightly (nonsignificant) longer ICT mean value and considerably
larger ($p<.04$) standard deviation than the CC pairings. Comparing
"single simulator" (CS and SC) results as though only a single simulator were used among pseudo piloted targets is also informative. In
this case, the ICT mean and standard deviation were significantly
larger ($p<.0005$ and $p<.002$) for CS compared to SC with the CS values
not significantly different from SS (the other "pilot active" pairing) but significantly different from CC. These results suggest that
controller issued speed requests result in greater variabilities for
environments of all simulators as opposed to all pseudo piloted targets and that mixing the two types of targets can produce very different results based on their sequential orders.

The distributed management condition provides further comparisons of S and C targets. The S flights had a CDTI while the C flights
obviously did not and were consequently still speed controlled through
controller request as before. Pilots managed their own speed control
to achieve the requested sequential order and to maintain safe spacing. The proper comparisons in this condition are between SS as the
"all CDTI simulator" environment, and CS or SC as a single CDTI simulator environment. As in centralized management, the mean ICT for SS

was significantly (p<.0005) longer than for SC but not for CS. On the other hand, the ICT standard deviation for SS was about 40% smaller (p<.05) than that for CS and 27% (nonsignificant) smaller than that for SC. In fact, for distributed management SS produced the smallest ICT standard deviation while the other pilot active pairing (CS) produced the largest. Combining the nonsignificantly different ICT values for the single simulator pairings (CS and SC) produces a standard deivation significantly larger than that for the all simulator (SS) pairing.

These observations from the distributed management condition again suggest that results may differ significantly when obtained from an "all simulator" CDTI environment or from a single simulator CDTI one.

Fig. 3 also suggests other differences between ATC simulation using none, one or multiple piloted simulators in dence traffic environments. In the distributed management conditions, the minimum observed ICT between two CDTI simulators (SS) was 43 seconds as contrasted with 37 seconds between one CDTI simulator and a pseudo piloted target preceding or following (CS or SC) and 23 seconds between two pseudo piloted targets. Approximately 10% of the single simulator-pseudo piloted ICT values and 25% of the pseudo-pseudo piloted ICT values were below the minimum "multiple simulator" ICT.

In the centralized management condition where all targets were speed controlled by controller request and pilots did not have CDTI, the minimum ICT for all four combinations of target pairings were below 20 seconds with no strong distinction between them for values less than 50 seconds. Above that time, the simulator active pairings (CS, SS) tended toward longer times than the pseudo piloted active pairs (SC, CC) again suggesting a strong dependence on whether a single simulator precedes or follows a pseudo piloted target.

Conclusions

The previous observations were obtained from a single task scenario of considerable difficulty and may or may not be generally applicable. However, it is clear that in the present investigation of the effect of CDTI on making curved approaches, interpretation of the results would differ considerably if based on a multiple or a single simulator environment. For example, combining the single simulator (CS, SC) ICT values produces means and standard deviation not significantly different for the two management conditions whereas comparing the "all simulator" (SS) ICT values shows a nonsignificant mean difference but a 50% smaller (p<.001) standard deviation in the distributed management case. The considerably smaller ICT variability for distributed management indicates a system in better control and not as prone to delays.

The differences in results obtained using none, one or multiple simulators can be attributed to the subtle but potent human piloting characteristics in simulators combined with the relative inflexibility and lack of fine control over pseudo piloted aircraft and the resultant air-to-air interactions. The differences were particularly apparent in the CDTI condition where pilots initiated their own flight adjustments via the CDTI displays which reflect the air-to-air interactions. Thus sensory-motor piloting skills and the opportunities for permitting realistic air-to-air interaction may strongly impact results of ATC studies which appear primarily to exercise decision-making and communication.

Multiple simulators would thus appear necessary in ATC simulations particularly those studying CDTI effects to provide the necessary environment in which the realistic latent air-air interactions which strongly affect system performance may manifest themselves. An unresolved question concerns the necessary number of (CDTI equipped) simulators. As the previous discussion noted, using a single CDTI equipped simulator would have seriously underestimated CDTI usefulness. The required number of simulators would probably increase with the number and seriousness of possible air-to-air interactions. For example, independent approaches to closely spaced parallel runways would produce important lateral interactions between the traffic streams as well as fore-and-aft interactions within each stream.

A minimum facility of three simulators might suffice for studies in which fore-aft interactions are likely as this provides data from the centered simulator, the most realistic position. Controllers, however, often merge, hold, interleave, etc., "strings" of, rather than, single aircraft and in studies of this nature, possibly six simulators (3 in each string) might be a minimum number necessary for realistic results from the interacting strongs.

The need for a low cost, multi simulator facility is implied for realistic, complex ATC experiments particularly those involving CDTI-distributed management in which new pilot responsibilities and new controller roles are being studied. The following sections discuss the development philosophy and prototyping of a low cost multi simulator designed to be modularly expandable and integratable into an existing computer mainframe-graphics facility.

MULTISIMULATOR FACILITY

Existing Facility - Centralized Processing

The three simulator ATC facility assembled in 1972 at NASA-ARC and used for a number of ATC-CDTI experiments is a centralized processing system.

Flight dynamics, path predictors, X, Y, Z positions, mode selec-

tions, etc., for each simulator are digitally computed by the mainframe computers using the pilot analog (aileron, elevator, throttle)
and discrete (switch selections) inputs from the simulator "cabs".
The computational results are passed to a separate graphics system
and displayed for each pilot on a CRT in the cab as in Fig. 2. The
pseudo piloted targets are also computed by the mainframe computer
and displayed with the simulator traffic on the controller's CRT. A
CRT text terminal and keyboard interfaced to the mainframe permits
appropriate information display for modification by the pseudo pilot
(S) of the preprogrammed targets. In this system, the simulator,
pseudo pilot and controller stations are essentially low level peripheral devices for information input and display output.

This centralized computation philosophy has several virtues and
drawbacks. Virtues include the absence of significant system integration problems in assembly, programming, processing or communication;
ease of fault insolation; relatively low maintenance and/or required
sophistication of maintenance; and compatability with other uses of
the facility. Drawbacks include the inflexible and absolute amount
of computational capacity-vs-speed (a crucial factor for real time
experimentation) and difficulty of modifying program structure. Ad-
ditions, deletions, changes or reordering of program elements in the
centralized computation system can tie up the entire facility. It is
also difficult to assess the impact of proposed program changes on
real time running speed without actually making them. These observations reflect characteristics of centralized processing rather than
criticisms of the excellent programming innovations and implementations in the existing NASA-ARC ATC simulation facility. Many of the
observations of centralized processing in fact are remarkably similar to those of centralized management.

The three simulators and the independent CRT displays completely absorbed the total capacity of the mainframe and the graphics system and for brief periods during experiments this capacity could be
exceeded depending on the instantaneous computation and display demands resulting in slower than real time running and loss of some of
the lesser important picture elements such as boundary lines. The
three independently computed flight dynamics and path predictors
consumed the major share of the real time computation capacity. Increasing the complexity of the displays or computations in order to
achieve more realistic conditions for pilots and controllers was not
possible let along supporting additional simulators. These computational and graphics bounds have potentially serious impacts on the
validity of experimental results and on the complexity of ATC problems, particularly in CDTI-distributed management, which must be
studied through simulation means.

Proposed Facility - Distributed Processing

A distributed processing phisolophy was adopted as the basis for

prototyping an alternative design for a multisimulator ATC facility
capable of supporting up to eight simulators and/or more sophisticated displays and computations. The basic decision was to utilize the
existing mainframe and graphics system but limit graphics to CDTI
displays and use electro-mechanical flight instruments in each simulator instead of an EADI (typically a higher graphics burden than
a CDTI). This spare capacity could then be used for adding and/or increasing the sophistication of CDTI displays. Similarly, the computation of flight dynamics for the simulators and the flight path predictors would be offloaded from the mainframe and distributed to each
simulator using an internal minicomputer. With each simulator computing its own flight dynamics, path predictor, navigation and various other functions, the central computational bottleneck is avoided.
The simulator facility is thus modularly designed and may be operated
with any number of simulators up to a limit set by considerations
other than computation capacity.

A maximum basic cycle time of 100 msec (0.1 sec) was set in order to minimize any detrimental effects of processing delays on pilot
control. All information linking pilot input to display via instruments or CDTI must be computed and transferred during the 100 ms period. The major system bottleneck (besides graphic capacity) then becomes the information transfer rate from a simulator to the mainframe/
graphics and back to the CDTI display with each simulator adding to
the total real time transfer burden. Thus a data communications system becomes an "invisible" but crucial design feature for shuttling
data from the simulators to the appropriate places. The basic design
of a modular multisimulator subfacility nested within the total facility is described in more detail in the following sections.

Design Overview

The complete ATC experimental facility consists of three major
systems as shown in Fig. 5.

1. Multiman System. Up to eight independent, minicomputer based,
piloted Flight Simulators, a supporting Host Computer which serves
primarily as a data concentrator and intermediary during experiments
and a Data Communications System.

2. Mainframe System. The supervisory, central concentrator controlling experiments and driving all computer graphics functions which
include at least one controller station and at least one pseudo pilot station. Stores selected data for later analyses and regulates
data flow to/from the Multiman System through a full-duplex, parallel, asynchronous, DMA data communications link. Data in the direction of the mainframe is periodically updated every 100 ms from the
simulators as forwarded by the Host and is used to generate CDTI's
and the controller's display. Data may be communicated from the mainframe to the multiman system over a low volume path and is primarily

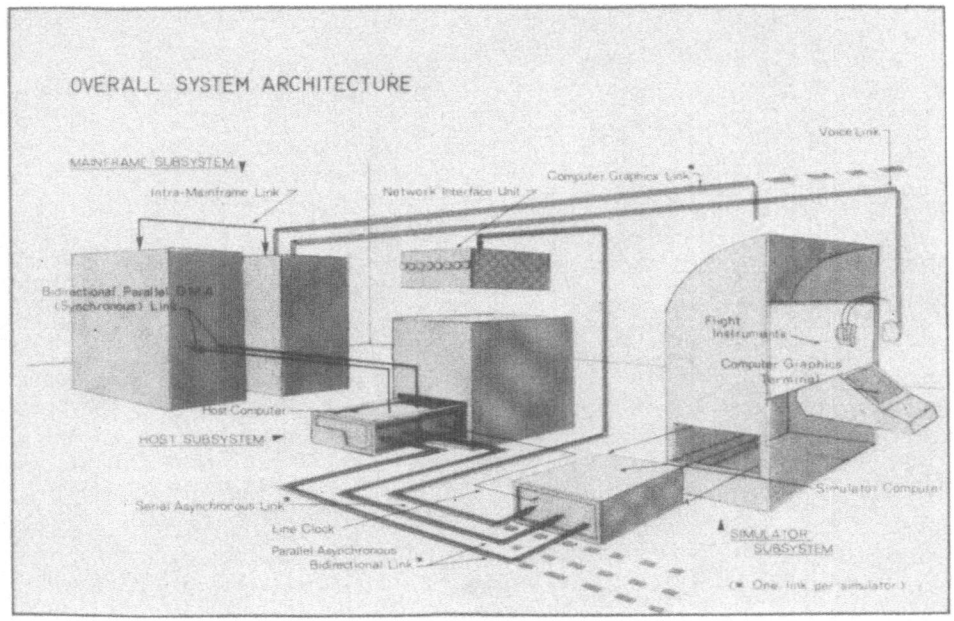

Fig. 5. Overview of the MultiSimulator ATC Facility.

used for supervisory control during experiments. The mainframe also
generates the pseudo targets and displays them on the controllers
display and CDTI's and in text format for the pseudo pilot.

3. <u>Voice System</u>. All stations (pilots, pseudo pilots, controllers)
are connected via a verbal communication network on which all communications can be recorded for later analyses. Neither the mainframe nor the voice systems will be discussed further in this paper.

<u>Multiman System</u>. The Multiman System shown in Fig. 6 is modular-
ly designed to accommodate up to eight <u>Flight Simulators</u> by the Host
Subsystem and Data Communications Systems. The Multiman System may be
operated as a facility without support by the Mainframe as long as
CRT graphics are not required, as for example, during program development, or during experiments not needing pilot CRT displays and in
which only limited data storage is required. This design feature eliminates any unnecessary burden on the Mainframe/graphics releasing
them for other experimenta, developmental or maintenance purposes.
Used in the stand along or "local" mode, the Multiman System can,
for example, support experiments using standard flight instruments.
An auxiliary benefit of the multiple independently programmable simulation is the ability to run subjects simultaneously rather than sequentially to increase experimental throughput.

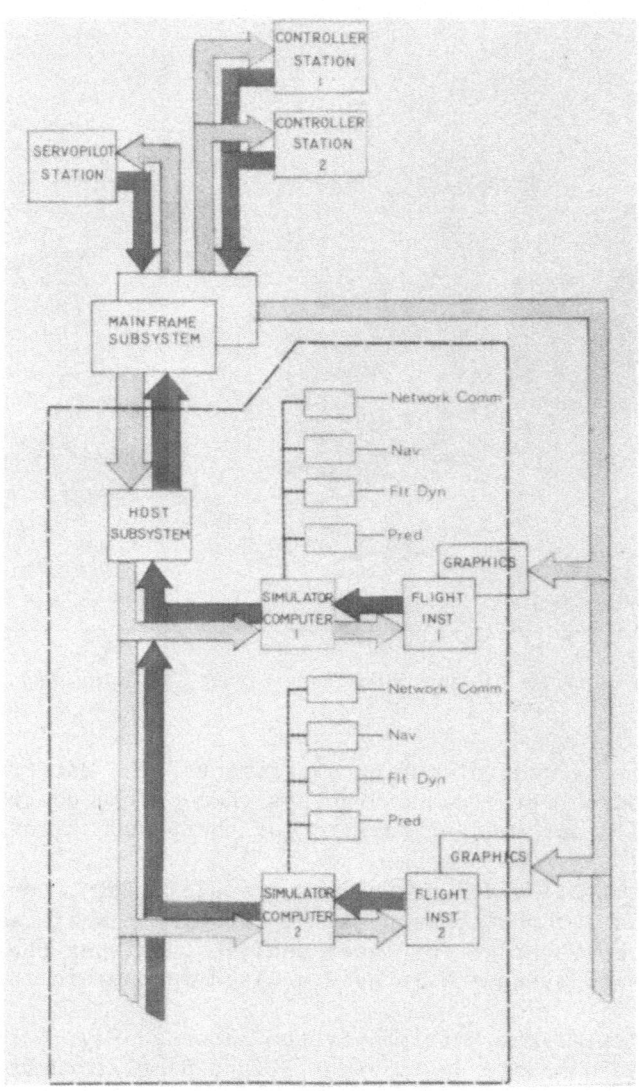

Fig. 6. The Multiman System.

Flight Simulator Subsystem

The Flight Simulator Subsystem consists of a commercially a-
vailable[1] instrument trainer, a minicomputer and an interfacing panel as shown in Fig. 7.

----

[1] Aviation Simulation Technology, Bedford, MA.

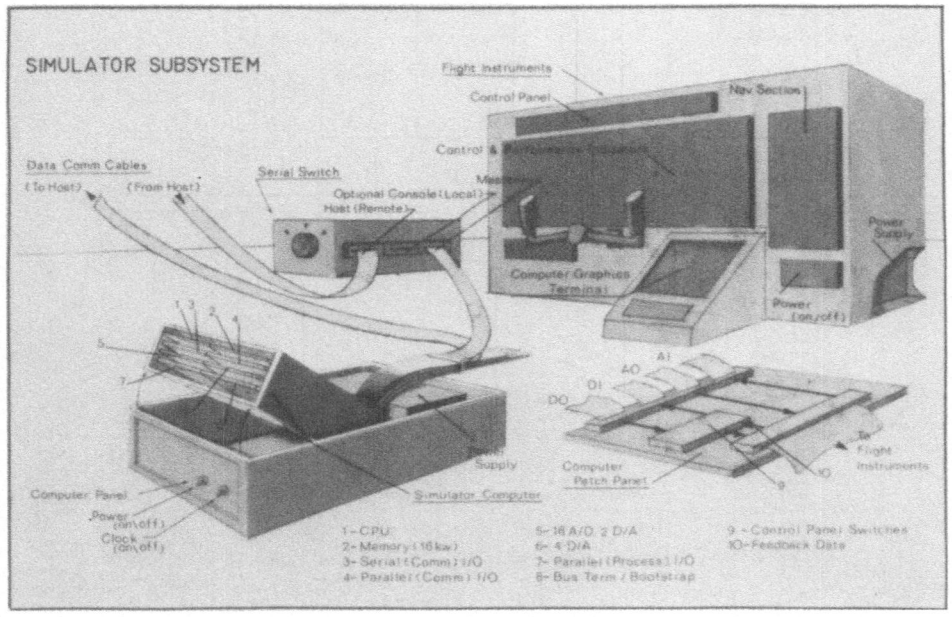

Fig. 7. Flight Simulator Subsystem.

Flight Panel. The instrument trainer (Flight Panel) provides a
high degree of visual and functional fidelity and can be operated as
a self contained unit (local mode) without the Simulator Computer and
is useful in this mode for training, etc. For developmental or experimental purposes, the Panel operates in remote mode supplying pilot
inputs (switch selections, aileron, elevator, throttle, etc.) to the
Simulator Computer which computes flight dynamics and outputs signals
for the electro-mechanical flight instruments. Compensated signals
drive phase-locked loops around each motor driven instrument using
circuitry supplied by the Panel manufacturer. It is possible, although
not presently configured, to input to the minicomputer the pilot inputs and quantities displayed on the flight and navigation instruments when the Panel is operated in the local mode. This would permit use of the Panel's internal flight dynamics and navigation sections allowing the minicomputer, for example, to compute and forward
to the Host the flight path predictor, X, Y, Z position etc.

Simulator Computer. The Simulator Computer is a commercially a-
vailable mini/micro computer structured around Digital Equipment Corporation's LSI-II as diagrammed in Fig. 8. The Simulator Computer
uses 16K words of MOS memory for on line experiment related software
which is downloaded from the Host at initial startup. Standard ven-
dor logic boards are used for CPU and RAM and a micro-coded floating
point chip is used for on line software arithmetic. The basic 100 ms

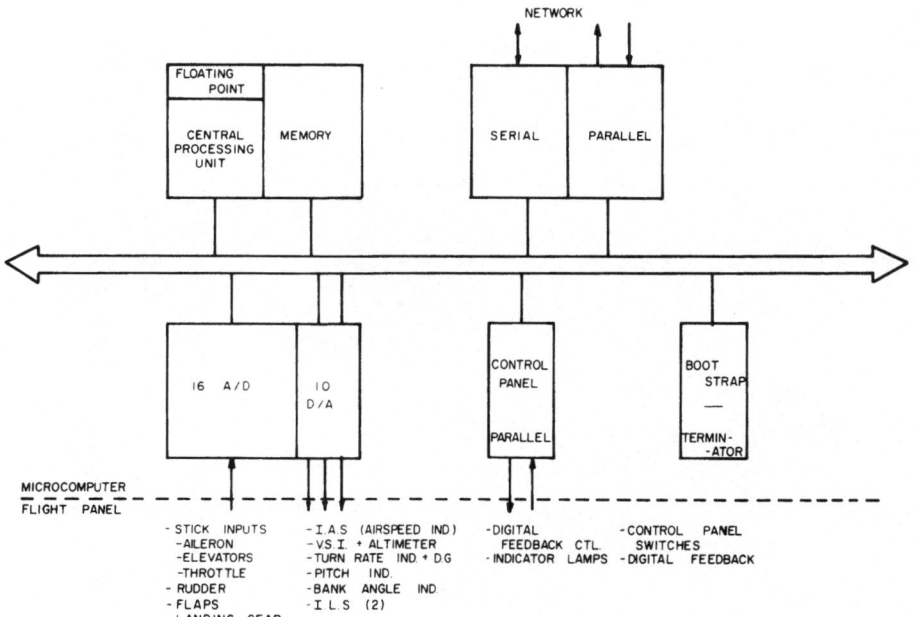

Fig. 8. Simulator Subsystem Bus Diagram.

software duty cycle is segmented by a standard 100 Hz or 60 Hz clock. Eight (12 bit) analog outputs and seven (12 bit) analog inputs are used in interfacing to the Panel. Sixteen discrete inputs for Control Panel operation and eight discrete output for status annunciation are required. Four additional discrete outputs are required for phase-encoded feedback circuit control and 12 inputs (plus interrupt) are needed for feedback circuit data input. These latter interfaces are software multiplexed with a single circuit on the Patch Panel. Resolution is expandable by Patch Panel-Local oscillator tuning and additional discrete inputs.

The Patch Panel interfaces the Simulator Computer and Flight Panel with phase-encoded analog feedback circuits and minimal hardware signal transducing.

Software Components. Fig. 9 shows the off line and on line program structures in A and B. The Simulator Computer software provides on line experiment flight controls. A specially-developed foreground/background operating system provides real time synchronous 100 ms framing for flight dynamics, predictor, ILS, and data communications to the Host. The background function provides control panel and data communications from the Host when the foreground is not running. Foreground functions require approximately 80 milliseconds as follows: flight dynamics (40), predictor, ILS, and data communications (40 ms about equally divided).

The flight dynamics are based on a simplified "Navion" model and are written in assembly language since the Simulator Computer does not support FORTRAN. A number of calculations and automatic control related macro-codes were developed to support modularization of the flight dynamics and other functions in conjunction with the LSI-II Computer.

A predictor is programmed to project the aircraft's position at six points into the near future. The present (but programmable) pilot selectable options are for a 15 or 30 second predictor. The predictor-model uses bank angle and airspeed to compute positions relative to the aircraft's absolute position. Wind effects are then added and the values are converted from floating point to integer for updating to the Host and the Mainframe computers.

At present, a curved approach deviation function is computed and displayed on the ILS for any of 5 approaches used in a previous experiment. Once an approach is selected by the pilot, the ILS Glide Slope

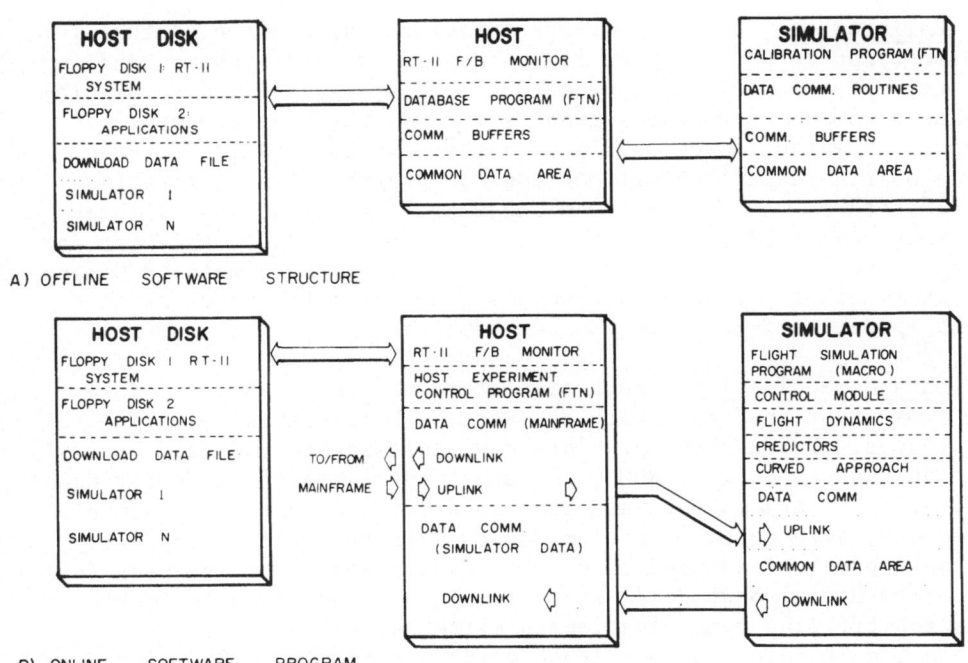

Fig. 9. Multiman System Software Structure Diagram.

and localized indicators are driven to indicate position deviations
from the ideal path maintained in the software model. A command function from the Host can initiate the aircraft at a given position and
altitude on an approach.

Data communications to the Host are activated on a "when-data-
is-ready" basis within the 100 ms real time software duty cycle. Control information, time (X, Y, Z) position, six scaled integer predictor points, two floating point values and simulator I.D. are trans-
mitted over the parallel asynchronous link to the Host in a "burst"
mode to limit interrupt processing impacts on the Host.

Data communications from the Host are handled with the Simulator
software in background mode when foreground functions are completed.
A background loop is entered to service uplink commands and local
control panel inputs. The uplink commands provide setting of simulator I.D. and clock frequency ("ticks"/100 ms cycle), predictor-peri-
od and scaling constants; (X, Y, Z) aircraft position, aircraft set
point information (altitude window, approach, release-when-ready),
and Control Panel Mask which permits the Host (or Mainframe through
the Host) to override the simulator-local pilot's control panel for
remote simulator control during experiments. The Control Panel permits selection of approach; position set pointing; selection of predictor ON/OFF and 15/30 sec. options; enable/disable data communications and other software control functions.

Calibration. Off line software provides calibration of the Simulator Computer software and Flight Panel hardware systems, and integration of the off line and on line software systems. Each simulator has a common data base segment stored on the Host Computer's
floppy disk which is accessible by either the Calibration Program
or the on line experiment flight control program. The calibration
program fetches the segment from the Host, uses it during calibration sequences then uploads the (tuned) segment back to the Host for
storage. When on line software is downloaded to the Simulator Computer, it receives the updated common segment from the Host and then
begins operations.

Data Communications. Each simulator has two (point-to-point)
data communications likns with the Host. A bidirectional, serial, a-
synchronous, RS-232C link provides download support (Serial, binary).
Power-up/start up LSI-11 ROM 'Absolute Leader' code is used for downloading. Applications software uses parallel, asynchronous channels
in two-way, alternate, 'burst' mode for data transfer. A specially
developed protocol is used to provide data transfer of 16-words (si-
mulator-to-Host), 8-words (Host-to-simulator) and variable length
words for off line communication in either direction. Transfers require a single interrupt per frame with an approximate 200 micro seconds average word transfer rate including all latencies and service
overhead.

Host Subsystem

The Host Subsystem shown in Pictorial and Functional form in Figs. 10 and 11 perform three major functions. The first function is the primary program development base for the Multiman System support-ed by mass storage on dual floppy disks and hardcopy by the teleprin-ter. The second function is the source and controller for downloading Simulator computers over 9600-band serial asynchronous lines (paral-lel line downloading is also supported). The third function is the data concentrator/distributor "store-and-forward" center between Main-frame and Simulator computers during on line experiments. Bidirectio-nal parallel DMA links between Host and Mainframe and "star" configu-ration asynchronous, parallel links between Host and Simulator support on line data communications.

The Host's software development base is Digital Equipment Corpo-ration's RT-11 (Foreground/Background) operating system, using FORTRAN. Off line simulator programs are FORTRAN-based ("stand alone FORTRAN") using MARCO-11-based data communications modules. On line simulator programs are MACRO-11-based using "FORTRAN-callable" subroutines for ease of conversion to/from FORTRAN and for remote debug.

A manual switching unit at the Host is used with a portable CRT for remote/local manual operation of a Simulator Computer and for

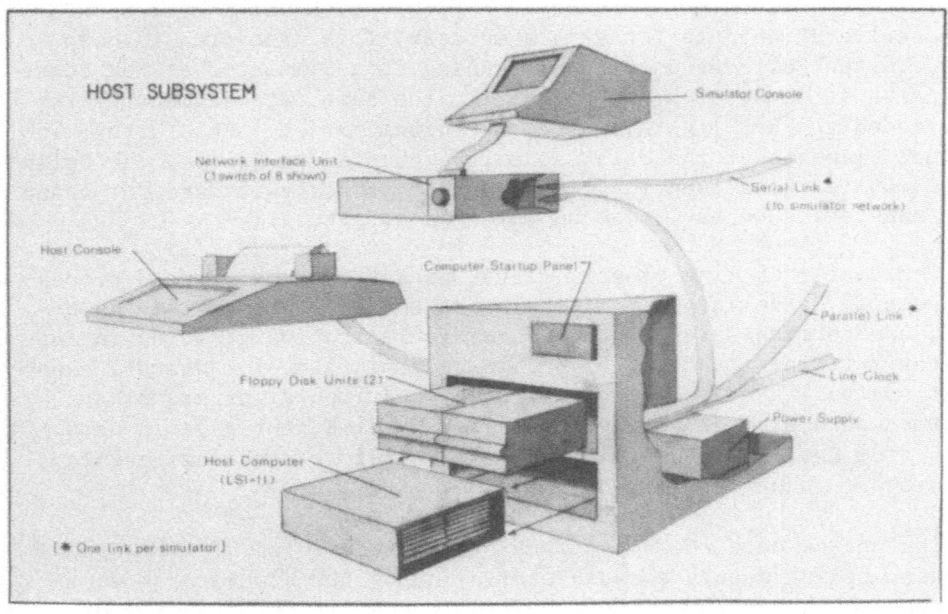

Fig. 10. Host Subsystem Pictorial.

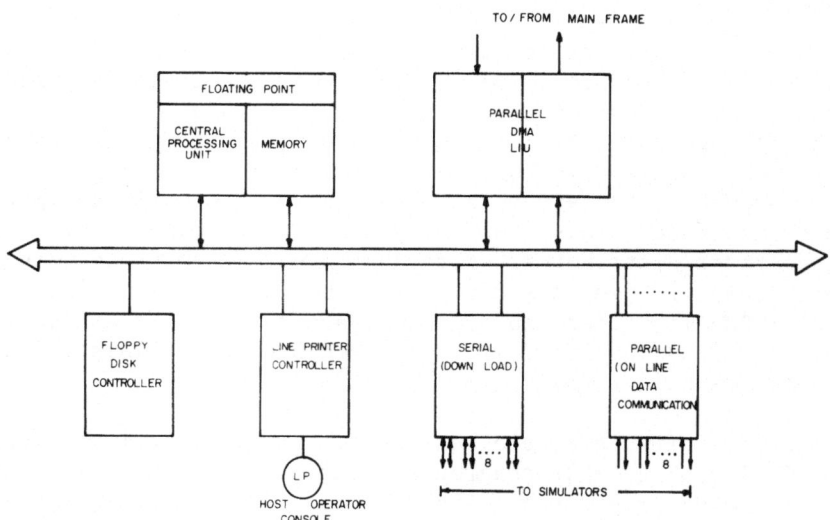

Fig. 11. Host Subsystem Functional Bus Diagram.

down load. A specially developed asynchronous driver permits down loading of binary data over serial lines. Higher-speed down loading over parallel lines is also used, although RT-11 "load image" alterations must be made for this case.

The Host maintains a database on floppy disk which corresponds to the calibration data for each simulator. This data area is down loaded to the Calibration program running in a Simulator at the start of a calibration session. Subsequently, the same data segment is re-down loaded to the Flight Simulation Program for on line startup; individual elements of the database may be changed on line by the Uplink Data Communications function. This dynamic property is used for changing parameters such as winds and turbulence factors.

The Host's on line program serves as a store-and-forward processor. Data is "pipelined" from Simulators to Mainframe on the "downlink", and commands are dispatched to Simulators (originating in the Host for Multiman link control and in the Mainframe for overall experiment control) over the "uplink". Some experimental programs have performed data storage on the Host's floppy disk, but gaps in data result from device latency limitations; therefore, on line Host-local data storage is not specified.

The uplink data communications structure provides for 8-word messages. Uplink messages cause overlaying of the Simulator's Control Panel Mask for remote Host control; in addition, a number of experiment control messages are used to alter parameters dynamically. The down link data communications structure provides for 16-word messages. The down link in the Host sees one 16-word message per simulator per 1/10 second; this amounts to 16 x 10 x 8 (simulators), or 1280 words/second over the parallel, asynchronous links, necessitating "burst mode" processing. The average latency in the Host is about 200-250 microsecond/word on the down link; this amounts to 3.2-4 millisecond/message, or about 26-32 millisecond/100 millisecond real time cycle (26-32%).

At initial startup, about 30 seconds are required to down load a Simulator Computer of which the actual data transfer requires about 20 seconds. Complete design and operating details of the Multiman Simulator System can be found in Kreifeldt (1979).

CONCLUSIONS

A modularly designed programmable flight simulator system has been prototyped as a means of providing an economical facility of up to eight simulators to interface with a mainframe/graphics system for ATC experimentation, particularly CDTI-distributed management in which pilot-pilot interaction can have a determining effect on system performance. Need for a multiman simulator facility was predicted on results from an earlier three simulator facility which indicated that experimental conclusions as to the utility of CDTI would be drastically altered if only a single simulator had been used. Future experiments are planned using the three simulator facility to determine further the appropriate level of simulator numerosity and level of fidelity.

ACKNOWLEDGEMENTS

Work described herein has been supported by NASA Grant NSG-2156 through the Man-Vehicle System Division of the NASA-Ames Research Center. The continuning support and suggestions of Thomas Wempe at NASA-ARC are most gratefully acknowledged. Particular thanks are due to Sandra Hart and many others at NASA-ARC without whose aid the experimental work could not have progressed as expeditiously.

REFERENCES

Bogdanoff, E., Brooks, H., Jasinski, F., Keys, L., Michael, A., Molnar, A., Proctor, G., Reeves, E. and Thorsell, B., 1960, Simulation: An Introduction of a New Technology, Report TM-499 System Development Corporation, Santa Monica, CA.
Buckley, E., House, K. and Rood, R., 1978, Development of a Performance Criterion for Air Traffic Control Personnel Research Through Air Traffic Control Simulation, Final Report. Report No. FAA-RD-78-71.

Geisler, M., 1960, Development of Man-Machine Simulation Technology, Report P-1945, RAND Corp., Santa Monica, CA.

Klass, P., 1980, Controller Numbers Keyed to Advances, Aviation Week and Space Technology, April 28, 61-63.

Kreifeldt, J., 1979, A Study of Pilot-Controller Interaction Under Various Simulated Advanced ATC and Navigation Concepts, Progress Report 12/1/78 - 8/31/79, Department of Engineering Design, Tufts University, Medford, MA.

Kreifeldt, J., 1980, Cockpit Displayed Traffic Information and Distributed Management in Air Traffic Control, Human Factors, 22 (6), 671-692.

Kreifeldt, J., Parkin, L. and Hart, S., 1977, ATC by Distributed Management in a MLS Environment, 13th Annual Conference on Manual Control, MIT Press, Cambridge, MA.

Parsons, H., 1972, "Man-Machine System Experiments", The Johns Hopkins Press, Baltimore and London.

Sackman, H., 1967, "Computers, System Science and Evolving Society", John Wiley & Sons, New York.

EVALUATION OF VARIOUS DESIGN ALTERNATIVES INFLUENCING DISCOMFORT IN A DRIVING SIMULATOR

John G. Casali

INTRODUCTION

The utilization of vehicular simulators is increasingly evident in research and design, training, and performance assessment. Numerous operator/vehicle interface simulations of airborne, land, and sea-going vehicles currently exist and are employed in a variety of applications. The advantages offered by these devices over the use of a full-scale vehicle are impressive, including increased safety, tighter experimental control, ease of sensing and recording operator performance, reduced data collection times, and lowered operational costs in some cases.

Simulator Sickness: A Significant Problem

Despite the ongoing expenditure of considerable effort and money toward the improvement of simulator technology, the use of many vehicular simulators, especially driving simulators, is hindered by a recurring malady often termed "simulator sickness". Subject drivers in both fixed-base and moving-base simulators have regularly exhibited symptoms characteristic of the simulator sickness syndrome, including: dizziness, disorientation, pallor, increased perspiration, altered heart and respiration rate, and even nausea and vomiting (Barrett and Nelson, 1966; Breda et al., 1972; Jex and Ringland, 1973; Testa, 1969). In some studies, driving simulator subjects have suffered acute symptoms which necessitated termination of the simulated driving task. In other cases, residual symp-

toms appeared following the simulator experience, often resulting
in malaise which persisted for several hours.

Despite its frequent occurrence, the incidence of sickness in
specific driving simulators is not often publicized by the investigators involved. The simulator sickness problem warrants significant
concern, however, and it is the responsibility of the simulator designers to consider the causes of sickness and eliminate them in the
design phase, if possible. Sickness adversely affects the validity
and generalizability of the simulator data. In fact, a nausea inducing simulator cannot be relied upon to yield valid and accurate
human response data, because a driver of the corresponding full-
scale vehicle does not experience sickness (Leonard and Wierwille,
1975). The instance of driver sickness in the simulator constitutes
an inappropriate extraneous variable, confounding the simulator data. It should also be realized that investigators who knowingly
place naive subjects in a sickness-inducing simulator without first
making them aware of the problem are in direct violation of accepted
ethical principles (American Psychological Association, 1973).

Simulator Sickness versus Motion Sickness

In 1957, Havron and Butler reported that 78% of questionnaire
respondents indicated that they had sustained some degree of "motion sickness" in response to a helicopter simulator. This appears
to be the first reference to such simulator experiences as motion
sickness. However, the generic term "motion sickness" should not be
used synonymously with "simulator sickness" for several reasons.
First, the term "motion" in "motion sickness" implies that physical
movement of the subject is necessary and solely responsible for any
sickness produced. In the case of fixed-base driving simulators,
however, the only feedback cues from control inputs that the subject
experiences are movement cues in the visual scene, auditory cues,
and possibly high-frequency vibration simulating drivetrain disturbances. In fixed-base devices, many of which have a history of
subject malaise, no translational or rotational movement of the subject is present, suggesting that illness is precipitated by factors
other than motion.

While it is true that some aspect of the motion system in moving-base simulators may contribute to nausea, it is doubtful that
motion is singularly accountable for the problem. Furthermore, after considering the number of simulator-produced stimuli that a
subject experiences visually and auditorially, as well as kinesthe-
tically and vestibularly, it is quite apparent that motion cues potentially play a minute role in causing malaise. Other characteristics of the simulator may contribute more significantly to simulator sickness. It is distinctly possible that the precipitating
causes of sickness are different for each individual simulator.

LITERATURE REVIEW

A literature review directed specifically at driving simulator sickness appears in Casali and Wierwille (1979). This review revealed that little definitive research had addressed either the etiology or the symptomatology of the sickness problem.

Fixed-Base Driving Simulator Sickness

Many older driving simulators and several devices currently under construction have no provision for physical movement of the driving subject. The prevalence of sickness among fixed-base simulators is well-known, with documentation appearing in Barrett and Nelson (1965); Barrett and Nelson (1966); Barrett and Thornton (1968b); Jex and Ringland (1973); Reason and Diaz (1971); and Testa (1969).

Several authors (e.g., Barrett and Thornton, 1968b) have hypothesized that the occurrence of sickness in a fixed-base simulator results from a sensory conflict between visual cues of apparent motion in the roadway scene and the absence of any corresponding physical motion cues. If this is indeed the case, the most obvious remedy is to include a motion system controlled by the same dynamics responsible for display movement. Similar conclusions pertaining to aircraft simulation have been offered by Havron and Butler (1957) and Miller and Goodson (1960), among others.

In a study by McLane and Wierwille (1975), the motion system of the moving-base Virginia Polytechnic Institute and State University (VPI&SU) simulator was deactivated, resulting in a temporary fixed-base simulation. Six degrees of vehicular motion were apparent in the computer-generated display. No evidence of subject uneasiness was observed in the study. These results suggest that the presence of a cue disparity between visual motion cues and lack of physical motion cues in fixed-base simulators is not sufficient for eliciting sickness. However, these results do not discount the conclusion of Reason and Brand (1975, p. 114) that it is "reasonable to assume that the essential stimulus for visually-induced sickness is the presence of a moving scene which, during real vehicle motion, would be accompanied by a stream of vestibular signals".

Barrett and Thornton (1968b, p. 307) add that "An interesting question is the degree of motion required to give the necessary body cues. Simple random vibration may be enough to eliminate the cue conflict (in fixed-base simulators), a possibility having considerable practical and economic import for the simulation art." However, several fixed-base devices which do incorporate simple random vibration are known to have serious subject illness problems (Jex and Ringland, 1973; Testa, 1969). Therefore, other factors besides the lack of physical motion may precipitate sickness in fixed-base simulators.

Research concerning the prediction and evaluation of fixed-base driving simulator sickness was reported by Testa (1969). The investigation did not directly assess simulator design influences on sickness, but did demonstrate that both self-report (questionnaire) and physiological measures were needed to properly identify a state of simulator sickness. A large percentage of Testa's subjects became ill in response to the simulated driving task. One prominent characteristic of the motion picture display system used in Testa's simulator was the extremely wide field-of-view, measuring approximately 2.62 radians (150 degrees) horizontally. Similar results occurred in the Havron and Butler (1957) helicopter simulator study, in which 28 of 36 subjects experienced illness. This simulator also incorporated a wide-angle scene of approximately 4.54 radians (260 degrees) azimuth by 1.31 radians (75 degrees) elevation.

It is generally accepted that wide-screen displays have a strong tendency to induce symptoms akin to those of motion sickness, especially when rapid movement is presented to viewers (Reason and Brand, 1975). This is perhaps one reason for the current trend away from the use of once-popular "cinerama" and "cinemascope" theatrical display systems in simulators. Several plausible explanations for the high incidence of sickness accompanying this type of display system are: (1) the viewer must attend to an abundance of detail presented throughout the wide field-of-view, resulting in disorientation, (2) both static and dynamic distortion of displayed objects occurs, in part due to the fact that the subject often views the screen from an angle unequal to that of the projector producing the image, (3) dynamic inaccuracies appear in the perceived angular and translational accelerations in the scene, and (4) a "strobing" phenomenon may result from the interaction of picture frame rate with the critical flicker fusion frequency of the viewer's peripheral vision.

Moving-Base Driving Simulator Sickness

The literature addressing moving-base simulator sickness is scant. With the exception of the Casali and Wierwille (1980) research, which will be discussed later in this paper, there appear to have been no other studies which attempted to determine specific causes of simulator sickness in moving-base driving simulators. Most studies involving driving simulators have not investigated simulator characteristics as independent variables, instead utilizing the simulator as a research tool in an applied nature. Casali and Wierwille (1979) examined this body of applied research for statements concerning malaise and for the characteristics of the driving simulators used. Also, interviews were conducted with several researchers who had driven various driving simulators, regarding the degree of discomfort they had experienced. Combining all available information, the design characteristics and simulator sickness histories of each existing driving simulator were compiled and tabled. From this table, which appears in Casali and Wierwille (1979), characteristics common to those simulators that had histories of nausea problems were determined. Among the moving-base simulators, three characteristics that appeared as potential contributors to simulator sickness were: (1) the manner in which translational motion was achieved, (2) the presence of lag or delay in the simulator response variables, and (3) the practice of enclosing the subject inside a cab. These three major potential contributors will be discussed individually.

Rotational simulation of translation versus true translation. The extent to which physical motion cues are accurately modeled directly affects the fidelity of a driving simulator. With each degree of freedom of movement added, the cost of the simulation increases considerably. Also, space limitations of the laboratory often restrict the number of motions included, as well as their associated excursion distances. Due to economic and space constraints, certain compromises have appeared in the motion systems of several driving simulators. The most prominent compromise is associated with the method of simulating lateral and longitudinal translation. The driver's platform of a simulator must traverse considerable distances if translational accelerations are sustained for any period of time, using the "normal" method of simulating these motions. This normal method is to translate the driver platform forward and backward for longitudinal translation and side-to-side for lateral translation. Several simulator designers have instead adopted the approach of using pitch and roll motion to approximate translation. By tilting (rotating) the subject in the roll axis, the lateral acceleration forces of cornering and lane-changing are simulated. Similarly, by rotating the subject in the pitch axis, longitudinal acceleration and braking forces are simulated. In either case, because of the rotation of the subject's vertical body axis with respect to the gravity vector, the subject supposedly experiences the sensation of lateral or longitudinal acceleration.

If the technique of simulating translational acceleration by rotation is truly a contributing factor to the incidence of simulator sickness, its influence may be explained in terms of a cue disparity theory. While the angular rotation does indeed produce a lateral or longitudinal component of acceleration to a seated subject, cue conflict may occur when the subject senses the rotational aspect of the motion, which is in this case an artifact. In other words, the possibility exists that the subject actually perceives the motion as rotational, when the motion the subject expects should be translational.

Delayed versus nondelayed dynamics. In a vehicle simulation that is closed-loop, the equations of motion of the vehicle dynamics must be solved on-line and in real time. The outputs of these equations provide the necessary signals to drive the display, instruments, and motion system. A problem associated with some driving simulators is that computational or response lags are introduced in addition to the normal vehicle dynamic responses. In these simulators the subject experiences delayed scene updating or delayed physical motion cues, or both.

The lag may be the result of any of the following: (1) a lack of computational speed, such as that due to serial processing in the computer that solves the vehicle dynamics equations, (2) delay in the response of servo systems used in the image generation process, such as those used to control the movement of a video camera over a terrain board, and (3) delay in the response of the hydraulic, mechanical, or electrical equipment used to move the platform (or the instruments) physically.

Regardless of their form, time lags in the feedback cues presented to the subject cause two problems. First, an apparent delay between the simulator's manual controls and driver feedback cues may cause the simulator to be difficult to handle. Inappropriate control-to-feedback delay places the additional burden on the operator of anticipating the vehicle's response and introducing lead compensation. Secondly, the delay is apt to contribute to operator discomfort. When delay occurs and is perceived by the subject, a cue disparity exists between actual feedback cues and expected feedback cues.

Enclosed versus open platform. The presence of a box-type cab has appeared on several simulators known to induce nausea. Usually no windows are built into these cabs, and the subject is enclosed by three walls and a roof, with the display constituting the front wall. Often the only light inside the cab is that emitted by the roadway scene. Interestingly, a lower incidence of illness has generally been reported with automobile body-cabbed and unenclosed simulators than in box-cabbed simulators.

The reason for the potential influence of enclosure on uneasiness are at best speculative. Nevertheless, two reasonable explanations have been formulated. First, the simple knowledge of being enclosed within a box-shaped structure may be discomforting to a subject. Individuals may experience claustrophobic reaction to enclosure, subsequently leading to symptoms of uneasiness. Also, the lack of any peripheral reference points other than the visual display, which appears to be suspended in dark space inside the cab, may be disorienting to the subject. Unenclosed and automobile body-cabbed simulators (with windows) do not have this problem, since room reference cues may be discernable, even in a darkened room.

Other contributors to simulator sickness. Less prominent potential causes, some of which are common to both fixed-base and moving-base devices, include: (1) lack of display collimation, (2) inappropriate or lack of streaming in the roadway periphery, (3) large lateral field-of-view (as mentioned earlier), (4) display distortion and poor resolution, (5) rapid luminance changes in the visual scene, (6) slow picture frame rates, and (7) poor coordination between visual

and kinesthetic-vestibular cues (other than delay, as mentioned previously)

INVESTIGATION OF MOVING-BASE SIMULATOR SICKNESS

An initial study performed by Casali and Wierwille (1980) is
discussed in the remainder of this paper. This research was undertaken with the objective of determining causes of sickness in moving-
base driving simulators. Obviously, it would be impossible to examine
all potential causes in a single research endeavor with the use of
one particular simulator as a research tool. Therefore, the selection
of independent variables for investigation required some degree of
judgment, based on the literature review results outlined above. In
this initial· study, the three major potential contributors were investigated as independent variables. The other less prominent causes
remain in need of future investigation.

Experimental Apparatus

A brief description of the fundamental apparatus used in the
Casali and Wierwille (1980) simulator sickness investigation follows.

Driving simulator. A highway driving simulator, located at VPI
&SU, was employed in the study (Fig. 1). This device provides realistic vehicle handling via a four-degree of freedom motion base (lateral translation, longitudinal translation, roll and yaw) which is accurately coordinated with a computer-generated roadway scene. This
particular simulator was easily adaptable to the study of simulator
sickness for two reasons: (1) in normal operation, the simulator had
never induced observable malaise in any of over 800 subjects, thereby
enabling the investigators to "degrade" the subsystems of the simulator in an effort to expose specific determinants of sickness, and (2)
the rapidly responding motion system, coupled with the analog/hybrid
computer-controlled dynamics, facilitated the simulation of feedback
delays and alternative motion system techniques characteristic of
problematic simulators.

Dependent measurement instruments. Eight dependent measures were
used to identify the state of simulator discomfort. Four physiological measures (pallor, respiration rate, forehead skin resistance, and
heart rate) were obtained using individual transducers.

Frequently accompanying the physiological symptoms exhibited by
simulator-sick subjects are degraded performance abilities of various
forms (Barrett and Thornton, 1968a; Miller and Goodson, 1960). A subject's ability to control the driving simulator was expected to decrease as a function of degraded simulator conditions, or malaise, or
both. Two measures of vehicle controllability affecting driver performance used were yaw (standard) deviation and steering wheel reversals/

Fig. 1. VPI&SU Driving simulator laboratory.

minute. In addition to assessing the performance of vehicle control, the potential influence of degraded simulator conditions on the performance of mental tasks was investigated. A pre-post simulator arithmetic test was used as the mental task measure. While no studies involving a driving simulator have addressed this problem, several research efforts have demonstrated that motion sickness is often accompanied by a decreased ability to perform mental tasks, such as arithmetic computations and estimation of elapsed time (Brand et al., 1968; Clark and Graybiel, 1961). The eighth and final dependent measure was a modified version of a self-report simulator reactions questionnaire used by Testa (1969).

Method

A between-subjects design was applied to the research, utilizing eight subjects in each of eight experimental conditions. The three independent variables, consisting of two levels each, are listed below. Abbreviations appear in parentheses.

1. Simulation of lateral acceleration (LAT): a. by true translation (T), normal method; b. by angular (roll) rotation (A). Replacement of normal translational simulation of lateral acceleration with angular (roll) rotation was performed by reprogramming the lateral-directional dynamics in half of the conditions. It should be noted that in condition A, the roll cue had to be attenuated to 35% of its true value because of the extreme roll excursions which would have resulted.

2. Simulator display and motion feedback dynamics (DEL): a. nondelayed (ND), normal method; b. delayed (D). In half of the experimental conditions, the normal (nondelayed) dynamics characteristic of a typical late-model, intermediate size sedan were used. In the remaining conditions the visual display and motion feedback systems were simultaneously delayed by 0.30 second over the normal vehicle response for a given steering input. This duration of delay appeared representative of feedback lags inherent in several sickness-inducing simulators.

3. Simulator driver platform (CAB): a. open (O), normal method, b. enclosed (E). The simulator operated unenclosed in half of the conditions. In the remaining conditions subjects were completely enclosed in a ventilated but windowless cab, which eliminated any visual perception of room reference cues. The narrow, box-like cab resembled those used on several other simulators, such as the Volkswagen (Lincke et al., 1973) and the General Motors Technical Center (Beinke and Williams, 1968) devices.

Prior to the driving task, baseline (pre-simulator) data were recorded for each subject. Next, the subject drove the simulator in a typical simulated driving task, consisting of straight roads, curves and moderate crosswinds. Experimental data were obtained during the latter part of the driving task. Finally, the subject expressed his or her subjective reactions to the simulator experience on the questionnaire.

Results

Following data collection, a single baseline-versus-experimental score was computed for each physiological measure and for the mental task measure for each subject. The questionnaire, yaw deviation, and steering reversals were each represented as a single score for each subject.

Multivariate analysis. To determine if the group of eight dependent measures was sensitive to changes in the simulator variables, the data were first subjected to a multivariate analysis of variance (MANOVA). Wilk's U criterion values were computed and then converted into exact F-ratios, using the standard formulae (Kramer, 1972). The group of measures were statistically significant for the method of simulating lateral acceleration (LAT), $F(8,49)=5.13$, $p=0.0001$; for the presence of delayed dynamics (DEL), $F(8,49)=3.76$, $p=0.0017$; and for the presence of an enclosure (CAB), $F(8.49)=2.44$, $p=0.0265$. In addition to the significant main effects, a significant two-way interaction (LAT by DEL) was found $F(8.49)=2.24$, $p=0.0403$.

Individual analyses. Subsequent to the MANOVA, a simple three-way between-subjects analysis of variance (ANOVA) was performed on each dependent measure. The intent of the individual ANOVA's was to determine which dependent measures were sensitive to the simulator (independent) variables. In maintaining a statistically conservative posture, only those independent effects within the domain of significance of the MANOVA were considered in the ANOVA investigations. Table I is a numerical summary of the significant effects revealed by the ANOVA's, $p<0.05$). Dependent measure means for each level of the simulator variables are also shown in Table I. In the case of the physiological measures, the means represent the experimental-minus-baseline difference score means.

The pallor ANOVA revealed that the increase in baseline to experimental pallor level was greater for simulation of lateral acceleration by angular rotation than for the normal method of lateral translation. Similar results were obtained for the respiration rate measure. However, in the case of both pallor and respiration rate, the main effect of LAT was restricted by a significant LAT by DEL interaction.

To examine the LAT by DEL interaction for pallor, a Newman-Keuls multiple comparison test was performed. It showed that whenever computational delay was introduced, regardless of the method of simulating lateral acceleration, the change in pallor increased significantly from the normal simulator configuration..As for respiration rate, the post hoc analysis revealed that the combination of angular rota-

Table I. Summary of significant sources of variance for each dependent measure.

Dependent Measure	Source of Variance	F	p	Factor Level	Mean Value
Pallor	LAT X DEL	6.51	0.0135	T/ND	1.27
				T/D	4.59
				R/ND	5.73
				R/D	4.06
	LAT	4.03	0.0495	T	2.93
				R	4.89
Respiration Rate	LAT X DEL	9.74	0.0028	T/ND	1.85
				T/D	1.26
				R/ND	2.23
				R/D	5.55
	LAT	13.89	0.0005	T	1.56
				R	3.89
	DEL	4.77	0.0331	ND	2.04
				D	3.41
	CAB	9.95	0.0026	O	1.74
				E	3.71
Forehead Skin Resistance	CAB	5.08	0.0281	O	5.67
				E	-1.81
Heart Rate	(none significant)	-	-	-	-
Mental Task	(none significant)	-	-	-	-
Reactions Questionnaire	(none significant)	-	-	-	-
Yaw Deviation	DEL	25.26	0.0001	ND	0.38
				D	0.68
Steering Reversals (per minute)	LAT	13.51	0.0005	T	87.33
				R	75.92

tion with delayed feedback dynamics (the most degraded simulator condition) had significantly greater increases in respiration rate from each of the other three less-degraded simulator configurations. Respiration rate was also found to be a sensitive measure for the main effects of delayed dynamics and enclosure of the subject.

Enclosed subjects reliably exhibited a decrease in skin resistance from baseline to experimental data-recording periods. This was indicative of increased forehead perspiration. No subjects in the open cab configuration demonstrated a decrease in forehead resistance. Because flow-through ventilation was provided, the increased perspiration occurring in the enclosed condition was probably not attributable to increased heat or humidity inside the cab.

The yaw deviation performance measure clearly indicated that delayed feedback dynamics degraded a subject's ability to maintain a steady vehicle heading. Yaw deviation values were convincingly higher ($p < 0.0001$) in conditions of delayed dynamics than in conditions of normal vehicle dynamics.

The method of simulating lateral acceleration was discovered to influence the number of steering reversals significantly. However, the results were initially surprising. The number of steering reversals is generally thought to increase as a function of driving task difficulty, which is in turn influenced by a number of factors, including the quality of the vehicle's handling. It appears then that the simulation of lateral acceleration by angular rotation would elicit more steering reversals than the normal translation method. Interestingly, the opposite effect was found to be true. One possible explanation for the lower number of steering reversals in response to angular rotation is that due to the "oversize" roll cue, subjects may quickly learn to refrain from making quick steering reversals which cause the simulator to move in successive rotational excursions. Continued rotational excursions, of the magnitude used in simulating a lateral acceleration cue, may be unpleasant to a subject.

While statistical significance was obtained for all of the physiological measures and simulator performance measures, it should be noted that no subjects experienced violent sickness in the simulator study reported above. However, the degraded simulator configurations were associated with symptoms pointing to the onset of malaise, including heightened pallor, and increased respiration rate and forehead perspiration. Vehicle controllability also suffered in the degraded configurations of delayed dynamics and angular (roll) simulation of lateral acceleration. The experiment purposely utilized attenuated roll motion and limited amounts of delay in these degraded configurations. Had full-size roll simulation of lateral acceleration and longer computational delays been introduced, acute observable sickness symptoms might have resulted.

CONCLUSIONS AND RECOMMENDATIONS

Several important conclusions concerning simulator design can be made on the basis of the empirical data and literature review results reported herein.

First, it is apparent that physical motion cues should be simulated with accurate movements in the same physical axis as in the full-scale vehicle. This is certainly feasible, using appropriately "washed-out" motion cues and reasonable excursion distances, and by taking advantage of the fact that the operator does perceive accelerations and decelerations but does not perceive them at low levels. Also, constant velocities are not perceived. The alternative techniques of simulating translational accelerations with rotational motions may reduce motion-base costs and conserve laboratory space, but these methods tend to be problematic in terms of both sickness inducement and undesirable handling.

The controlling systems for the simulator response variables should be free of any computational delays or servo lags to insure that the manual control inputs are coordinated with the visual and physical motion feedback cues. If not, subject uneasiness may result and vehicle controllability may degrade. Delayed feedback in response to steering inputs burdens the driving subject with the task of introducing compensating lead to control the vehicle. The increased workload and constant attentional demand placed on the subject may heighten the stress level, further contributing to malaise.

On both fixed-base and moving-base simulators, the necessity of a box-type cab is questionable. As previously discussed, the presence of an enclosure over the driver platform was sufficient for significantly increased forehead perspiration and respiration rate. Furthermore, in most cases it is doubtful that a windowless, box-type cab enhances the fidelity of a driving simulator.

Certain display characteristics which potentially contribute to simulator sickness are in need of future investigation. It is well-known that overly wide field-of-view displays, such as "cinerama" systems, induce illness, perhaps for the reasons stated in the introduction. It is of utmost importance that the roadway scene be geometrically accurate throughout the field-of-view, with no perceptible static or dynamic distortion of objects or movements occurring in the display. Tradeoffs between the level of detail presented and potential subject disorientation also should be considered in the selection of a roadway scene.

In a general sense, to avoid simulator sickness, it is very important to provide the subject with "interface cues" that are sufficient in number, accurately represented, and properly coordinated. Distortions in time, space, or response will tend to increase the chances of inducing simulator sickness. Such distortions must be minimized or preferably, eliminated.

ACKNOWLEDGMENT

The author wishes to express his thanks to Dr. W.W. Wierwille, who designed and constructed the VPI&SU driving simulator. His steadfast assistance and support were essential to the research described above.

REFERENCES

American Psychological Association, 1973, Ethical principles in the conduct of research with human participants, American Psychological Association, Inc., Washington, D.C.

Barrett, G.V. and Nelson, D.D., 1965, Human factors evaluation of a driving simulator; summary of human factors evaluation, Goodyear Engineering Report No. 12400, Volume XV.

Barrett, G.V. and Nelson, D.D., 1966, Human factors evaluation of a driving simulator; summary of virtual image display studies, Goodyear Engineering Report No. 12400, Volume XX.

Barrett, G.V. and Thornton, C.L., 1968a, Relationship between perceptual style and driver reaction to an emergency situation, Journal of Applied Psychology, 52 (2), 169-176.

Barrett, G.V. and Thornton, C.L., 1968b, Relationship between perceptual style and simulator sickness, Journal of Applied Psychology, 52 (4), 304-308.

Beinke, R.E. and Williams, J.K., 1968, Driving simulator, Paper presented at the General Motors Corporation Automotive Safety Seminar, Milford, Michigan, July 11-12.

Brand, J.J., Colquhoun, W.P. and Perry, W.L.M., 1968, Side effects of l-hyoscine and cyclizine studied by objective tests, Aerospace Medicine, 39, 999-1002.

Breda, W.M., Kirkpatrick, M. and Shaffer, C.L., 1972, A study of route guidance techniques, final report, North American Rockwell, Columbus, Ohio, Report No. DOT-FH-11-7708.

Casali, J.G. and Wierwille, W.W., 1979, A driving simulator overview concerning simulator sickness, Virginia Polytechnic Institute and State University, Human Factors Laboratory, Report No. HFL-79-13, Blacksburg, Virginia.

Casali, J.G. and Wierwille, W.W., 1980, Investigation of the effects of various design alternatives on moving-base driving simulator discomfort, Human Factors, 22 (6), 741-756.

Clark, B. and Graybiel, A., 1961, Human performance during adaptation to stress in the Pensacola slow rotation room, Aerospace Medicine, 32, 93-106.

Havron, M.D. and Butler, L.F., 1957, Evaluation of training effectiveness of the 2-FH-2 helicopter flight trainer research tool, U.S. Naval Training Device Center, Report No. NAVTRADEVCEN 1915-00-1.

Jex, H.R. and Ringland, R.F., 1973, Notes on car simulator state of the art, motion, and visual requirements, Systems Technology, Inc., STI Paper No. 2039-2, Hawthorne, California.

Kramer, C.Y., 1972, "A first course in methods of multivariate analysis", C.Y. Kramer, Publisher, Blacksburg, Virginia.

Leonard, J.J. and Wierwille, W.W., 1975, Human performance validation of simulators: theory and experimental verification, Proceedings of the Human Factors Society 19th Annual Meeting, Dallas, Texas, October 14-16, 446-456.

Lincke, W., Richter, B. and Schmidt, R., 1973, Simulation and measurement of driver vehicle handling performance, Paper presented at the SAE Automobile Engineering Meeting, Detroit, Michigan, SAE Paper No. 730489.

McLane, R.C. and Wierwille, W.W., 1975, The influence of motion and audio cues on driver performance in an automobile simulator, Human Factors, 17 (5), 488-501.

Miller, J.W. and Goodson, J.E., 1960, Motion sickness in a helicopter simulator, Aerospace Medicine, 31, 204-211.

Reason, J.T. and Brand, J.J., 1975, "Motion sickness", Academic Press, London.

Reason, J.T. and Diaz, E., 1971, Simulator sickness in passive observers, Flying Personnel Research Committee Report No. 1310.

Testa, C.J., 1969, The prediction and evaluation of simulator illness symptomatology, Unpublished doctoral dissertation, UCLA, Los Angeles, California.

VIDEO DISC TECHNOLOGY: A NEW APPROACH TO THE DESIGN OF

TRAINING DEVICES

Steven Levin and J. Dexter Fletcher

INTRODUCTION

The productivity limits of our current training technology may
have been reached. Massive efforts to improve "stand-up" lectures,
printed training materials, and "hands-on" laboratory experience
have begun to yield too few significant returns. We need revolutio-
nary new techniques that will break through the constraints posed by
existing training technologies and allows us to beat the current
tradeoffs that must be made among costs, quantity, and quality. Video disc technology is one of the most promising sources of these
new techniques.

Video disc technology is creating opportunities for new kinds
and forms of training devices at surprisingly low cost. At the heart
of this technology is the capability to access tens of thousands of
color images, including stereo sound, in seconds of fractions there-
of. When coupled with the now uniquitous microprocessor, video disc
technology offers new opportunities for training applications as well
as new issues for the training specialist.

In this paper we share some of our ideas and experiences in the
application of video discs to training. Specifically, we will examine
four new ideas for training applications which use video disc technology. We then consider issues relevant to these applications and their
implications for training. We begin a brief review of the characteristics and capabilities of the optical video disc.

Optical Video Disc Antecedents

The functional capabilities of video discs have long been noted

as desirable, and there have been various attempts to provide these
capabilities in earlier instructional systems. In 1966, an IBM 1500
Instructional System was installed in a public school setting for
use by investigators at Stanford University (Suppes, 1966). This system offered random access to 1,024 16-millimeter still photographs,
about 30 minutes of recorded audio information that could be divided
into addressable segments of arbitrary length, and a small number of
digitally stored graphic characters. The photographic and audio information was stored on tape and the speed with which random access
could be accomplished left something to be desired.

In the 1968-69 school year, Stanford began using a digitized
audio capability in its DEC PDP-10 based instructional system (Fletcher and Atkinson, 1972). Speed of random access to the audio messages was never a problem on this system, but storage was expensive.
It took 36,000 bits to store one second of audio information. Graphic, photographic, and/or video information was out of the question
because the student terminals for this system were teletypewriters--
although valiant efforts by Lorton (1973) to present graphics on
Model 33 teletypewriters should not go unmentioned. In 1963 the University of Illinois PLATO project developed a video scanner that provided rapid (less than one second) random access to 128 still photographs. In order to reduce costs, the project later developed a capability for rapid random access to 256 still photographs stored on
microfiche (Stifle, 1971). Around 1972-73 the Brigham Young Univer-
sity/MITRE TICCIT project pioneered the use of computer-controlled
interactive television, and issues of computer-controlled video information began to arise in instructional design (Bunderson, 1974).

Disc storage of video information has been available for some
time as sports referees and umpires must occasionally acknowledge.
These discs provide digital storage and a read-write capability of
relatively low capacity although improvements in magnetic pick-up
heads and magnetic oxide technology have increased this capacity by
a factor of about four since the discs first appeared. Currently,
they store about 1200 frames. These discs incorporate most of the
features available in video discs--with the quite notable exceptions
of low cost and large storage capacity. Slow motion, still frame,
and reverse motion are all available, but there seem to be no examples of these discs being used in instruction.

Optical Video Disc Technology

A video disc is similar to an audio record except that each
side of a 12-inch disc contains 30 minutes of television. Since television is the rapid presentation of pictures at a rate of 30 images
per second, each side of a video disc contains 54,000 still pictures
in color.

The player for a video disc is much like a turnable except that

under computer control the player can rapidly find any particular
part of the video disc. Specifically, any one of the 54,000 images,
or any part of the 30 minutes of television on a video disc, can be
located typically in a fraction of a second. A video disc provides
a combination of moving and still images, any part of which can be
quickly located.

Video discs are made like audio records using original materials
that can be movies, video tapes, pages of text, tables of numbers,
graphs, charts, maps, drawings, diagrams, or photographs. From the
original materials, a master disc is produced that in turn is used
to "press" multiple copies speedily and at low cost.

Two types of video discs and video disc players are now available. The players are characterized by the market they are targeted
for: general consumer usage and industrial/educational applications.
Various potential manufacturers such as Sony, RCA, JVC, IBM, Xerox,
Zenith, ARDEV (Altantic Richfield Development) to name a few, are
waiting in the wings, but only three are now marketing video disc
players.

- Magnavox, a wholly owned subsidiary of Phillips, began selling a
  consumer model player for about $800 in December 1978.
- Discovision (DVA), under license by MCA, began general sales of an
  industrial player for about $3,000 in June 1979.
- Thompson-CSF began sales of an industrial player for about $3,500
  early in 1980.

Some of the aforementioned companies have announced marketing plans
for their video disc systems. Sony has said that they will begin
U.S. wide sales of their industrial player in late 1980. Pioneer and
RCA have announced marketing dates of third quarter 1980 and early
1981 respectively for consumer video disc systems.

The differences between consumer and industrial systems are significant as Table I shows. For instructional settings, the industrial
players incorporate microprocessor control for fast random access,
pre-programmed branching, and frame selection. Microprocessors have
also been installed in the less expensive consumer players to provide
random access, branching, and frame selection. However, the servo-
mechanism in these players requires from 15 to 20 seconds to locate
a frame in the worst case. Maximum access times for the industrial
players are under four seconds. The consumer players lend themselves
to instructional application, but the functional capabilities of the
industrial players may be sufficiently greater that their higher
cost is justified. With experience, we should discover what various
video disc features buy in terms of instructional achievement and
will thereby be able to recommend commercial or industrial players
depending on instructional settings.

Table I. Comparison of Consumer With Industrial Video Discs

Features	Consumer Players	Industrial/Educational Players
Cost	$800	$2500-$3500
Still frame	Manual or automatic	Manual or programmed[*]
Frame random access	Manual Visual identification	Manual or programmed Visual, keyboard selection, or programmed
Frame-by-frame "stepping"	Manual	Manual or programmed
Variable speed motion (forward or reverse)	Manual	Manual or programmed
Two discrete sound channels	Yes	Yes

[*]Programmed control may be through an on-board video disc player microprocessor or an external computer.

The three video disc players discussed above are all optical, laser-based systems. They use a 12-inch disc with a spiral track. The track is pitted with oblong depressions or micropits about 1 micron deep that vary in accordance with the audio or video information they represent. During playback the disc spins at 1,800 revolutions per minute while these micropits modulate a low power helium-neon laser focused on the track and thereby generate a signal that is processed and passed to a standard video monitor (i.e., to the antenna terminals of a TV set).

One video frame is stored on each track and there are 54,000 tracks per disc. Video, audio, and still photographic information can all be intermingled on these discs. These video disc systems effectively provide rapid access to 30 minutes of video information, 30 minutes of analogue audio information, 54,00 still photographs, well in excess of 400 hours of digitized audio information, 30 minutes of motion picture information, or various combinations of the above. The point to be made is that video discs provide rapid random access to a lot of information which can be inexpensively stored and replicated. Table II presents a summary comparison of the three video discs discussed in this section.

There are at least two important differences between the DVA and the Thompson-CSF video disc systems. First, the Thompson disc is "transmissive" and the DVA disc is "reflective". Light from a laser

Table II. Summary Comparison of Three Video Disc Systems

	Magnavox/Phillips	Discovision-DVA	Thompson-CSF
Cost	$800	$3,000	$3,500
Market	Consumer	Industrial/Educational	Industrial/Educational
Cost of Master	$1,500[a]	$1,500[a]	$1,500[a]
Cost of Copies	$5-$10[c]	$5-$10[c]	$18[b]
Technology	Optical-reflective (two-sided aluminum disc)	Optical-reflective (One-sided aluminum disc)	Optical-transmissive (two-sided plastic disc)
Standard	NTSC	NTSC	NTSC, PAL/SECAM

[a] Cost per side

[b] Cost includes $1.00 for protective plastic cover

[c] Depends on quantity.

passes through the Thompson CSF disc and is picked up on the other side. The DVA disc reflects light back up so that light is sent from and received on the same side of the disc. Three implications of this difference in technical approach are: (1) The Thompson CSF disc is "floppy", it can be rolled up in magazines and newspapers, and the DVA disc is rigid; (2) by adjusting the focus of the laser, either side of the Thompson CSF disc can be read without physically turning the disc over. If both sides of the DVA disc are to be used, it must be physically turned over-- or there must be a light source for each side; and (3) the Thompson CSF disc is more sensitive to dust and dirt than the DVA disc.

The second important difference between the two discs is that the Thompson CSF disc systems observe European PAL/SECAM television standards as well as U.S. standard NTSC. The DVA disc observes only American NTSC television standards. When used in PAL/SECAM mode, the Thompson CSF disc therefore provides more lines per display. The major instructional implication of this difference seems to be that because of the better resolution with PAL/SECAM standards, programs that display large amounts of text may be better suited to the Thompson disc. On the other hand, if instruction designers want to take advantage of the millions of already purchased television sets in American homes, schools, and industries, they may be well advised to use NTSC encoded video discs.

Second Generation Disc Technology

Three second generation developments in video disc technology appear to be particularly notable for instructional applications.

First is the anticipated appearance of direct read after write, or DRAW, technology. The important aspects of the DRAW disc is that it stores $10^{10}-10^{12}$ bits of digital information on each side of the disc. This development is being undertaken by both RCA and Phillips. In the near term, video disc systems using DRAW technology are likely to be very expensive. However, DRAW disc systems may be better suited for the small volume experimental development that is characteristic of many instructional settings.

Second is the development of still frame sound. All currently available video disc systems only provide sound when the disc is played at 30 frames per second. Unfortunately, this detracts considerably from the disc's instructional value during slide-type presentations. DVA and Sony have both indicated that some form of compressed audio providing 3 to 30 seconds of sound per frame is under development.

Third, ARDEV has been developing an optical film based digital disc system that will use up to nine parallel reading heads. The use of multiple reading heads will enable track to track tradeoffs of video bandwidth with compressed sound and optically stored digital data. In addition, the ARDEV video disc system is planned to include a keyboard and full capability microprocessor.

Video Disc Versus Video Tape

A brief comparison of videotape capabilities with those of video discs may be in order at this point. After all, videotapes represent a rapidly maturing technology, and they can offer many of the features of video discs such as random access, variable speed play, and reverse motion. Three points appear to be salient:

1. Tapes can be edited and current video discs cannot. This difference is likely to be removed with the appearance of the DRAW technology discussed above. DRAW technology will, of course, provide editable video discs quickly and relatively inexpensively.
2. Random access to arbitrary points on a video disc is much faster than similar access using videotape. Random access on industrial video discs is less than 4 seconds compared with 20-100 seconds on videotape. Moreover, accuracy of random access on videotape leaves something to be desired.
   Still frames are possible using either video disc or videotape, but it is exceedingly expensive and cumbersome on videotape and single stepping through a series of still frames is essentially impossible. A far more comprehensive and detailed comparison of

videotape and video disc has been prepared by Heuston (1978).

APPLICATIONS OF VIDEO DISC TECHNOLOGY TO TRAINING

One of the most exciting aspects of video disc technology is that is makes possible an entirely new set of training experiences. This section describes four such new applications ranging from new kinds of training movies to low-cost simulators.

Interactive Movies

Interactive Movies, illustrated in Fig. 1, translate movie viewing into an active participatory process. In effect, the viewer becomes the director and controls many features of the movie. A sampling of feature controls available to the viewer is the following:
1. Perspective. The movie can be seen from different directions. In effect, the viewer can "walk around" ongoing action in the movie or view it from above or below.
2. Detail. The viewer can "zoom in" to see selected, detailed aspects of the ongoing action or can "back off" to gain more perspective on the action and simultaneous activity elsewhere.

Fig. 1. Illustration of an Interactive Movie.

3. Level of instruction. In some cases, the ongoing action may be too rich in detail or it may include too much irrelevant detail. The viewer can hear more or less about the ongoing process by so instructing the Interactive Movie system.

4. Level of abstraction. In some instances the viewer may wish to see the process being described in an entirely different form. For example, the viewer might choose to see an animated line drawing of an engine's operation to get a clearer understanding of what is going on. In some cases, elements shown in the line drawings may be invisible in the ongoing action—for instance, electrons or force fields can be shown.

5. Speed. Viewers can, of course, view the ongoing action at a wide, continuous range of speed—including reverse action and no action (still frame).

6. Plot. Viewers can change the "plot" to see the results of different decisions made at selected times during the movie.

A typical application for Interactive Movies would be in training (and aiding) equipment technicians. The technician could not only see how a particular part is located and installed from several points of view (e.g. top versus bottom) but could interactively control how detailed a description is either seen or heard regarding that maintenance activity.

Several Interactive Movie video discs have been completed using hand to hand combat (i.e., karate) as the subject area. These discs let the viewer not only control playing a particular karate move forward and backward at any rate, but also include multiple views and closeup views following every move from four different positions. In progress is an Interactive Movie that will focus on a particular equipment maintenance task.

Surrogate Travel

Surrogate travel, illustrated in Fig. 2, forms a new approach to locale familiarization and low cost trainers. The basic principle is simple. On video discs are stored up to 108,000 images showing discontinuous motion along a large number of paths in an area. Under microprocessor control, the user accesses different sections of the disc, simulating movement over the selected path.

The user sees with photographic realism the area of interest. Unlike a travel movie, the user is able to both choose the path and control the speed of advance through the area using simple controls. The video disc frames the viewer sees originate as filmed views to what one would actually see in the area. To allow coverage of very large areas, the frames are taken at periodic intervals that may range from every foot inside a building, to every ten feet down a city street, to hundreds of feet in a large open area (e.g., a harbor).

Fig. 2. Illustration of Surrogate Travel.

The rate of frame playback, which is the number of times each
frame is displayed before the next frame is shown, determines the
apparent speed of travel. Free choice in what routes may be taken is
obtained by filming all possible paths in the area as well as all
possible turns through all intersections. While it might first appear that this would be a time consuming and expensive technology,
it is in fact relatively efficient and inexpensive because of the
design of special equipment and procedures for doing the filming.

Demonstrations of this technology have been developed for building interiors (MIT, National Gallery of Art), a small town (Aspen,
Colorado), and industrial facility (nuclear power plant), a weapon
site, and San Francisco Harbor. Plans are underway to produce a prototype video mapping library of broader scope for selected areas
worldwide.

To provide training in reading and understanding maps, the pho-
tograph-based Surrogate Travel is linked to different sorts of maps
of the area. In effect, the viewer can travel across a map, can fly
into it getting greater and greater detail from what can by presented
by standard map symbology, and then "fall through" the map to see

photographically what the map depicts. In addition, the viewer can
switch among different types of maps (e.g., topographic, infrared,
etc.) to develop an understanding of how different map symbologies
and representations interact.

In addition to ground level travel, including the inside and
outside of buildings, aerial flight experience can be produced and
used for flight training and airport familiarization. Similarly, other forms of travel experience, such as anchorage piloting and low
level nap of the earth flying, are also easily accommodated.

Electronic Libraries

Electronic libraries, illustrated in Fig. 3, in the form of Spatial Data Management Systems (SDMS) provide students and instructors
quick and easy access to an assortment of multi-source and multi-
media information (Levin, 1980). Users literally "fly over" information and select what they want by simply pointing. Spatiality is used
to group materials into lesson plans so that different information
spaces represent course concepts, additional instruction, and assessment procedures.

Fig. 3. Illustration of an Electronic Library.

Stored on a video disc are tens of thousands of frames consisting of photographs, diagrams, charts, text, movies, spoken speeches,
music, graphs, etc. The pages can be organized, reassembled, segmented, and/or duplicated in accordance with the users's need and
growing sophistication with the subject matter. The pages can be annotated, highlighted, drawn-on, underlined, etc., at the user's con-
venience and pleasure.

For the instructor, the SDMS provides ready access to a wealth
of material which might otherwise be unaccessible. Instructors can
access the SDMS to create their own information spaces (i.e., courses
or lectures) and subsequently present such materials to large audi-
ences in single locations via large screen television projection or
to multiple locations through cable distribution systems.

Students can independently use the SDMS for self paced instruction by either working through previously designed information spaces
or by browsing on their own. When students and instructors are in remote locations, offsite instruction is facilitated by linking two or
more SDMSs together using regular telephone lines. In this manner,
a student or instructor can literally fly the other to a topic of interest, sharing at geographically remote sites a large library of information.

The same video materials can be used for hundreds of different
users. The only thing that must be changed from user to user is the
magnetic storage medium which serves as the user's private librarian
for the video disc.

The U.S. Marine Corps Education Center is using the video disc
and SDMS technology to create the electronic library which we have
discussed. The low cost replication characteristics of the video
disc make this medium an attractive alternative for the distribution of previously archived information.

Simulation

Video disc technology can be used very effectively for producing
low-cost visual simulators. An example of this, illustrated in Fig.
4, is the development of a tank gunnery trainer (Thomas et al.,
1980). In this low-cost trainer, a gunner is being taught to locate,
track, and fire at enemy tanks. Instructional sequences consisting
of both the visuals seen by the gunner and the constantly changing
problem information needed to provide instructional feedback are accessed from a video disc.

The video disc provides rapid access to many problem sets as
well as surprisingly high fidelity display of what is normally seen
by tank gunners. The trainers can be linked together to provide in-
tra-tank training, for tank crews, or inter-tank training for tank

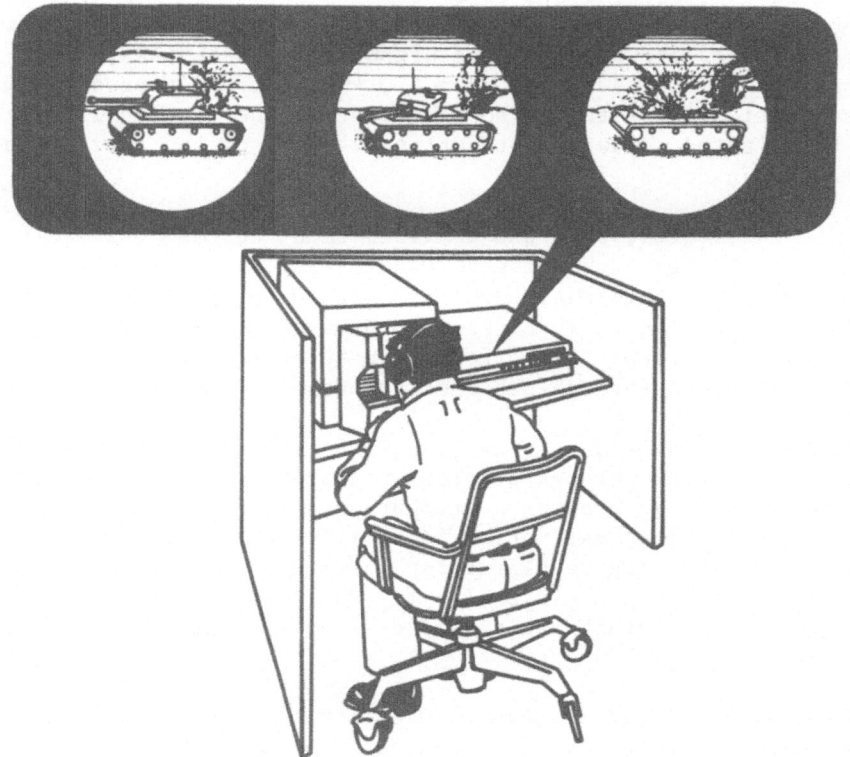

Fig. 4. Illustration of Visual Simulator.

platoons. Shoot-offs and "quick-fire" exercises are presented to increase motivation. All sighting devices and sight reticles are included in the trainer. Computer graphics overlaid on the video sequences are used to show trajectory and burst-on-target information.
Daytime, nighttime, smoke, and dust sequences are all included. The
device captures the fun and excitement of arcade games in a job-re-
levant, militarily-essential training activity.

ISSUES

     Development and use of training applications employing the vi-
deo disc such as those discussed in the previous section raise a set
of issues to be considered by the training community. A few of these
issues are briefly discussed below:
1. What are the authoring requirements in producing video disc based
   training systems? Probably the most important requirement is the
   need to work freely with the capability of storing and almost in-
   stantaneously accessing over 50,000 frames of information. Our experience is that many video disc first-timers make only incremen-

tal use of the technology because they do not take advantage of
its random access characteristics. Missing from the video disc
milieu are inexpensive authoring systems which would enable authors to put together and try out new video disc program concepts.
Bunderson and Campbell (1980) have prepared a more complete discussion of this issue.

2. What is the role of digitized sound and synthesized speech in video disc applications? Currently, several groups are experimenting
with a number of means to overcome the lack of still sound in video disc systems. Future video disc systems will almost certainly
incorporate still sound capabilities. However, synthesized speech
may have an important place in video disc applications given the
rapid, random access capabilities of video discs. For example, in
the Surrogate Travel application discussed earlier, the controlling
computer also generates narrative descriptions that accompany the
viewer's travel. Speech synthesis is used as the audio channel
because it is not reasonable to prestore all things that could be
said for all possible routes that the viewer might take.

3. How does video disc technology affect the accessibility of training devices? A lot. With the video disc, the training device designer gets a comparatively low cost (potentially less than
$1,000) system that brings together an array of media (slides,
text, graphs, movies, sound) in a compact and robust form. In some
instances, very expensive computer graphics based systems may be
replaced by inexpensive video disc players.

4. How "smart" a video disc training device is needed? Smart, yes,
but not brilliant. The four applications discussed earlier all
use video disc players controlled by small, inexpensive microcom-
puters. Future video disc players may incorporate sufficiently capable microprocessor systems to allow the necessary functions to
be completely on board. While it is true that the players available today include a small microprocessor which can support branch-
ed instruction--usage in this mode is far removed from the real potential of the video disc as a training device.

5. How can training take advantage of the "super realism" offered by
video discs? In the Surrogate Travel and low cost trainer applications, the viewer sees the real place (e.g., a real tank and,
if appropriate, a real explosion). This is all made possible by
storing a facsimile version of events and materials on the video
disc. Thus, instead of seeing a simplification (although at times
this may be beneficial), the user interacts with the object or
process as it will be encountered in the work place.

6. Are video disc training devices inherently "fun"? Stated in another form, since television is fun, will video disc training devices find greater acceptance? Without attempting a full analysis
we state our experience: Users enjoy working with these systems.
We cannot say if this is because the systems are well designed,
because they include color, sound, and motion, because they are
a familiar medium (television), or because of something else.
Nonetheless, the overwhelming impression we and others have

gained from working with video disc systems is that users like them.

The list of issues above is in no way complete. What we have attempted to do is alert the reader to some of the questions which have arisen through our own experiences. We feel that the area of video disc based training devices is extremely promising and deserves considerable attention by the training community.

REFERENCES

Bunderson, C.V., 1974, The design and production of learner-controlled courseware for the TICCIT system: A progress report, International Journal of Man-Machine Studies, 6, 479-491.
Bunderson, C.V. and Campbell, J.O., 1980, Videodisc Training Delivery Systems and Videodisc Authoring/Production Systems: Hardware, Software, and Procedures, U.S. Army Research Institute, Alexandria, VA, in press.
Fletcher, J.D. and Atkinson, R.C., 1972, An evaluation of the Stanford CAI Program in initial reading, Journal of Educational Psychology, 63, 597-602.
Heuston, D.H., 1978, A Comparison Between the Videotape and Videodisc as Educational Devices, WICAT, Inc., Orem, UT.
Levin, S., 1980, Video disc-based spatial data management, in: Proceedings of the AFIPS 1980 Office Automation Conference, American Federation of Information Processing Societies, Inc., Washington, D.C.
Lorton, P.V., 1973, Computer-based instruction in spelling: An investigation of optimal strategies for presenting instructional material, Unpublished doctoral dissertation, Stanford University.
Stifle, J., 1971, The PLATO IV Architecture, University of Illinois, Computer-based Education Research Laboratory, CERL Report X-20, Urbana-Champaign, ILL.
Suppes, P., 1966, The uses of computers in education, Scientific American, 215, 206-220.
Thomas, J.O., Madni, A. and Weltman, G., 1980, Systems Requirements Analysis and Prototype Design of a Low-Cost Portable Simulator for Performance Training, Perceptronics, Inc., Annual Technical Report PATR-1085-80-2, Woodland Hills, CA.

CONTRIBUTORS

Wilhelm Berheide,
        Forschungsinstitut für Anthropotechnik,
        Königstrasse 2,
        5307 Wachtberg-Werthhoven,
        W.-Germany

John G. Casali,
        Human Factors Laboratory,
        Dept. Industrial Engineering and Operations Research,
        Virginia Polytechnic Institute and State University,
        Blacksburg, Virginia 24061,
        U.S.A.

Gerald P. Chubb,
        Crew Systems Effectiveness Branch,
        Human Engineering Division,
        Air Force Aerospace Medical Research Laboratory,
        Wright-Patterson Air Force Base, Ohio 45433,
        U.S.A.

Bernhard Doering,
        Forschungsinstitut für Anthropotechnik,
        Königstrasse 2,
        5307 Wachtberg-Werthhoven,
        W.-Germany

J. Dexter Fletcher,
        U.S. Army Res. Inst. for the Behavioral and Social Sciences,
        5001 Eisenhower Avenue,
        Alexandria, Virginia 22333,
        U.S.A.

Klaus-Peter Gärtner,
        Forschungsinstitut für Anthropotechnik,
        Königstrasse 2,
        5307 Wachtberg-Werthhoven,
        W.-Germany

Frank E. Gomer,
    Engineering Psychology Department,
    McDonnel Douglas Astronautics Company,
    St. Louis Division,
    Box 516,
    Saint Louis, Missouri 63166,
    U.S.A.

Klaus-Peter Holzhausen,
    Forschungsinstitut für Anthropotechnik,
    Königstrasse 2,
    5307 Wachtberg-Werthhoven,
    W.-Germany

Richard van der Horst,
    Institute for Perception TNO,
    P.O. Box 23,
    3769 ZG Soesterberg,
    The Netherlands

Jürgen Kaster,
    Forschungsinstitut für Anthropotechnik,
    Königstrasse 2,
    5307 Wachtberg-Werthhoven,
    W.-Germany

Karl-Friedrich Kraiss,
    Forschungsinstitut für Anthropotechnik,
    Königstrasse 2,
    5307 Wachtberg- Werthhoven,
    W.-Germany

John G. Kreifeldt,
    Department of Engineering Design,
    Tufts University,
    Medford, Massachusetts 02155,
    U.S.A.

Norman Lane,
    Naval Air Development Center,
    Warminster, Pennsylvania 18974,
    U.S.A.

Steve Levin,
    U.S. Army Res. Inst. for the Behavioral and Social Sciences,
    5001 Eisenhower Avenue,
    Alexandria, Virginia 22333,
    U.S.A.

Wouter A. Lotens,
     Institute for Perception TNO,
     P.O. Box 23,
     3769 ZG Soesterberg,
     The Netherlands

Jan Moraal,
     Institute for Perception TNO,
     P.O. Box 23,
     3769 ZG Soesterberg,
     The Netherlands

Ben Ostrofsky,
     Department of Industrial Engineering,
     University of Houston,
     Central Campus,
     Houston, Texas 77004,
     U.S.A.

Dennis L. Price,
     Transportation Safety Projects Office,
     167 Whittemore Hall,
     Virginia Polytechnic Institute and State University,
     Blacksburg, Virginia 24061,
     U.S.A.

Hans Radke,
     Deutsche Forschungs- und Versuchsanstalt für
     Luft- und Raumfahrt e.V. (DFVLR),
     Postfach 3267,
     3300 Braunschweig,
     W.-Germany

Günter Rau,
     Helmholtz Institute for Biomedical Engineering,
     Goethestrasse 27-29,
     5100 Aachen,
     W.-Germany

Fred V. Schick,
     Deutsche Forschungs- und Versuchsanstalt für
     Luft- und Raumfahrt e.V. (DFVLR),
     Postfach 3267,
     3300 Braunschweig,
     W.-Germany

Carol A. Simpson,
     Psycho-Linguistic Research Associates
     2055 Sterling Avenue
     Menlo Park, California 94025,
     U.S.A.

Melvin I. Strieb,
      Analytics Inc.
      Willow Grove, Pennsylvania,
      U.S.A.

Donald A. Topmiller,
      Crew Systems Effectiveness Branch,
      Human Engineering Division,
      Air Force Aerospace Medical Research Laboratory,
      Wright-Patterson Air Force Base, Ohio 45433,
      U.S.A.

Manfred Voss,
      Fraunhofer-Institut für Informations- und
      Datenverarbeitung - IITB,
      Sebastian-Kneipp-Strasse 12-14,
      7500 Karlsruhe 1,
      W.-Germany

Heino Widdel,
      Forschungsinstitut für Anthropotechnik,
      Königstrasse 2,
      5703 Wachtberg-Werthhoven,
      W.-Germany

Walter W. Wierwille,
      Human Factors Laboratory,
      Dept. Industrial Engineering and Operations Research,
      Virginia Polytechnic Institute and State University,
      Blacksburg, Virginia 24061,
      U.S.A.

Robert C. Williges,
      Human Factors Laboratory,
      Dept. Industrial Engineering and Operations Research,
      Virginia Polytechnic Institute and State University,
      Blacksburg, Virginia 24061,
      U.S.A.

INDEX